Molecular, Genetic, and Cellular Advances in Cerebrovascular Diseases

Editors

Hua Su

Center for Cerebrovascular Research
Department of Anesthesia and Perioperative Care
University of California, San Francisco, USA

Michael T Lawton

Department of Neurosurgery
University of California, San Francisco, USA

World Scientific

NEW JERSEY • LONDON • SINGAPORE • BEIJING • SHANGHAI • HONG KONG • TAIPEI • CHENNAI • TOKYO

Published by

World Scientific Publishing Co. Pte. Ltd.
5 Toh Tuck Link, Singapore 596224
USA office: 27 Warren Street, Suite 401-402, Hackensack, NJ 07601
UK office: 57 Shelton Street, Covent Garden, London WC2H 9HE

Library of Congress Cataloging-in-Publication Data
Names: Su, Hua (Of the University of California, San Francisco), editor. |
Lawton, Michael T., editor.
Title: Molecular, genetic, and cellular advances in cerebrovascular diseases /
[edited by] Hua Su, Michael T. Lawton.
Description: New Jersey : World Scientific, 2017. | Includes bibliographical references and index.
Identifiers: LCCN 2017026793 | ISBN 9789814723299 (hardcover : alk. paper)
Subjects: | MESH: Cerebrovascular Disorders--therapy |
Cerebrovascular Disorders--diagnostic imaging
Classification: LCC RC388.5 | NLM WL 355 | DDC 616.8/1075--dc23
LC record available at https://lccn.loc.gov/2017026793

British Library Cataloguing-in-Publication Data
A catalogue record for this book is available from the British Library.

For any available supplementary material, please visit
http://www.worldscientific.com/worldscibooks/10.1142/9793#t=suppl

Typeset by Stallion Press
Email: enquiries@stallionpress.com

Printed in Singapore

Preface

On behalf of my co-editor, Hua Su, and the distinguished authors of these contributions, I am proud to present this book, **Molecular, Genetic, and Cellular Advances in Cerebrovascular Diseases**. The cerebrovascular diseases have long been of interest to clinicians who care for afflicted patients, but this domain has only recently attracted the attention of basic scientists, whose knowledge and insights have truly advanced the field. We need to look no further than brain arteriovenous malformations (AVMs) to find an example of how collaboration with scientists has changed our thinking about AVM pathogenesis. These lesions were considered congenital lesions that result from errors in vasculogenesis or angiogenesis during fetal development, in large part because these explanations originated from clinicians who had limited research activity or insights into underlying genetic or molecular mechanisms. However, study of patients with the rare disease hereditary hemorrhagic telangiectasia (HHT) has identified the genetic basis for AVMs in these patients, which has then led to models of brain AVM in transgenic mice that facilitate the search for the true pathogenesis and novel therapies for patients. Our reinterpretation of AVM pathogenesis and pathophysiology allows us to explain quirks of clinical practice like recurrent AVMs in children after previous complete resections and *de novo* AVMs that form in a variety of other clinical settings. Today, we stand poised to rewrite the story of AVMs and generate pharmacological therapies for patients in whom surgery is too risky or radiosurgery is ineffective. These advances originate from collaborations between neurosurgeons, neurologists, geneticists, molecular scientists, neuroradiologists, and neuroanesthesiologists, and this book is a reflection of this collaborative effort.

Molecular, Genetic, and Cellular Advances in Cerebrovascular Diseases begins with an in-depth review of the brain's structure and techniques to image it. The book continues with discussions of stroke and neural repair after injury. Attention then turns to AVMs, their current management, and experimental breakthroughs in the laboratory. Brain aneurysms are 10 times more common than AVMs, and this lesion's biology and hemodynamics are reviewed. Dural arterio-venous fistulas (DAVFs) are recognized as acquired lesions, and this pathogenesis has been critical in rethinking the pathogenesis of brain AVMs and other cerebro-vascular disease. DAVFs and their management are also included in the book. Finally, the book finishes with reviews of interventional and stem cell therapies, which hold tremendous promise for innovative therapies in the near future.

Editing a book is an awkward experience because contributors do so much of the work of composing the chapters, and yet the editors are named on the cover. Dr. Su and I would like to acknowledge the efforts of our authors and thank them for sharing their insights. We are proud of this collection of topics and hope that you learn from the content, and also feel inspired to build collaborative research efforts that will help further unravel the mysteries of cerebrovascular diseases.

Michael T. Lawton, MD

Contents

Imaging in Cerebrovascular Disease

1

Christopher P. Hess

Department of Radiology and Biomedical Imaging, University of California, San Francisco, San Francisco, California, USA

The past two decades have seen tremendous advances in our understanding of the anatomy and physiology of the central nervous system and its supporting vasculature. The lesion studies, anatomic dissection, and light microscopy techniques of the past have been largely supplanted over the last three decades by modern neuroimaging, which now uniquely permits non-invasive and *in vivo* visualization of the brain's structure and physiology in the individual patient. These tools serve as the cornerstone for the detection, characterization, and monitoring of all cerebrovascular abnormalities, for open surgical, endovascular and radiosurgical treatment planning, and for follow-up after clinical management. Imaging is also essential for patient selection in clinical trials and, in some cases, can serve as a powerful marker to predict patient outcomes.

Cerebrovascular imaging has evolved at a rapid pace, to a point where there are now multiple modalities in routine use, each with its own dizzying array of specialized techniques that characterize specific anatomic or physiologic features of disease. Computed tomography (CT), magnetic resonance imaging (MRI), and digital subtraction catheter angiography (DSA) serve as the three core imaging modalities for vascular imaging in current practice. Each provides a complementary lens through which to view the biological complexity of cerebrovascular disease; in some cases, more than one modality is necessary.

This chapter reviews the current state of the art in cerebrovascular imaging and describes the fundamental strengths and limitations of the three key modalities. The primary goal is to provide the reader with a general, "30,000-foot" overview of the techniques used in current clinical practice to image vessels, blood flow and the brain's microcirculation. The first section discusses the original and most commonly employed imaging tool to evaluate the intracranial vascular system: angiographic imaging of the vessel lumen. The sections that follow outline techniques for dynamic (time-resolved) angiography, and methods for quantifying brain physiology from the microcirculation. The final section describes the importance of brain parenchymal imaging for the evaluation of cerebrovascular disease.

1.1 Imaging of the Arterial and Venous Lumen

The most frequently used methods for imaging of cerebrovascular system are based upon visualization of blood within the vascular lumen using X-ray, CT, or MR angiography.

1.1.1 *CT angiography (CTA/CTV)*

Because differences in the intrinsic density of the blood pool and the vessel wall are visually imperceptible, luminal contrast is generated with CT via bolus injection of dense iodinated contrast and rapid imaging of the head during transit of the contrast bolus through the cerebral circulation. Images are acquired as intravascular contrast medium extends into patent and opacified cerebral vessels. Modern multi-detector CT (MDCT) enables coverage of the entire head and neck within 4–8 seconds, and first-pass transit for the contrast bonus from the arterial to the venous circulation takes place in less than 10 seconds. By deliberately timing the acquisition of CT images with respect to the anticipated location of the contrast bolus within the circulation, images can be obtained to emphasize the arterial and/or venous system (Figure 1). CT is also most often performed prior to any contrast injection in order to identify the presence of acute hemorrhage and other structural abnormalities that might require immediate patient care, for a full exam time of less than 5 minutes.

The high density of injected iodinated contrast relative to other brain and neck soft tissues allows CT to reveal the patent vascular lumen within the cervical and intracranial circulation with very high spatial resolution (typical MDCT voxel sizes are 0.5–1 mm). This resolution equips CT arteriography (CTA) with the ability to characterize stenotic disease with extremely high accuracy, to identify

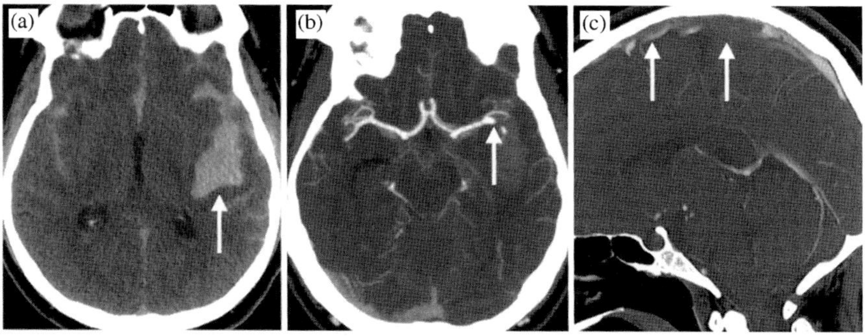

Figure 1. CT, CTA, and CTV. Axial images from CT, CTA, and CTV obtained using MDCT without contrast (a) and in the arterial (b) and venous (c) phases following bolus injection of contrast, respectively. The non-contrast CT image in (a) shows diffuse subarachnoid hemorrhage, with high density blood concentrated in the left sylvian fissure (arrow), due to rupture of a left middle cerebral artery (MCA) aneurysm (arrow) shown in the corresponding CTA (b). CTV obtained in a different patient (c) outlines clot within the superior sagittal sinus (arrows), the result of dural venous sinus thrombosis.

contour irregularities and aneurysms in arteries of the circle of Willis and cortical branches, and to detect arterial occlusions in major vessels. To better study the venous system, high-resolution CT venography (CTV) is performed in an analogous fashion to CTA, but with delayed timing of MDCT acquisition relative to the bolus injection (usually around 45 seconds).

An important limitation of CTA arises in the setting of occlusive or severe stenotic arterial disease. In these low-flow situations, the injected contrast bolus may not opacify the lumen of arteries upstream of the site of disease during the time of CT acquisition. This limitation becomes particularly important in the setting of arterial thrombus in stroke, as it precludes accurate determination of the length of the thrombus (an important consideration for planning endovascular treatment). To overcome this shortfall, many centers repeat CT at a later venous phase, with sufficient delay to permit contrast to extend into the vessel of interest if it remains patent.

Given its ability to rapidly acquire images, widespread availability in hospitals and emergency departments, and relatively lower cost compared to other modalities, CT/CTA is the most regularly used technique in patients with emergent neurologic disorders. Most neurosurgical emergencies, including large vessel occlusions (LVOs), ruptured aneurysms, hydrocephalus, herniation, and acute hemorrhage, are readily diagnosed using a combination of CT and CTA and/or CTV. However, although vascular contrast is high with bolus CTA, CT suffers from poor soft tissue contrast that can make subtle lesions in the brain parenchyma difficult or impossible to detect. Evaluation of vessels around the bone may also be limited by beam-hardening artifacts. The sensitivity of CT is thus limited for many other disorders. Furthermore, CT/CTA requires radiation for imaging, at a time when medical radiation to patients is gaining attention in the scientific literature and even in the mainstream media (Brenner & Hall, 2007; Smith-Bindman *et al.*, 2009).

1.1.2 *MR angiography* (*MRA/MRV*)

Like CT, MR angiography can be achieved using bolus injection of contrast media (in the case of MRI, typically a gadolinium chelate) for contrast-enhanced MRA (CE-MRA). Because acquisition of high-resolution MR images requires more time than CT, however, more careful synchronization is required between the acquisition of imaging data and the temporal passage of the bolus through the craniocervical circulation. Optimal timing for acquisition during peak arterial or venous enhancement (for MRA or MRV, respectively) is typically estimated empirically from a separate timing or "tracking" bolus experiment before angiographic imaging to assess first-pass contrast kinetics. Elliptical centric MR acquisition is then used during a second bolus injection to generate luminal images with optimal arterial or venous contrast. With this technique, the second MR acquisition is modified to acquire data containing contrast information early in the scan, during the peak arterial or venous phase of the second administered contrast bolus. Data containing information about edges in the image are acquired later during the scan, a process that takes approximately 2 minutes in total. Using bolus timing and elliptic-centric acquisition, it is possible to perform CE-MRA with very high spatial resolution, usually 0.5 mm in plane with 1–2 mm slice thickness using parallel imaging at 3T. Although this spatial resolution falls short of what can be achieved with CTA, it is more than sufficient to monitor changes in coiled cerebral aneurysms stents (Wallace *et al.*, 2007), without the downside of repeated radiation exposure and artifacts that come with serial CTA.

Using time of flight (TOF) MR angiography, it is also possible to obtain excellent images of the vascular lumen in the intracranial circulation without the requirement for intravenous contrast. The ability to study vessels without contrast using TOF MRA and MRV makes these the dominant techniques used in practice for routine screening of patients with suspected arterial or venous disease. Both exploit the principle of flow-related enhancement, in which the MRI signal from stationary tissue is suppressed and that within the vascular lumen is enhanced by sensitizing the signal to flowing blood. The contrast and spatial resolution of TOF

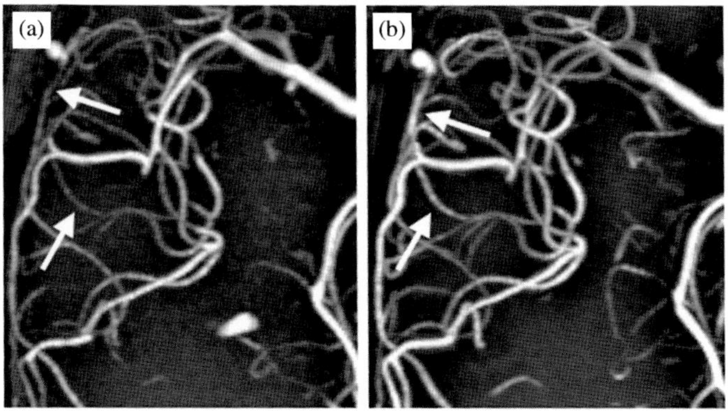

Figure 2. 3T versus 7T time of flight MRA. Moving from 3T (a) to 7T (b) field strength permits better visualization of cortical MCA branches (arrows), as shown on these axial MIP images from a time-of-flight MRA obtained in the same subject at both field strengths. The improved conspicuity of cortical arterial branches reflects the combined effects of greater background suppression and stronger signal from flow-related enhancement at 7T.

MRA and MRV can be further enhanced by using higher field strength, 3T and even 7T, MRI (Figure 2). It is important to note, however, that this method for generating luminal contrast is fundamentally limited when blood flow is slow or turbulent. TOF techniques suffer from significant limitations in this setting, including over-estimation of the degree of narrowing in stenotic disease and failure to accurately depict luminal diameter in complex flow, such as within the carotid bifurcation, giant intracranial aneurysms or arteriovenous malformations (AVMs). In these cases, gadolinium-enhanced CE-MRA is preferred.

In addition to its angiographic capabilities, MRI has higher soft tissue contrast than any other modality, and thus provides the best possible evaluation of the brain parenchyma. In contrast to CT and DSA, it also requires no ionizing radiation. However, MRI requires longer imaging times than CT, often 15–30 minutes for optimal diagnostic quality, which may delay treatment. MRI scanners are also less accessible than CT scanners, and examinations are more costly. Furthermore, patients with severe claustrophobia cannot safely undergo MRI without anesthesia, and MRI is contraindicated in patients with certain implanted devices.

1.1.3 *Catheter-based digital subtraction angiography*

The use of catheter angiography has gradually declined over the last three decades, as the accuracy of non-invasive CT and MRI has improved significantly. Nonetheless, given its inherently high spatial and temporal resolution, it remains the gold standard for evaluation in many vascular disorders. DSA, which differs from the old "cut film" angiography techniques of the past in its temporal subtraction of images to remove bony structures and depict only intravascular contrast, now serves as the *de facto* standard to evaluate lesion angioarchitecture and understand the dynamics of arterial blood supply to and venous drainage from complicated lesions. DSA also uniquely permits endovascular treatment, including the removal of acute occlusive thrombi in stroke and transcatheter treatment of aneurysms, fistulas, and tumor blood supply.

There are a number of disadvantages to DSA. It is an invasive technique that requires direct arterial or venous cannulation, catheter and wire manipulation for vessel selection, and injection of relatively high volumes of contrast media. Large post-procedural groin hematomas, internal bleeding related to vessel puncture, and significant vascular injury requiring surgical repair are extremely rare in experienced hands. However, even with the best technique, there is a small but finite risk of causing transient ischemic attack, stroke, or extremity ischemia when thrombus forms at the tip of the catheter, plaque is dislodged from aortic or other arterial atheroma, or small volumes of air are inadvertently injected. According to a cooperative statement issued by the Society of Interventional Radiology, the American Society of Interventional and Therapeutic Neuroradiology, and the American Society of Neuroradiology, the reported range of reversible neurological deficits with DSA ranges from 0 to 2.3%, and that of permanent neurological deficit, 0 to 5% (Citron *et al.*, 2003). A second drawback of DSA is its requirement for significantly higher doses of radiation than CTA in most cases. Although dose can be reduced using high-sensitivity detectors, image noise reduction techniques, and low frame-rate acquisition, recent estimates for effective radiation dose are 2–7 times higher for cranial DSA than CTA, averaging 2.71 mSv for DSA (by comparison, averaging 0.67 mSv for CTA) (Manninen *et al.*, 2012). Effective DSA doses are even higher in patients undergoing catheter-directed interventions.

Other disadvantages of DSA are its requirement for relatively larger contrast volumes and the inherent limitations that come with planar radiography. The former is especially problematic in patients with diminished renal function at baseline, as large-volume injected arterial contrast may rarely precipitate acute renal failure. Finally, most cerebral angiography is performed using biplane units, permitting only projections in two planes to be obtained during injections and thereby making it impossible to accurately separate overlapping vessels. The 3D structure of the vascular system and vascular lesions is not depicted with biplane angiography. Newer angiographic systems capable of tomographic reconstruction and 3D rendering of vessels by rotating the angiography unit's imaging heads around the patient are becoming more commonly available in clinical practice. These hybrid

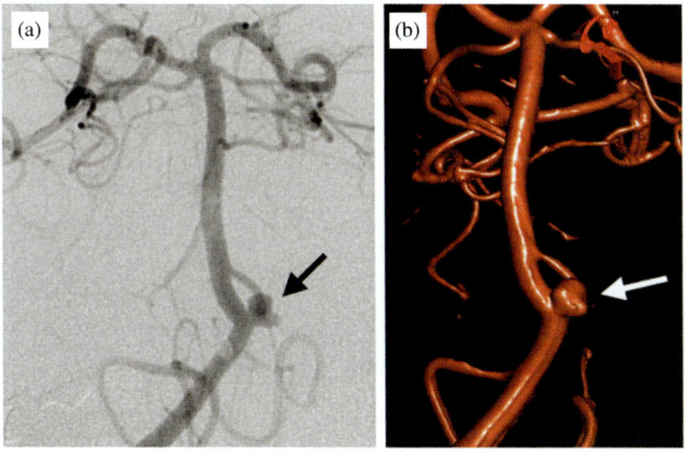

Figure 3. High-resolution projection and tomographic DSA. Standard anteroposterior planar (a) and 3D tomographic reconstruction (b) of an unruptured, bilobed proximal basilar aneurysm obtained using DSA. The latter are reconstructed from multiple planar projections acquired as the imaging camera is rotated around the patient during catheter injection of contrast. The precise morphology of the aneurysm, including size and dimensions of the neck, can be measured from the 3D reconstructions, which can be useful for endovascular or surgical treatment planning.

DSA-tomographic techniques are capable of providing the most accurate, highest resolution 3D renderings of vascular abnormalities currently possible with any modality (Figure 3). They also enable 3D, CT-like images to be obtained in real time during angiographic procedures.

1.2 Dynamic Imaging of Blood Flow

The contrast-enhanced cross-sectional CTA/CTV and CE-MRA techniques discussed to this point aim to acquire a snapshot of the vascular system at a single point in time following the injection of a bolus of contrast. This strategy permits reconstruction of high-quality images of the arterial and/or venous system and identification of a variety of cerebrovascular disorders. In reality, however, it is possible to obtain a host of additional information by studying the temporal passage of the contrast bolus through the circulation. At the macroscopic scale of the level of cranial and cervical vessels, blood flow through the circulation is spatially and temporally complex. Enlargements and narrowings of arteries, networks of collateral arteries, and the presence of shunts between arteries and veins introduce hemodynamic nuances that cannot be adequately described by angiographic snapshots from a single time instant.

As discussed above, DSA is inherently a time-resolved technique, in which serial planar projections are obtained rapidly during the injection of a single vessel, allowing real-time visualization of the transit of contrast through the intracranial circulation in planar projections. Viewing the temporal dynamics of the flow of contrast facilitates ready detection of arterial occlusions and slow flow, collateral pathways, and arteriovenous shunts. DSA thus remains the gold standard for imaging complex intracranial blood flow; its high spatial and temporal resolution make it the most sensitive technique for detecting subtle vascular abnormalities and understanding their impact on blood flow.

Post-processing techniques for quantitative flow evaluation using DSA may provide additional insight into disorders in which normal blood flow is disrupted. So-called quantitative DSA (QDSA) techniques, based upon temporal analysis of injected contrast agents, are now being applied to study the timing of blood flow in various vascular disorders. Because the X-ray attenuation at each pixel on planar DSA projections is related to the concentration of injected contrast media within that pixel, the X-ray attenuation within each pixel varies as the injected contrast passes through the circulation, and over time encodes information about the nature of blood flow in that region. A number of interesting flow parameters, such as time to peak contrast and area under contrast curves, can be derived to illustrate dominant feeding vessels to AVMs (Figure 4) and to measure the changes in flow that come about as a result of catheter-directed interventions.

Extending the same concepts of rapid acquisition of serial snapshots in time to tomographic reconstruction with CT, multiphase CTA techniques have been developed that exploit the rapid acquisition capability of MDCT scanners to obtain temporally consecutive 3D volumes through the brain after bolus contrast administration. For example, Menon *et al.* developed a technique for three-phase CT (peak arterial, equilibrium/peak venous, and late venous) with whole-head coverage and relatively low patient radiation dose (Menon *et al.*, 2015). Although 3 phases is considerably lower than the 20–50 phases typically captured with each DSA injection, multiphase CTA permits visualization of time-resolved images of pial surface collateral arteries in patients with stroke, showing the promise of this technique as a tool for selecting patients presenting with acute stroke for intervention and potentially for predicting their clinical outcome after LVO.

Time-resolved imaging of 3D volumes is also possible using gadolinium-bolus MRA. As noted above, however, there is an inherent tradeoff between spatial and temporal resolution with all MR techniques. High spatial resolution requires

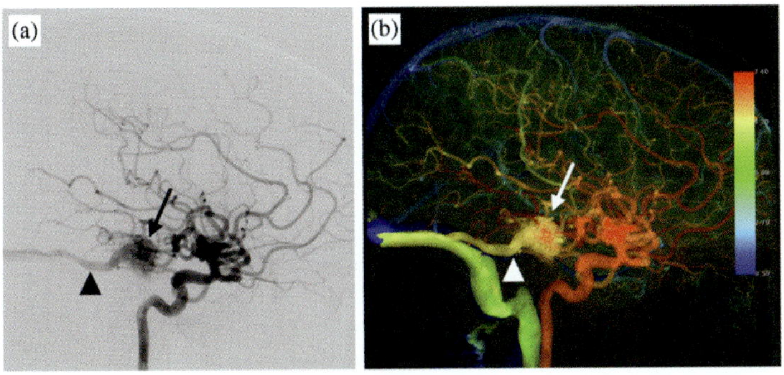

Figure 4. Lateral planar DSA (a) and lateral color-encoded flow map (b) obtained from DSA time series in a patient with AVM. The lateral projection in (a) shows the nidus of the AVM (arrow) as well as an early draining vein (b, arrowhead) as a result of the arteriovenous shunting. The flow map in (b) is colored according to the timing of contrast arrival, with arteries first (red), and the nidus (arrow) and draining vein (arrowhead) later (and thus colored yellow). Images courtesy of Daniel L. Cooke, Department of Radiology and Biomedical Imaging, University of California, San Francisco.

longer imaging times, and high temporal resolution requires short imaging times. However, there is intrinsic redundancy in dynamic data, in that much of the anatomic detail remains the same across the time dimension, and the difference between individual time points reflects only changes in vascular signal intensity due to passage of the contrast. Time-resolved MRA techniques such as Time-Resolved Imaging of Contrast KineticS (TRICKS) and Time-resolved angiography WIth Stochastic Trajectories (TWIST) exploit this intrinsic temporal redundancy to overcome the normal conflict between spatial and temporal resolution, using so-called "view sharing" reconstruction (Hennig *et al.*, 1997; Korosec *et al.*, 1996). These methods lack the high temporal resolution of biplane DSA and are not currently capable of achieving sufficient spatial resolution to accurately define the angioarchitecture of vascular abnormalities like DSA. Nonetheless, dynamic MRA has found application in brain and spine AVM and AVF, dural sinus thrombosis, and other vascular anomalies (Griffiths *et al.*, 2000; Meckel *et al.*, 2008; Vargas *et al.*, 2010). Figure 5 shows an example of how dynamic MRA can show early venous drainage in the setting of AVM-related arteriovenous shunting.

Adding to its capacity to track the passage of a contrast bolus through the intracranial circulation, MRI can furthermore quantify the velocity and direction of intraluminal blood flow using phase-contrast and 4D-flow MRA techniques. Obviating the need for gadolinium contrast like conventional time-of-flight MRA, these approaches rely upon sensitizing the MR acquisition to moving protons within

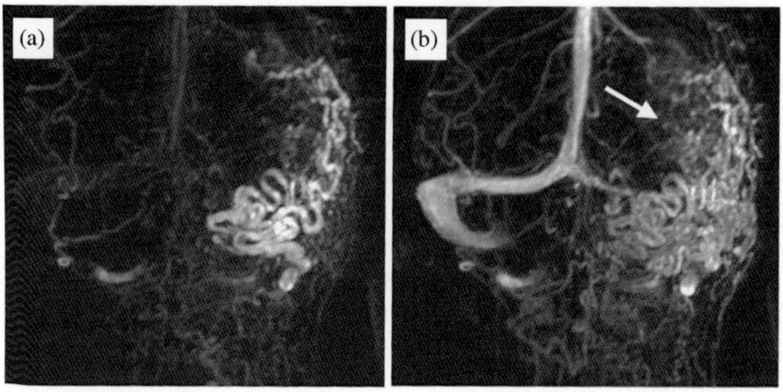

Figure 5. Dynamic MRA of the brain using view-sharing. Coronal projections of dynamic contrast-enhanced 3D TRICKS MRA obtained in a patient with a large left hemispheric AVM. Images are shown in the arterial (a) and venous (b) phases following intravenous gadolinium administration. Note later filling of the venous component of the AVM nidus (arrow, as an example).

a fixed range of velocities. By gating the acquisition to the cardiac cycle, phase-contrast MRA interrogates the time-dependent velocity of flow during cardiac systole and diastole within individual vessels. Such measurements have been used, for example, to characterize the severity of vascular stenoses or feeding artery flow rate and volume to the AVM nidus and shear stress in aneurysms (Boussel *et al.*, 2009). 4D-flow MRA, in contrast, refers to the more recent extension of 2D phase-contrast MRA methods to resolve all three dimensions of space as well as the dimension of time along the cardiac cycle (3D + time = 4D). This feat is made possible using multichannel coils, highly parallel acquisition techniques, and sparse sampling that have only recently been introduced clinically for MRI (Bammer, *et al.*, 2007; Wetzel *et al.*, 2007). Rather than characterize velocity along a single or few directions, 4D-flow MRA also measures velocity along multiple directions and thereby permits measurement of voxel-wise flow rate and direction and visualization of flow within the vessel lumen. Figure 6 shows the quality of images that can be obtained with modern 4D-flow reconstructions to illustrate the complexity of flow within aneurysmal vessels.

1.3 Intracranial Vessel Wall Imaging

Luminal imaging enables the identification of vascular occlusions, aneurysms, and abnormalities in vascular caliber. However, these techniques lack the ability to characterize disorders centered within the vessel wall, such as plaque structure in intracranial atherosclerotic disease (ICAD), inflammatory or infectious vasculitis,

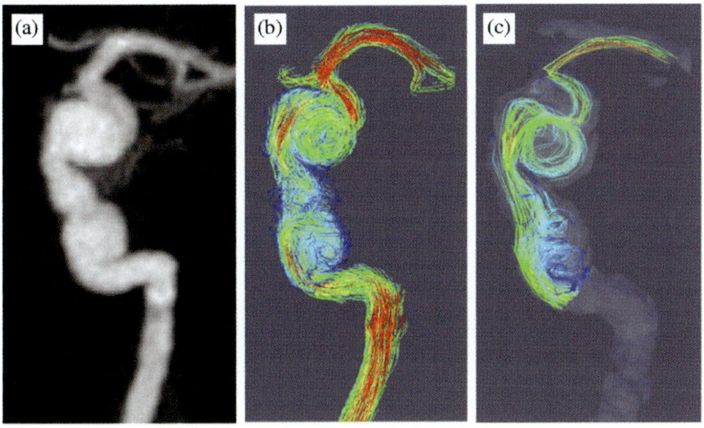

Figure 6. 4D flow MRA in a fusiform dysplastic internal carotid artery aneurysm. The targeted anteroposterior maximum intensity projection in (a) shows the morphology of the aneurysm. For visualization of flow within the aneurysm, the images in (b) and (c) show 3D flow streamlines and pathlines computed from the 4D flow data, respectively. The former illustrates the direction and rate of flow within each point in space at a single time point during the cardiac cycle (in this case, peak systole). The latter depict the theoretical path of flowing blood through the entirety of the cardiac cycle, as calculated by following the course of flow starting at the origin of each of the pathlines. Images courtesy of David Saloner, Department of Radiology and Biomedical Imaging, University of California, San Francisco.

and other vasculopathies. In these scenarios, the only clue that a disorder of the vessel wall is present may be abnormal cerebrospinal fluid (CSF) obtained with lumbar puncture, focal or diffuse disruption in luminal caliber (such as with dissection or primary CNS vasculitis), or the observation of the end effects of altered hemodynamics in the brain, such as infarcts in the territory of involved vessels.

To visualize the vessel wall outside of its lumen, it is necessary to overcome a number of significant technical challenges. Arteries in the circle of Willis are normally only 2–5 mm in diameter, with a combined intimal, medial, and adventitial thickness of 700 μ or less. Furthermore, intracranial arteries are surrounded by CSF, which pulsates at a lower velocity than blood but with the same periodicity, driven by the cardiac cycle. Accurate visualization of the vessel wall additionally requires inherently high contrast between the vessel wall, the pulsating blood that it contains, and the circulating CSF that surrounds it. This is possible at present only using MRI vessel wall imaging (VWI).

The first MRI attempts to target the vessel wall used 2D T1 black-blood, CSF-suppressed techniques (2D T1 FLAIR), with large matrix sizes to achieve

0.5–0.6 mm in-plane spatial resolution with relatively thick 2–3 mm slices oriented in multiple planes around vessels of interest (Schwartz *et al.*, 2009). Long imaging times limited the application of these initial approaches to special cases. However, the more recent implementation of variable-flip angle, fast-spin echo 3D T1 FLAIR techniques (Busse *et al.*, 2006) on MR scanners equipped with multichannel head coils has inspired a new generation of 3D VWI techniques capable of achieving isotropic high spatial resolution (<1 × 1 × 1 mm) within clinically feasible scan times (<8 minutes) (Alexander *et al.*, 2008; Qiao *et al.*, 2011). Images acquired using these methods can be reformatted in any plane, including planes specifically targeted to the axis of the vessel of interest. In this fashion, abnormal mural thickening and/or remodeling are evident in vasculitis, acute stroke, ICAD, and in reversible cerebral vasoconstriction syndrome (RCVS) (Mandell *et al.*, 2012; Xu *et al.*, 2010). With sufficient spatial resolution, it is similarly possible to characterize the internal architecture of plaque in atherosclerosis or mural thrombus lining the walls of larger aneurysms.

The use of T1-weighted MR for VWI also opens the door to using intravenous gadolinium chelates to evaluate for pathological vessel wall enhancement, which has now been found in a number of vascular disorders. Although the reasons for mural enhancement are not entirely understood at this point, putative mechanisms that have been posited include contrast leakage from the lumen through abnormally permeable endothelial cells into the arterial wall, and leakage from disrupted vasa vasorum. Concentric or eccentric arterial enhancement is a consistent feature of inflammatory CNS vasculitis, at the site of arterial occlusion in ischemic stroke, within ruptured aneurysms, and in some atherosclerotic plaques.

By adding high-resolution T2-weighted imaging for multiparametric VWI, the resulting images can be used to accurately differentiate between ICAD, RCVS, and inflammatory vasculitis (Mossa-Basha *et al.*, 2015). VWI may ultimately prove useful as a tool to evaluate for active inflammation within the vessel wall, signaling lesions at greater risk of causing brain injury and therefore defining treatment with anti-inflammatory regimens (Figure 7). Using 7T MRI techniques, the spatial resolution of VWI is further increasing (van der Kolk *et al.*, 2013).

1.4 Imaging of Parenchymal Physiology

Visualizing the lumen, flow, and even the vessel walls of the intracranial vasculature paints a rich but incomplete picture of cerebrovascular disease, as each of these lacks the ability to describe the impact of the disease on the brain's microcirculation or overall hemodynamic status. A number of important cerebrovascular

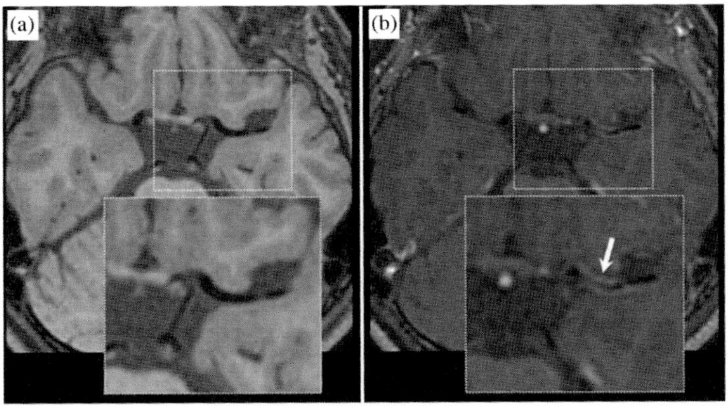

Figure 7. Intracranial VWI with MRI. High-resolution 3D T1 images with inset images focused on the left MCA, acquired before (a) and following (b) the administration of gadolinium. Images show subtle enhancement (arrow, b) of the left MCA in a child presenting with acute left lateral lenticulostriate territory infarcts. Concentric mural enhancement is a feature of transient cerebral arteriopathy of childhood, an inflammatory cause for stroke in this age group.

disorders are angiographically occult in that DSA or cross-sectional vascular imaging techniques are normal even in the face of severe disease. Even when overt abnormalities such as LVOs are evident on catheter, CT, or MR angiography, the impact of these abnormalities on the brain's local perfusion and function and their potential for causing injury to the brain cannot be reliably discerned with these techniques. These tasks lie in the domain of physiologic imaging, using diffusion MRI and PET, CT, and MR perfusion (MRP) imaging.

MR diffusion-weighted imaging (DWI), which measures the average rate of water movement within tissue, or apparent diffusion, is highly sensitive to irreversible ischemic cell death. Within minutes of arterial occlusion, apparent diffusion decreases within infarcted brain tissue (Figure 8). Although the origin of reduced diffusion in stroke has not been clearly defined, it is thought that the dominant contributor is hypoxic failure of the transmembrane Na+/K+ ATPase within the neuronal cell wall. With the failure of this ion pump, normal neuronal osmotic balance is disrupted, and cellular swelling ensues. The consequent reduction in extracellular volume reduces the rate of diffusion. Interestingly, the pattern of cellular swelling has recently been elucidated, and it appears that affected neurons adopt a beaded morphology rather than uniformly across the cell membrane (Budde & Frank, 2010). Importantly, apparent diffusion in infarction varies as a

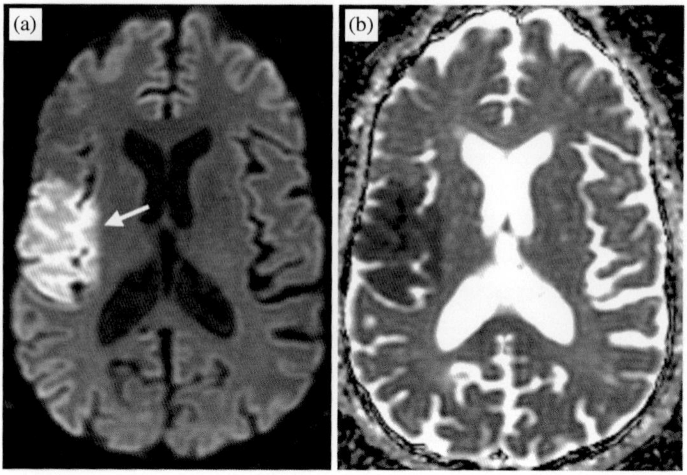

Figure 8. Diffusion-weighted images in ischemic stroke. Diffusion-weighted MR (a) and associated apparent diffusion coefficient map (ADC) (b) in a patient with right MCA infarction. The infarct appears as hyperintense on the diffusion-weighted image (arrow) and hypointense on the corresponding ADC map, as is typical of infarctions in the first 4 days of the injury. Diffusion MRI shows the core of the infarction, the zone of irreversible cell death.

function of the time from the initial onset of ischemia, initially becoming lower and lower until peaking in the 2–4-day window, normalizing compared to non-ischemic tissue at 7–10 days, and then gradually increasing with time until it is higher than normal brain parenchyma (Lansberg *et al.*, 2001).

As a group, perfusion imaging techniques are based upon the serial measurement of an endogenously or exogenously administered tracer over time in order to quantitatively characterize the localized delivery of the tracer to the brain. The acquired measurements are used to estimate hemodynamic parameters within tissue by fitting the acquired data to a mathematical model of how the tracer is distributed within or passes through that tissue. Both the parameters of the infusion technique and the expected pharmacokinetics of the tracer must be included in the model. In the former case, the rate and volume of the contrast bolus are used. In the latter, variables in the model include diffusibility from the intravascular to extravascular space, the volume of distribution of the tracer within tissue, and the equilibrium half-life of the tracer.

Once fit to a time series of data, hemodynamic parameters such as cerebral blood volume (CBV), cerebral blood flow (CBF), mean transit time (MTT), and bolus arrival time are derived from model-based fitting. CBV defines the volume of blood within blood vessels in a given region of the brain and is commonly measured in milliliters per 100 g of tissue. CBF is the rate of delivery of blood to that region and is measured in units of milliliter per minute per 100 g of tissue. MTT refers to the average period of time that blood spends within a given region of brain tissue as it passes from the arterial to the venous circulation, and is commonly measured in seconds. Finally, bolus arrival time refers to the time that it takes for the contrast bolus to arrive at a given region of the brain, also commonly measured in seconds. Many tracers and modalities can be used to obtain these measurements, ranging from ^{15}O-labeled radiotracers with PET, iodinated contrast with CT perfusion, or gadolinium chelates with MR perfusion (MRP) techniques such as dynamic susceptibility contrast (DSC) MRI.

Arterial spin labeling perfusion (ASL perfusion), an alternative method for MRP that is gaining popularity in the clinical setting, also merits discussion. The tracer with ASL is endogenous, consisting of protons in blood that are non-invasively labeled or "tagged" with T1 signal in the upper neck. When these labeled protons enter the brain, they can be recognized by their unique magnetic signature. By subtracting images obtained with and without labeling, the local concentration of labeled protons can be measured and used to calculate CBF. The main drawback of ASL is its inherently low signal-to-noise ratio (SNR), as labeled blood protons comprise only about 1% of the static tissue signal. Also, the technique only permits reliable measurement of CBF, and not CBV or MTT (Diebler, *et al.*, 2008).

The clinical utility of perfusion can best be understood through its application to stroke, and the concept of "ischemic penumbra" (Symon, 2007). When CBF falls below a critical threshold, there is onset of irreversible cellular damage within a short period of time. However, brain tissue that is subjected to milder CBF reductions experiences failure of electrical activity that remains reversible upon the restoration of normal perfusion. The CBF thresholds for irreversible injury depend upon time, as demonstrated in the seminal experiments of Jones *et al.* (Jones *et al.*, 1981). Following the abrupt occlusion of a large artery, irreversible injury can thus be expected after a short time in areas of the brain that experience the greatest degree of CBF reduction. In other areas for which CBF is reduced to a lesser degree, or sustained by blood flow from collateral vessels, the effects of tissue oligemia may be entirely reversible. It is this tissue at risk of infarction that is considered the primary impetus for chemical thrombolysis or mechanical thrombectomy to restore blood flow in the setting of acute LVO.

Because of the widespread availability of MDCT, the most commonly applied imaging method in acute stroke to assess ischemic penumbra has been CTP (Campbell *et al.*, 2012; Konstas *et al.*, 2009a; 2009b). In current practice, brain

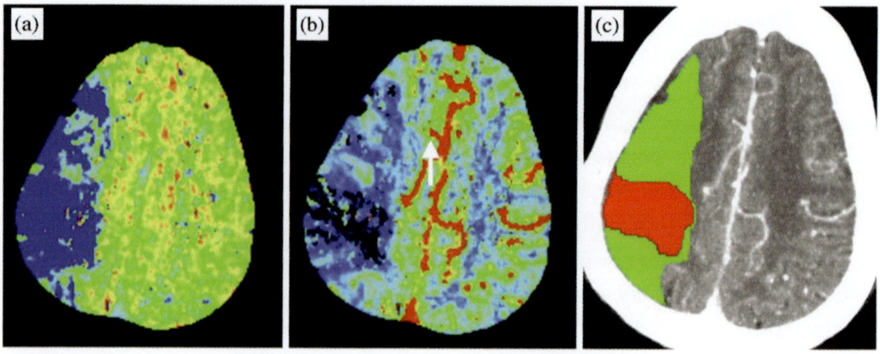

Figure 9. CTP in acute stroke. Selected axial processed CTP images in a patient with acute right MCA territory infarction showing calculated maps of (a) MTT, (b) CBF and (c) estimated infarction volume.. Brain regions with prolonged mean transit and low CBF are shown in blue. Estimates of infarction core and surrounding ischemic penumbra derived from these maps are depicted in red and green, respectively.

regions for which CBF falls below a critical threshold are considered to be irreversibly damaged, representing a core of infarcted tissue that would have no benefit from reperfusion. The surrounding zone of ischemic tissue at risk of infarction due to LVO is calculated from maps of MTT (Figure 9). Although the presence of prolonged transit time does not correspond directly to viable tissue suffering from reversible ischemia, MTT can be accurately quantified using CTP and indirectly correlates to CBF decreases that do result in reversible ischemia. CTP, however, yields inherently noisy measurements, and its accuracy for determining core infarction has been called into question (Schaefer *et al.*, 2015).

As cerebral perfusion pressure is varied, blood supply to the brain is kept relatively constant through a process of compensatory arterial vasodilatation or vasoconstriction that appropriately increases or decreases CBV in order to maintain constant CBF. The effects of chronic arterial narrowing, as with moyamoya or atherosclerotic carotid stenosis, can be overcome to a point through these normal autoregulatory mechanisms. However, when arteries are maximally dilated, further decreases in perfusion pressure place the patient at the risk of developing infarction. The capacity to increase CBV in response to a hemodynamic challenge is referred to as cerebrovascular reserve and can be measured in the individual patient by evaluating for brain perfusion changes when in the setting of a vasodilatory challenge, such as injected acetazolamide or inhaled carbon dioxide (Pillai & Mikulis, 2015). Such methods are increasingly important to guide the institution of therapy in chronic steno-occlusive disease (Gupta *et al.*, 2012).

Blood oxygen level-dependent (BOLD) MRI serves as a complementary technique for assessing cerebrovascular reactivity that should be mentioned. The BOLD effect refers to the increase in MRI T2* signal that occurs as a result of decreases in the deoxy- to oxyhemoglobin ratio within vessels, which occurs when CBF increases within functionally active areas of the brain in response to metabolic demand. Changes in the magnitude of the BOLD signal in the face of a vasodilatory challenge also reflect cerebrovascular reserve, in that brain regions that exhibit no increase in BOLD signal are most likely supplied by blood vessels that are already maximally dilated and therefore offer very limited or no capacity for cerebrovascular reserve (Pillai & Mikulis, 2015).

To conclude this section, it should be noted that techniques have been developed more recently to quantitatively assess oxygen consumption in the brain. The rate of oxygen consumption is referred to as the cerebral metabolic rate of oxygen ($CMRO_2$), an important direct marker for brain function and metabolism. Oxygen is not produced endogenously in the brain and must be delivered locally through the blood supply. Measurements of oxygen extraction fraction (OEF), reflecting the ratio of oxygen consumption to delivery, have traditionally used ^{15}O PET, which suffers from low spatial resolution and the requirement for close proximity to a cyclotron given the short half-life of ^{15}O tracers. MR techniques have been developed over the last several years to measure OEF (Oja *et al.*, 1999), which can be used in the calculation of local $CMRO_2$ (Bolar *et al.*, 2011). However, these techniques require further validation and comparison with PET before they can be applied clinically.

1.5 Anatomic Brain Imaging

Equally important to imaging the vessels and revealing the physiology within the brain is detecting the effects of cerebrovascular disease on the brain parenchyma. Vasogenic edema, mass effect, and hemorrhage are examples of important clinical endpoints that may be occult on angiographic and physiologic images, yet critical to management of disease. Furthermore, multiple primary vascular disorders, including cerebral cavernous malformations, capillary telangiectasias, lacunar infarcts, small-vessel vasculitis, and thrombotic microangiopathy, are depicted well only on anatomic images of the brain. These involve small vessels beyond the resolution of conventional angiography and cause only subtle changes in underlying hemodynamic function.

As noted previously, combined CT/CTA (sometimes with CTP) is used at most institutions for patients with suspected acute cerebrovascular disease. Unenhanced CT provides sufficient anatomic contrast to identify most acute neurological emergencies, such as hemorrhage, hydrocephalus, or herniation. CTA further increases sensitivity to LVO, ruptured aneurysms, arteritis, dural sinus thrombosis, and brain AVM. CTP adds interpretive value regarding the hemodynamic effects of LVO on brain perfusion. However, when diagnostic evaluation is elective or less urgent, MRI remains the modality of choice.

For optimal anatomic imaging of vascular disease, multicontrast anatomic MRI including T1, T2 and/or T2 fluid-sensitive inversion recovery (FLAIR), diffusion and susceptibility sensitive sequences are far more sensitive to parenchymal abnormalities than CT. T1, High-quality T2 and T2 FLAIR images are used to identify vascular masses, as well as any associated mass effect and vasogenic edema. Even when the underlying vascular lesion is occult on these anatomic images, these findings often provide clues as to the vascular origin of disease. Given its sensitivity to acutely ischemic tissue, as discussed above, diffusion imaging is also routinely performed in both axial and coronal planes. Finally, a susceptibility-sensitive, T2*-weighted sequence, such as T2* gradient echo or susceptibility weighted imaging (SWI), is included in order to increase sensitivity to the presence of blood products within the brain parenchyma or extra-axial spaces (Haacke *et al.*, 2009, Mittal *et al.*, 2009).

The presence of blood products on MRI serves as an important clue as to the presence of a vascular lesion on susceptibility sensitive images. Acute clot can be identified as blooming susceptibility on T2* MRI within occluded arteries. Furthermore, even minute quantities of extravasated blood are visible with current T2*-weighted imaging. Cerebral amyloid angiopathy (CAA) provides an excellent example for the application of these techniques. In this condition, abnormal aggregations of amyloid protein build up in the walls of small arteries in the brain, rendering them prone to leakage and increasing the risk of lobar hemorrhage. Silent microhemorrhages and pial siderosis from CAA are readily detected using susceptibility sensitive techniques and serve an important role in the antemortem diagnosis of this disorder (Knudsen *et al.*, 2001; Linn *et al.*, 2010). Similarly, it has recently been noted pathologically that some treatment-naïve brain AVMs exhibit surrounding blood products as evidence of old hemorrhage (EOOH), even when patients have no clinical history of symptomatic AVM rupture (Guo *et al.*, 2012). T2*-weighted imaging is sensitive and can detect this "silent" microhemorrhage, which is thought to predict the risk of future hemorrhage, and therefore it can be useful to suggest the need for more invasive treatment. Figure 10 shows examples of CAA and EOOH on susceptibility-sensitive MRI.

Because deoxyhemoglobin is a paramagnetic substance, deoxygenated blood within venous structures stands out on susceptibility-sensitive images, especially SWI. As a result, parenchymal veins become more conspicuous when they are abnormally engorged or contain blood with a greater concentration of deoxyhemoglobin.

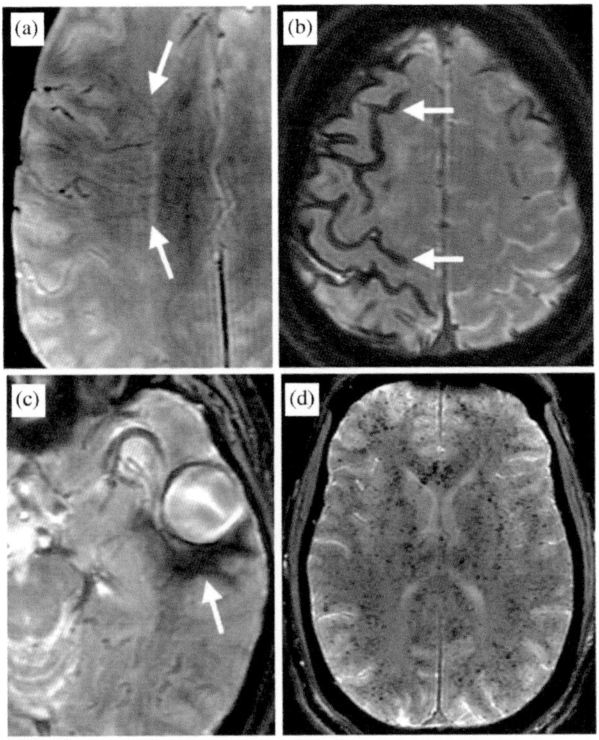

Figure 10. Susceptibility sensitive MRI techniques in cerebrovascular disease. T2*-weighted images in four different patients. Right MCA infarction (same patient as in Figure 9) associated with prominent venous structures draining the territory of the infarction, consistent with elevated OEF in (a). In (b), multiple gyral surfaces have low signal consistent with pial staining (siderosis), in this patient due to CAA. The low signal around an AVM in (c) reflects the sequela of remote hemorrhage. Finally, multiple punctate foci of low T2* scattered throughout the brain in (d) in a patient with thrombotic microangiopathy.

The presence of an arteriovenous fistula, for example, may render veins within the involved venous territory more prominent on SWI. As another example, the "brush sign" on SWI as recently described by Morita *et al.*, 2008) refers to asymmetric prominence of the deep medullary venous system draining brain areas affected by ischemic stroke (Figure 10). This observation suggests increased deoxyhemoglobin within the areas of involved perfusion and may portend hemorrhagic transformation after thrombolytic therapy (Terasawa *et al.*, 2014).

As a final note on the subject of anatomic brain imaging, a few additional notes regarding the use of intravenous contrast media should be mentioned. The use of iodinated contrast in CTA and CTP, and of gadolinium chelates for some types of

MRA and MRP acquisition, have been discussed at previous points in this chapter. However, contrast for both CT and MRI carries additional value in its ability to detect disruption of the normal blood–brain barrier. Contrast media can be useful in dating ischemic stroke, as the typical onset of parenchymal enhancement is in the 5–7-day range. Severe venous hypertension, as present in some dural fistulas, may also lead to parenchymal enhancement. Ferumoxytol, a colloid containing supermagnetic iron oxide, also shortens T1 and can be used in lieu of gadolinium as an exogenous contrast agent. In fact, the degree of T1 shortening with this agent is stronger than gadolinium chelates, making the agent excellent for depiction of vascular lesions (Dósa, *et al.*, 2011). After initial injection, ferumoxytol remains in the vascular system for a much longer period than gadolinium and heightens sensitivity to small vascular lesions. Although initially confined to the vascular compartment, delayed ferumoxytol T2* imaging can also take advantage of intracellular uptake of the nanoparticles to image inflammation (Storey & Arbini, 2014). This method for imaging inflammation has been applied to vascular abnormalities such as aneurysms and AVMs, under the premise that active inflammation within these lesions may predispose to their instability and increase the risk of rupture (Hasan *et al.*, 2012).

1.6 Summary

A number of powerful imaging tools are now at the disposal of the clinician and scientist for the evaluation of vascular lesions. Together, multiple modalities, especially CT, MR, and DSA, are capable of interrogating the vessel lumen and wall, visualizing the dynamic flow of blood through the arterial and venous cerebral circulation, and quantifying the physiology of blood delivery to the brain. Far from "one size fits all," the performance of these different techniques depends on the individual patient, the underlying vascular abnormality in question, and the potential impact of the lesion on the brain. In many cases, more than one modality may be necessary for comprehensive evaluation. A number of technical advances in image acquisition and analysis on the near horizon will not only enhance the accuracy of existing techniques, but also add quantitative and predictive power for the evaluation of cerebrovascular disease and the selection of appropriate treatments.

Acknowledgments

This research is supported by a grant from the U.S. National Institutes of Health (U54 NS065705).

References

Alexander, A.L., Buswell, H.R., *et al.* (1998). Intracranial black-blood MR angiography with high-resolution 3D fast spin echo. *Magn Reson Med* **40**(2): 298–310.

Bammer, R., Hope, T.A., *et al.* (2007). Time-resolved 3D quantitative flow MRI of the major intracranial vessels: Initial experience and comparative evaluation at 1.5T and 3.0T in combination with parallel imaging. *Magn Reson Med* **57**(1): 127–140.

Bolar, D.S., Rosen, B.R., *et al.* (2011). QUantitative Imaging of eXtraction of oxygen and TIssue consumption (QUIXOTIC) using venular-targeted velocity-selective spin labeling. *Magn Reson Med* **66**(6): 1550–1562.

Boussel, L., Rayz, V., *et al.* (2009). Phase-contrast magnetic resonance imaging measurements in intracranial aneurysms *in vivo* of flow patterns, velocity fields, and wall shear stress: Comparison with computational fluid dynamics. *Magn Reson Med* **61**(2): 409–417.

Brenner, D.J., Hall, E.J. (2007). Computed tomography — an increasing source of radiation exposure. *N Engl J Med* **357**(22): 2277–2284.

Budde, M.D., Frank, J.A. (2010). Neurite beading is sufficient to decrease the apparent diffusion coefficient after ischemic stroke. *Proc Nat Acad Sci USA* **107**(32): 14472–14477.

Busse, R.F., Hariharan, H., *et al.* (2006). Fast spin echo sequences with very long echo trains: Design of variable refocusing flip angle schedules and generation of clinical T2 contrast. *Magn Reson Med* **55**(5): 1030–1037.

Campbell, B.C., Christensen, S., *et al.* (2012). Comparison of computed tomography perfusion and magnetic resonance imaging perfusion-diffusion mismatch in ischemic stroke. *Stroke* **43**(10): 2648–2653.

Citron, S.J., Wallace, R.C., *et al.* (2003). Quality improvement guidelines for adult diagnostic neuroangiography. Cooperative study between ASITN, ASNR, and SIR. *J Vasc Interv Radiol* **14**(9 Pt 2): S257–S262.

Diebler, A.R., Pollack, J.M., *et al.* (2008). Arterial spin-labeling in routine clinical practice, part 1: Technique and artifacts. *AJNR Am J Neuroradiol* **29**(7): 1228–1234.

Dósa, E., Tuladhar, S., *et al.* (2011). MRI using ferumoxytol improves the visualization of central nervous system vascular malformations. *Stroke* **42**(6): 1581–1588.

Griffiths, P.D., Hoggard, N., *et al.* (2000). Brain arteriovenous malformations: Assessment with dynamic MR digital subtraction angiography. *AJNR Am J Neuroradiol* **21**(10): 1892–1899.

Guo, Y., Saunders, T., *et al.* (2012). Silent intralesional microhemorrhage as a risk factor for brain arteriovenous malformation rupture. *Stroke* **43**(5): 1240–1246.

Gupta, A., Chazen, J.L., *et al.* (2012). Cerebrovascular reserve and stroke risk in patients with carotid stenosis or occlusion: A systematic review and meta-analysis. *Stroke* **43**(11): 2884–2891.

Haacke, E.M., Mittal, S., *et al.* (2009). Susceptibility-weighted imaging: Technical aspects and clinical applications, part 1. *AJNR Am J Neuroradiol* **30**(1): 19–30.

Hasan, D.M., Amans, M., *et al.* (2012). Ferumoxytol-enhanced MRI to image inflammation within human brain arteriovenous malformations: A pilot investigation. *Transl Stroke Res* **3**(Suppl 1): 166–173.

Hennig, J., Scheffler, K., *et al.* (1997). Time-resolved projection angiography after bolus injection of contrast agent. *Magn Reson Med* **37**(3): 341–345.

Jones, T.H., Morawetz, R.B., *et al.* (1981). Thresholds of focal cerebral ischemia in awake monkeys. *J Neurosurg* **54**(6): 773–782.

Knudsen, K.A., Rosand, J., *et al.* (2001). Clinical diagnosis of cerebral amyloid angiopathy: Validation of the Boston criteria. *Neurology* **56**(4): 537–539.

Konstas, A.A., Goldmakher, G.V., *et al.* (2009a). Theoretic basis and technical implementations of CT perfusion in acute ischemic stroke, Part 1: Theoretic basis. *AJNR Am J Neuroradiol* **30**(4): 662–668.

Konstas, A.A., Goldmakher, G.V., *et al.* (2009b). Theoretic basis and technical implementations of CT perfusion in acute ischemic stroke, Part 2: Technical implementations. *AJNR Am J Neuroradiol* **30**(5): 885–892.

Korosec, F.R., Frayne, R., *et al.* (1996). Time-resolved contrast-enhanced 3D MR angiography. *Magn Reson Med* **36**(3): 345–351.

Lansberg, M.G., Thijs, V.N., *et al.* (2001). Evolution of apparent diffusion coefficient, diffusion-weighted and T2-weighted signal intensity of acute stroke. *AJNR Am J Neuroradiol* **22**(4): 637–644.

Linn, J., Halpin, A., *et al.* (2010). Prevalence of superficial siderosis in patients with cerebral amyloid angiopathy. *Neurology* **74**(17): 1346–1350.

Mandell, D.M., Matouk, C.C., *et al.* (2012). Vessel wall MRI to differentiate between reversible cerebral vasoconstriction syndrome and central nervous system vasculitis: Preliminary results. *Stroke* **43**(3): 860–862.

Manninen, A.L., Isokangas, J.M., *et al.* (2012). A comparison of radiation exposure between diagnostic CTA and DSA examinations of the cranial and cervicocerebral vessels. *AJNR Am J Neuroradiol* **33**(11): 2038–2042.

Meckel, S., Glucker, T.M., *et al.* (2008). Display of dural sinuses with time-resolved, contrast-enhanced three-dimensional MR venography. *Cerebrovasc Dis* **25**(3): 217–224.

Menon B.K., d'Esterre, C.D., *et al.* (2015). Multiphase CT angiography: A new tool for the imaging triage of patients with acute ischemic stroke. *Radiology* **275**(2): 510–520.

Mittal, S., Wu, Z., *et al.* (2009). Susceptibility-weighted imaging: Technical aspects and clinical applications, part 2. *AJNR Am J Neuroradiol* **30**(2): 232–252.

Morita, N., Harada, M., *et al.* (2008). Ischemic findings of T2*-weighted 3-tesla MRI in acute stroke patients. *Cerebrovasc Dis* **26**(4): 367–375.

Mossa-Basha, M., Hwang, W.D., *et al.* (2015). Multicontrast high-resolution vessel wall magnetic resonance imaging and its value in differentiating intracranial vasculopathic processes. *Stroke* **46**(6): 1567–1573.

Oja, J.M., Gillen, J.S., *et al.* (1999). Determination of oxygen extraction ratios by magnetic resonance imaging. *J Cereb Blood Flow Metab* **19**(12): 1289–1295.

Pillai, J.J., Mikulis, D.J. (2015). Cerebrovascular reactivity mapping: An evolving standard for clinical functional imaging. *AJNR Am J Neuroradiol* **36**(1): 7–13.

Qiao, Y., Steinman, D.A., *et al.* (2011). Intracranial arterial wall imaging using three-dimensional high isotropic resolution black blood MRI at 3.0 Tesla. *J Magn Reson Imaging* **34**(1): 22–30.

Schaefer, P.W., Souza, L., *et al.* (2015). Limited reliability of computed tomographic perfusion acute infarct volume measurements compared to diffusion-weighted in anterior circulation stroke. *Stroke* **46**(2): 419–424.

Smith-Bindman, R., Lipson, J., *et al.* (2009). Radiation dose associated with common computed tomography examinations and the associated lifetime attributable risk of cancer. *Arch Intern Med* **169**(22): 2078–2086.

Storey, P., Arbini, A.A. (2014). Bone marrow uptake of ferumoxytol: A preliminary study in healthy human subjects. *J Magn Reson Imaging* **39**(6): 1401–1410.

Swartz, R.H., Bhuta, S.S., *et al.* (2009). Intracranial arterial wall imaging using high-resolution 3-Tesla contrast-enhanced MRI. *Neurology* **72**(7): 627–634.

Symon, L. (2007). The ischemic penumbra–the beginning, Chapter 1. In: Donnan, G.A., Baron, J.C., *et al.*, (eds). *The Ischemic Penumbra: Pathophysiology, Imaging and Therapy.* New York: Informa Healthcare.

Terasawa, Y., Yamamoto, N., *et al.* (2014). Brush sign on 3-T T2*-weighted MRI as a potential predictor of hemorrhagic transformation after tissue plasminogen activator therapy. *Stroke* **45**(1): 274–276.

van der Kolk, A.G., Hendrikse, J., *et al.* (2013). Multi-sequence whole-brain intracranial vessel wall imaging at 7.0 tesla. *Eur Radiol* **23**(11): 2996–3004.

Vargas, M.I., Nguyen, D., *et al.* (2010). Dynamic MR angiography (MRA) of spinal vascular diseases at 3T. *Eur Radiol* **20**(10): 2491–2495.

Wallace, R.C., Karis, J.P., *et al.* (2007). Noninvasive imaging of treated cerebral aneuryms, part I: MR angiographic follow-up of coiled aneurysms. *AJNR Am J Neuroradiol* **28**(6): 1001–1008.

Wetzel, S., Meckel, S., *et al.* (2007). *In vivo* assessment and visualization of intracranial arterial hemodynamics with flow-sensitized 4D MR imaging at 3T. *AJNR Am J Neuroradiol* **28**(3): 433–438.

Xu, W.H., Li, M.L., *et al.* (2010). *In vivo* high-resolution MR imaging of symptomatic and asymptomatic middle cerebral artery atherosclerotic stenosis. *Atherosclerosis* **212**(2): 507–511.

Cell Mechanisms and Clinical Targets in Stroke

2

*Jing Lan**,†, *Elga Esposito**, *MingMing Ning**, *Xunming Ji**,†, *and Eng H. Lo**

**Neuroprotection Research Laboratory, Departments of Radiology and Neurology, Massachusetts General Hospital and Harvard Medical School, Boston, Massachusetts, USA*
†*Department of Neurology, Xuanwu Hospital, Capital Medical University, Beijing, China*

The pursuit of neuroprotection in stroke has been a challenging undertaking, as many clinical trials have not achieved clear positive results. Recent studies now suggest that "pure" neuroprotection, i.e., preventing neuronal death alone, may not be sufficient. In the context of the neurovascular unit, rescuing function and restoring homeostasis across all neuronal, glial, and vascular compartments will be required. In this chapter, we summarize translational challenges in neuroprotection, survey recent concepts in cell–cell signaling, and briefly explore examples of clinical targets that might be addressed with reverse translation.

2.1 Neuroprotection

Stroke continues to be a major cause of mortality and morbidity, and with a rapidly aging population, it is certain that the incidence of stroke will increase worldwide. Because stroke and cerebrovascular diseases also comprise a major contributing factor to cognitive impairment and dementia, the cumulative cost and burden to society would be astronomical (Gorelick *et al.*, 2011).

Neuroprotection has been an elusive goal not only for stroke but for almost all central nervous system (CNS) disorders. The initial major pursuit for neuroprotection after stroke was based on the concept of excitotoxicity (Lo *et al.*, 2003). After cerebral ischemia, loss of blood supply, glucose and oxygen disrupts brain metabolism, leading to neuronal depolarization, uncontrolled release of the excitatory neurotransmitter glutamate, and overactivation of NMDA, AMPA, and metabotropic glutamate receptors, which ultimately result in ionic imbalance and neuronal death (Lipton, 1999). Thus, blockade of glutamate receptors was thought to be a rational target. Unfortunately, many clinical trials testing glutamate receptor antagonist did not achieve positive outcomes.

Another seemingly reasonable target in neuroprotection was based on the overall idea of oxidative stress. After stroke, energetic loss, and ischemic injury perturb mitochondrial function, reactive oxygen and nitrogen species are generated, and these damaging radicals are then thought to damage the neuron from within (Chan, 2001). The development of various free radical scavengers was therefore pursued; ultimately, however, these drugs did not work in randomized clinical trials for stroke patients (Shuaib *et al.*, 2007).

More recent studies, at least in some experimental model systems, indicate that cerebral ischemia may also lead to an upregulation of various programmed cell death pathways, e.g., apoptosis, autophagy, and necroptosis (Yuan & Yankner, 2000). In neuronal cell culture models of oxygen–glucose deprivation and animal models of focal cerebral ischemia, blockade of these molecular cell death mechanisms appears to significantly reduce neuronal injury. However, whether these neuroprotection approaches will work in human stroke patients remains to be proven.

The lack of effective neuroprotection has been widely analyzed and debated. In the early trials testing glutamate receptor antagonists, the treatment time windows might have been too long, and patient populations possibly included many "non-responders" as well as suboptimal stroke etiologies. It has also been suggested that some of the neuroprotective compounds that were tested did not effectively penetrate the blood–brain barrier (Dehouck *et al.*, 2002), and more recently, that the statistical approaches and primary endpoints typically used (dichotomized chi-squared analyses of 3-month modified Rankin scores) might not have been sensitive enough (Bath *et al.*, 2012).

2.2 Reperfusion

The lack of effective translation nevertheless does not mean that basic molecular and cellular mechanisms are incorrect or irrelevant (Fisher, 2013; Lo, 2014).

In carefully selected patients with acute ischemic stroke, rapid thrombolysis with tissue plasminogen activator is a proven FDA-approved approach because reversing the initial insult by restoring blood flow is an "upstream" method with large effects in the right patients (Campbell *et al.*, 2015). However, with therapies that seek to block "downstream" molecular mechanisms of cell death or neuroinflammation, these effects may be more nuanced and subtle, and hence more difficult to detect. The search for widely applicable and effective treatments for diverse patient populations remains an urgent and unmet need.

Recent clinical trials now suggest that mechanical reperfusion with thrombectomy devices may also be effective in carefully selected patient populations (Yarbrough *et al.*, 2015). However, some analyses suggest that there remains a 20–30% mismatch between effective arterial opening and significant improvement in neurological outcome scores (Grotta & Hacke, 2015). Furthermore, these patients are already in a tertiary-care hospital setting, and the 60-to-90-minute time window before successful recanalization represents an opportunity to infuse drugs (Menon *et al.*, 2015). Perhaps most importantly, these advances in mechanical thrombectomy allow a combination of controlled reperfusion and infusion of a putative neuroprotectant.

2.3 Neurovascular Unit

Although the field typically refers to "neuroprotection," it is implicitly understood that the goal is to save not just neurons but all cell types in the brain, including glial and vascular cells. Throughout the past decade or so, the concept of the neurovascular unit has provided a conceptual framework for dissecting mechanisms and pursuing this goal (Lo *et al.*, 2003; 2004). Saving neurons from cell death will preserve the brain's ability to fire action potentials. However, in order for the entire brain to function, neurons also need to work in coordination with many other cell types. Proper neurotransmission requires a balance between neuronal release and glial reuptake mechanisms. Blood–brain barrier function requires crosstalk among the cerebral endothelium, pericytes, and astrocytic endfeet. Renewal and homeostasis in white matter require crosstalk between endothelial cells and oligodendrocyte precursors. Even the revolutionary functional MRI method itself is predicated on crosstalk between neuronal activity and hemodynamic response. Similarly, dysfunction and disease are also mediated by crosstalk among all cell types in the brain. Hence, the pathophysiology of stroke (and perhaps all CNS disorders) cannot be effectively investigated without careful consideration of the entire neurovascular unit (Guo & Lo, 2009).

Recent attempts to search for stroke therapeutics have begun to take the neurovascular unit concept into consideration. For example, the SUR1-targeting compound glibenclamide may affect both neurons and astrocytes, thus potentially achieving neuroprotection as well as ameliorating edema (Simard *et al.*, 2014). Another potential approach involves the use of the homeostatic blood protease activated protein C (APC). New data now suggest that modified forms of APC may have simultaneous multifactorial beneficial actions on endothelium, astrocytes, and neurons, thus preserving the entire neurovascular unit after stroke (Griffin *et al.*, 2015). Although there are no guarantees, the ability to rescue multiple cell types in the neurovascular unit may increase the odds of finding effective therapies for stroke.

2.4 Biphasic Penumbra

The ischemic penumbra provides an important conceptual target in stroke (Heiss & Graf, 1994). In theory, low levels of perfusion may still be sustained by collateral supply in areas surrounding the severely damaged cores of a focal stroke. However, these penumbral areas are known to be "meta-stable," and without proper treatment most penumbral tissue will collapse and convert into irreversibly damaged core (von Bornstadt *et al.*, 2015). Diffusion–perfusion mismatch MRI and combinations of PET scans designed to measured blood flow and oxygen/glucose metabolism have all provided support for this penumbral concept. Viable tissue can be detected in stroke patients during early-stage post-onset, and some areas can be rescued with reperfusion. Indeed, some clinical trials have sought to use imaging techniques as an inclusion criterion to find patients with salvageable penumbra.

Recent findings suggest, however, that the penumbra may be more complex than initially thought. Injury and repair mechanisms may be triggered simultaneously, and it is now suspected that the penumbra not only dies but also actively tries to recover (Lo, 2008). The mammalian brain possesses remarkable endogenous mechanisms for compensation, plasticity, and remodeling; therefore, many mediators in stroke pathophysiology may display biphasic properties, i.e., damaging under some conditions but beneficial under other conditions. For example, overactivation of NMDA receptors kills neurons via excitotoxicity during acute stroke, but properly regulated NMDA signaling contributes to recovery when patients transition to rehabilitation (Lo, 2008). The same may be true for other cell death mechanisms such as oxidative stress and programmed cell death. Uncontrolled generation of high levels of radicals is damaging, but homeostatic levels of some radical species are required for angiogenesis and endothelial progenitor cell function

(Ushio-Fukai, 2006). Programmed cell death pathways are also essential for maintaining functional cell subsets, and deletion of autophagy genes induces neurodegeneration in some model systems (Komatsu *et al.*, 2006). Even neuroinflammation displays biphasic properties: M1-like microglia and macrophages are damaging, whereas M2-like subsets promote tissue cleanup and remodeling (Hu *et al.*, 2015).

Hence, targeting the penumbra in stroke, and perhaps more broadly the treatment of any CNS disorder, will require careful attention to complex and shifting substrates and signals as the brain transitions from initial injury to subsequent modes of neurovascular recovery (Lo, 2010). Blocking every targ*et al*one without nuance may no longer be effective.

2.5 Cofactors and Comorbidities

The concept of the neurovascular unit emphasizes the need to investigate crosstalk among neuronal, glial, and vascular compartments after stroke (Lo *et al.*, 2003). Because recent studies now acknowledge that the brain is not an "isolated organ," an "extended" neurovascular unit has been proposed (Lo, 2014). Beyond the CNS, interactions with peripheral responses from immune, hormonal, and cardiac systems may also be critically important. The spleen, bone marrow, and circulating immune cells are all activated after stroke (Iadecola & Anrather, 2011). Changes in the brain are correlated with changes in the heart (Ishikawa *et al.*, 2013).

Within this context of central–peripheral crosstalk, the complex issue of how to account for the many cofactors and comorbidities often associated with stroke patients has been raised. GWAS studies suggest that many stroke genes represent cardiovascular pathways (Markus & Bevan, 2014). Hypertension and diabetes have long been known to be significant factors in stroke (Mozzaffarian *et al.*, 2016). In fact, many of these modifying factors not only alter stroke risk, but also significantly contribute to the acute process of brain injury and the delayed process of brain repair. It has been suggested, for example, that diabetic vascular systems suffer from impaired angiogenic programs that may influence stroke recovery (Ergul *et al.*, 2015).

From the experimental side, model systems will have to be adapted to incorporate these important modifying factors. From the clinical side, stroke trials may have to more carefully identify and separate patient cohorts depending on how the mechanisms being targeted are affected by various risk factors. Ultimately, a systems biology approach would be helpful. For example, mapping the transcriptome of reactive astrocytes should offer insight into how deleterious versus pro-recovery gliosis can be regulated for therapeutic benefit (Zamanian *et al.*, 2012). Recent attempts to map the brain vasculome should also provide a rich database for dissecting new mechanisms and stroke targets in susceptible patient populations (Guo *et al.*, 2012, 2016).

2.6 Translational Opportunities

In spite of the many challenges, translation in stroke is still an exciting goal. An enormous clinical need exists, and there are now many opportunities to test novel approaches. Combination therapies can be revisited in the context of carefully selected patients undergoing mechanical thrombectomy. Pharmacologic rescue, together with reperfusion, should yield greater odds for success. Finally, stroke genetics and remote conditioning are currently offering new and hopeful insights for translational studies.

2.6.1 *Stroke genetics*

This emerging field focuses on the potential roles of endogenous signals that may assist CNS remodeling during stroke recovery. In experimental models, reactive astrocytes release multiple factors that promote recovery, including HMGB1 that attracts circulating endothelial progenitors for angiogenesis (Hayakawa *et al.*, 2012), and thrombospondin-1 that promotes synaptic plasticity (Christopherson *et al.*, 2005; Liauw *et al.*, 2008). Another pro-recovery candidate is the growth factor BDNF that can be expressed in both reactive astrocytes (Miyamoto *et al.*, 2015) and the brain vasculome (Guo *et al.*, 2012). Recent investigations into human BDNF signals are facilitating inquiry into the possibility that variations in BDNF-related genetics can potentially separate responders and non-responders in stroke recovery trials. The BDNF Val66Met polymorphism was associated with the 90-day poor outcome of ischemic stroke in Chinese patients (Zhao *et al.*, 2013), and 2-week and 1-year poor outcomes in Korean patients (Kim *et al.*, 2012). However, a different study was unable to detect any significant association between Val66Met polymorphism and 3-month stroke recovery in a Swedish population (Stanne *et al.*, 2014). Nevertheless, the BDNF polymorphism can be applied not only to stroke prognosis but also to treatment and rehabilitation decisions. Even though patients are able to adapt their walking pattern after practice, regardless of whether they have the implicated BDNF polymorphisms or not, the presence of the polymorphism may still influence the rate of improvement (Helm *et al.*, 2016). The BDNF genotype may also individualize the type and magnitude of aerobic exercise to enhance rehabilitation in stroke survivors (Mang *et al.*, 2013). Further studies are required to carefully dissect the role of these BDNF signals in stroke recovery patients, but these emerging data provide proof-of-principle that translational opportunities exist in linking genetic data with both experimental studies and clinical trials.

2.6.2 *Remote conditioning*

Another promising opportunity involves accumulating observations regarding remote conditioning. Hougaard *et al.* (2014) tested if remote preconditioning (rPerC) can treat acute ischemic stroke together with rtPA. Although there was no significance in penumbral salvage, tissue survival analysis suggested that rPerC was neuroprotective. Besides acute ischemic stroke, remote conditioning has also been tested on recurrent stroke in intracranial arterial stenosis (IAS). Meng *et al.* found that bilateral arm ischemic preconditioning (BAIPC) for 300 consecutive days potentially reduces the incidence of recurrent stroke and accelerates recovery after stroke, not only in patients under 80 years of age (Meng *et al.*, 2012) but also in octo- and nonagenarians (Meng *et al.*, 2015).

Because remote conditioning has shown promise for the treatment and prevention of ischemic stroke, clinical applications are currently being considered. For remote conditioning in acute ischemic stroke, more than 60 hospitals in China are recruiting patients to confirm if BAIPC will safely reduce the ischemic stroke recurrence rate (Hou *et al.*, 2016); the EPIC-sICAS trial (Wei *et al.*, 2016) is examining the effect of rPerC on the recovery of nerve function and collateral circulation in stroke patients. For remote preconditioning in penumbral salvage, another study on rPerC (Pico *et al.*, 2016), which has important design differences compared with the Hougaard study (Hougaard *et al.*, 2014), is recruiting patients from five French hospitals to see if rPerC can be neuroprotective in a different, defined patient cohort. Results from these trials will determine the translational relevance of these new approaches.

2.7 Summary

Stroke represents a challenging translational problem in part because: (1) it triggers rapidly evolving injury mechanisms with sometimes narrow time windows; (2) it involves many cell types within as well as outside the CNS; (3) it is influenced by multiple cofactors and comorbidities; and (4) it involves sick patients with extremely heterogeneous etiologies and backgrounds. Ultimately, effective translation will require rigorous attempts to define and link mechanisms across all cell, animal, and human model systems (Lopez *et al.*, 2015; Ning *et al.*, 2011; 2013).

References

Bath, P.M., Lees, K.R., *et al.* (2012). Statistical analysis of the primary outcome in acute stroke trials. *Stroke* **43**(4): 1171–1178.

Campbell, B.C., Meretoja, A., *et al.* (2015). Twenty-year history of the evolution of stroke thrombolysis with intravenous alteplase to reduce long-term disability. *Stroke* **46**(8): 2341–2346.

Chan, P.H. (2001). Reactive oxygen radicals in signaling and damage in the ischemic brain. *J Cereb Blood Flow Metab* **21**(1): 2–14.

Christopherson, K.S., Ullian, E.M., *et al.* (2005). Thrombospondins are astrocyte-secreted proteins that promote CNS synaptogenesis. *Cell* **120**(3): 421–433.

Dehouck, M.P., Cecchelli, R., *et al.* (2002). *In vitro* blood-brain barrier permeability and cerebral endothelial cell uptake of the neuroprotective nitrone compound NXY-059 in normoxic, hypoxic and ischemic conditions. *Brain Res* **955**(1–2): 229–235.

Ergul, A., Valenzuela, J.P., *et al.* (2015). Cellular connections, microenvironment and brain angiogenesis in diabetes: Lost communication signals in the post-stroke period. *Brain Res* **1623**: 81–96.

Fisher, M. (2013). The spectrum of translational stroke research. *Neurol Res* **35**(5): 443–447.

Gorelick, P.B., Scuteri, A., *et al.* (2011). Vascular contributions to cognitive impairment and dementia: A statement for healthcare professionals from the American Heart Association/American Stroke Association. *Stroke* **42**(9): 2672–2713.

Griffin, J.H., Zlokovic, B.V., *et al.* (2015). Activated protein C: Biased for translation. *Blood* **125**(19): 2898–2907.

Grotta, J.C., Hacke, W. (2015). Stroke neurologist's perspective on the new endovascular trials. *Stroke* **46**(6): 1447–1452.

Guo, S., Lo, E.H. (2009). Dysfunctional cell–cell signaling in the neurovascular unit as a paradigm for central nervous system disease. *Stroke* **40**(3 Suppl): S4–S7.

Guo, S., Lok, J., *et al.* (2016). Effects of controlled cortical impact on the mouse brain vasculome. *J Neurotrauma* **33**(14): 1303–1316.

Guo, S., Zhou, Y., *et al.* (2012). The vasculome of the mouse brain. *PLoS One* **7**(12): e52665.

Hayakawa, K., Pham, L.D., *et al.* (2012). Astrocytic high-mobility group box 1 promotes endothelial progenitor cell-mediated neurovascular remodeling during stroke recovery. *Proc Natl Acad Sci USA* **109**(19): 7505–7510.

Heiss, W.D., Graf, R. (1994). The ischemic penumbra. *Curr Opin Neurol* **7**(1): 11–19.

Helm, E.E., Tyrell, C.M., *et al.* (2016). The presence of a single-nucleotide polymorphism in the BDNF gene affects the rate of locomotor adaptation after stroke. *Exp Brain Res* **234**(2): 341–351.

Hou, C., Duan, J., *et al.* (2016). Remote limb ischemic conditioning treatment for intracranial atherosclerotic stenosis patients. *Int J Stroke* **11**(7): 831–838.

Hougaard, K.D., Hjort, N., *et al.* (2014). Remote ischemic preconditioning as an adjunct therapy to thrombolysis in patients with acute ischemic stroke: A randomized trial. *Stroke* **45**(1): 159–167.

Hu, X., Leak, R.K., *et al.* (2015). Microglial and macrophage polarization-new prospects for brain repair. *Nat Rev Neurol* **11**(1): 56–64.

Iadecola, C., Anrather, J. (2011). The immunology of stroke: From mechanisms to translation. *Nat Med* **17**(7): 796–808.

Ishikawa, H., Tajiri, N., *et al.* (2013). Ischemic stroke brain sends indirect cell death signals to the heart. *Stroke* **44**(11): 3175–3182.

Kim, J.M., Stewart, R., *et al.* (2012). Associations of BDNF genotype and promoter methylation with acute and long-term stroke outcomes in an East Asian cohort. *PLoS One* **7**(12): e51280.

Komatsu, M., Waguri, S., *et al.* (2006). Loss of autophagy in the central nervous system causes neurodegeneration in mice. *Nature* **441**(7095): 880–884.

Liauw, J., Hoang, S., *et al.* (2008). Thrombospondins 1 and 2 are necessary for synaptic plasticity and functional recovery after stroke. *J Cereb Blood Flow Metab* **28**(10): 1722–1732.

Lipton, P. (1999). Ischemic cell death in brain neurons. *Physiol Rev* **79**(4): 1431–1568.

Lo, E.H. (2008). A new penumbra: Transitioning from injury into repair after stroke. *Nat Med* **14**(5): 497–500.

Lo, E.H. (2010). Degeneration and repair in central nervous system disease. *Nat Med* **16**(11): 1205–1209.

Lo, E.H. (2014). 2013 Thomas Willis Award Lecture: Causation and collaboration for stroke research. *Stroke* **45**(1): 305–308.

Lo, E.H., Broderick, J. P., *et al.* (2004). tPA and proteolysis in the neurovascular unit. *Stroke* **35**(2): 354–356.

Lo, E.H., Dalkara, T., *et al.* (2003). Mechanisms, challenges and opportunities in stroke. *Nat Rev Neurosci* **4**(5): 399–415.

Lopez, M.F., Krastins, B., *et al.* (2015). Proteomic signatures of serum albumin-bound proteins from stroke patients with and without endovascular closure of PFO are significantly different and suggest a novel mechanism for cholesterol efflux. *Clin Proteomics* **12**(1): 2.

Mang, C.S., Campbell, K.L., *et al.* (2013). Promoting neuroplasticity for motor rehabilitation after stroke: Considering the effects of aerobic exercise and genetic variation on brain-derived neurotrophic factor. *Phys Ther* **93**(12): 1707–1716.

Markus, H.S., Bevan, S. (2014). Mechanisms and treatment of ischaemic stroke — insights from genetic associations. *Nat Rev Neurol* **10**(12): 723–730.

Meng, R., Asmaro, K., *et al.* (2012). Upper limb ischemic preconditioning prevents recurrent stroke in intracranial arterial stenosis. *Neurology* **79**(18): 1853–1861.

Meng, R., Ding, Y., *et al.* (2015). Ischemic conditioning is safe and effective for octo- and nonagenarians in stroke prevention and treatment. *Neurotherapeutics* **12**(3): 667–677.

Menon, B.K., Campbell, B.C., *et al.* (2015). Role of imaging in current acute ischemic stroke workflow for endovascular therapy. *Stroke* **46**(6): 1453–1461.

Miyamoto, N., Maki, T., *et al.* (2015). Astrocytes promote oligodendrogenesis after white matter damage via brain-derived neurotrophic factor. *J Neurosci* **35**(41): 14002–14008.

Mozaffarian, D., Benjamin, E.J., *et al.* (2016). Heart disease and stroke statistics — 2016 update: A report from the American Heart Association. *Circulation* **133**(4): e38–360.

Ning, M., Lo, E.H., *et al.* (2013). The brain's heart — therapeutic opportunities for patent foramen ovale (PFO) and neurovascular disease. *Pharmacol Ther,* **139**(2): 111–123.

Ning, M., Sarracino, D.A., *et al.* (2011). Proteomic temporal profile of human brain endothelium after oxidative stress. *Stroke* **42**(1): 37–43.

Pico, F., Rosso, C., *et al.* (2016). A multicenter, randomized trial on neuroprotection with remote ischemic per-conditioning during acute ischemic stroke: The REmote iSchemic Conditioning in acUtE BRAin INfarction study protocol. *Int J Stroke* **11**(8): 938–943.

Shuaib, A., Lees, K.R., *et al.* (2007). NXY-059 for the treatment of acute ischemic stroke. *N Engl J Med* **357**(6): 562–571.

Simard, J.M., Sheth, K.N., *et al.* (2014). Glibenclamide in cerebral ischemia and stroke. *Neurocrit Care* **20**(2): 319–333.

Stanne, T.M., Tjarnlund-Wolf, A., *et al.* (2014). Genetic variation at the BDNF locus: Evidence for association with long-term outcome after ischemic stroke. *PLoS One* **9**(12): e114156.

Ushio-Fukai, M. (2006). Redox signaling in angiogenesis: role of NADPH oxidase. *Cardiovasc Res* **71**(2): 226–235.

von Bornstadt, D., Houben, T., *et al.* (2015). Supply-demand mismatch transients in susceptible peri-infarct hot zones explain the origins of spreading injury depolarizations. *Neuron* **85**(5): 1117–1131.

Wei, M., Huo, K., *et al.* (2016). The design and rationale of a clinical trial evaluating limb postconditioning in young patients with intracranial arterial stenosis. *J Stroke Cerebrovasc Dis* **25**(10): 2506–2512.

Yarbrough, C.K., Ong, C.J., *et al.* (2015). Endovascular thrombectomy for anterior circulation stroke: Systematic review and meta-Analysis. *Stroke* **46**(11): 3177–3183.

Yuan, J., Yankner, B.A. (2000). Apoptosis in the nervous system. *Nature* **407**(6805): 802–809.

Zamanian, J.L., Xu, L., *et al.* (2012). Genomic analysis of reactive astrogliosis. *J Neurosci* **32**(18): 6391–6410.

Zhao, J., Wu, H., *et al.* (2013). Brain-derived neurotrophic factor G196A polymorphism predicts 90-day outcome of ischemic stroke in Chinese: A novel finding. *Brain Res* **1537**: 312–318.

Neural Repair for Cerebrovascular Diseases 3

Steven C. Cramer

Department of Neurology, Anatomy & Neurobiology, and Department of Physical Medicine & Rehabilitation, University of California, Irvine, California, USA

Neural repair refers to restoration of the central nervous system (CNS) structure or function after an insult such as stroke. This repair can be spontaneous or therapeutically induced. Repair contrasts with therapeutic strategies that aim to prevent cerebrovascular disease, and with approaches that aim to limit extent of injury from a new stroke such as acute reperfusion or neuroprotection. Neural repair is focused on regrowth, restoration, rewiring, and rehabilitation.

The burden of disability after stroke remains large. In middle- and high-income countries around the world, stroke is the leading neurological cause of lost disability-adjusted life years (Johnston & Hauser, 2008). An estimated 6,600,000 Americans adults have had a symptomatic stroke, with a prevalence that increases with age, and 13,000,000 people in the United States have had a silent stroke (Writing Group Members *et al.*, 2016). Each year, 795,000 people experience a stroke, of which 610,000 are first-ever symptomatic stroke. The mean survival after stroke is 6–7 years, with approximately 85% of patients living past the first year of stroke (Lloyd-Jones *et al.*, 2010). Most patients survive their stroke and live with enduring disability for years to come.

3.1 Current Therapies

Current therapies for a new stroke reduce disability in only a subset of patients. The only drug approved to treat acute stroke remains tissue plasminogen activator (tPA) (Hacke *et al.*, 2008; National Institue of Neurological Disorders (NINDS) & Stroke rt-PA Stroke Study Group, 1995). A limited fraction of patients receive this medicine (Adeoye *et al.*, 2011; Reed *et al.*, 2001), in large part due to the narrow time window for safe drug administration; according to recent estimates, approximately 5% of stroke patients receive this medication (Adeoye *et al.*, 2011). Moreover, of those so treated, half or more have significant long-term disability (Hacke *et al.*, 2008; NINDS & Stroke rt-PA Stroke Study Group, 1995). Even fewer patients receive acute endovascular reperfusion therapies (Goyal *et al.*, 2016). Because most repair-based approaches have a time window measured in days rather than hours, any repair-based approach that obtains regulatory approval will have the potential to help a much larger proportion of patients affected by stroke.

3.2 Spontaneous Recovery from Stroke

Stroke triggers several cellular and molecular cascades that facilitate processes, including spontaneous repair and behavioral recovery (Overmann & Carmichael, 2014). Animal studies have enhanced the understanding of these mechanisms (Arai & Lo, 2009; Carmichael, 2006; Cramer & Chopp, 2000; Li & Carmichael, 2006; Murphy & Corbett, 2009; Nudo, 2011; Wieloch & Nikolich, 2006; Zhang & Chopp, 2009). Experimental stroke models depict altered expression for numerous genes (Li & Carmichael, 2006), which is related to subsequent increases in levels of several growth factors (Finklestein *et al.*, 1990; Kawamata *et al.*, 1997a), growth of synapses and dendrites (Jones *et al.*, 1996; Jones & Schallert, 1992), axonal remodeling and angiogenesis (Ding *et al.*, 2008; Hermann & Chopp, 2012; Seevinck *et al.*, 2010; Teng *et al.*, 2008), and enhanced brain excitability mediated by alterations in glutamate and gamma-aminobutyric acid (GABA) receptor subtypes (Clarkson *et al.*, 2010, 2011; Que *et al.*, 1999; Redecker *et al.*, 2002). These events are often concentrated in perilesional tissue but are not confined there; indeed, many are also present in the contralesional hemisphere (Dijkhuizen *et al.*, 2012; Jones *et al.*, 1996; Kawamata *et al.*, 1997b), in ipsilesional areas connected to the lesioned area (Dancause *et al.*, 2005), and even downstream in the spinal cord (Sist *et al.*, 2014).

Brain responses to injury can be organized into three recovery epochs. The first occurs during the initial hours after stroke onset and represents an opportunity to salvage threatened tissue, e.g., via reperfusion or neuroprotection.

The second commences days to weeks following stroke and corresponds to neural repair. Mechanisms of spontaneous recovery are most robust during this time (Cramer, 2008a). The third denotes a chronic phase of neural repair when the brain is relatively stable with regards to endogenous repair-related events, but modifications in brain structure and function are still possible with specific interventions. These three epochs delineate periods of neuroprotection and repair, and have clinical implications with regards to delivery of restorative therapies.

While animal studies are able to study mechanisms of recovery using any number of invasive techniques that enable a range of molecular and histological assessments, studies in human subjects are generally restricted to non-invasive methods. The most commonly used methods include structural and functional magnetic resonance imaging (fMRI), positron emission tomography (PET), single photon emission computed tomography (SPECT), electroencephalography (EEG), magnetoencephalography (MEG), transcranial magnetic stimulation (TMS), and near infrared spectroscopy (NIRS); these provide a systems-level perspective on neural repair (Cramer, 2008a). These studies show modulation in local and distant cortical and subcortical activity, changes in interactions between hemispheres, shifts in cortical representational maps, and alterations in brain connectivity (Cassidy & Cramer, 2016; Cramer, 2008a; Grefkes & Ward, 2014; Stiner *et al.*, 2014). The exact contribution that such findings make to behavioral recovery can be difficult to estimate across the human stroke population, where there are many sources of heterogeneity that include pre-stroke status, site/size of stroke injury, and post-stroke therapies. In general, restitution of normal patterns of brain function is associated with the best outcomes (Ward *et al.*, 2003). Consistent with this, post-stroke modulations in brain structure and function, as above, may be of benefit to patients with more severe injury (Bradnam *et al.*, 2012; Heiss & Thiel, 2006; Johansen-Berg *et al.*, 2002; Lotze *et al.*, 2006; Netz *et al.*, 1997; Turton *et al.*, 1996) but prove detrimental in patients with less severe injury. The same pattern of CNS reorganization can be likened to a crutch, improving function among more impaired subjects and hindering function in less impaired individuals.

Diaschisis is known to occur after stroke and has received limited attention, but it may be particularly important in relation to recovery after stroke. As one of several theories of functional recovery originally postulated by von Monakow in the early 20th century, diaschisis refers to reduced function in brain areas that are spared injury but are functionally connected to sites of injury (von Monakow, 1969; Feeney & Baron, 1986). Both animals (Buchkremer-Ratzmann *et al.*, 1996; Carmichael *et al.*, 2004; Redecker *et al.*, 2002) and humans (Bonilha *et al.*, 2014; Nhan *et al.*, 2004; Pappata *et al.*, 2000; Sobesky *et al.*, 2005; Volz *et al.*, 2016) demonstrate diaschisis with changes in cerebral blood flow, metabolism, and neurotransmitter activity in areas distant from the lesion. A reversal of diaschisis is suggestive of recovery, but additional work is needed to better understand the timing of diaschisis and the process of diaschisis reversal relative to time post-stroke.

3.3 Therapies to Promote Neural Repair

Many categories of restorative therapy are under study in relation to promoting neural repair (Cumberland Consensus Working Group *et al.*, 2009; Cramer, 2008b; Knecht *et al.*, 2011; Zhang & Chopp, 2009). Several have reached the point where they are being studied in human subjects, though in general most are in early phase trials. The current emphasis is on the study of single agents, a key step to understanding therapeutic efficacy, but combinations of therapy will likely be examined over time. Some restorative therapies are started in the early days after stroke onset, aiming to amplify innate repair mechanisms. Other restorative therapies are initiated months after stroke onset, at a time when most spontaneous behavioral recovery has occurred, and so have as the goal induction of new neural plasticity.

3.3.1 *Growth factors*

Growth factors are appealing as a restorative agent because of their established role in normal development and because they play a critical role in spontaneous neural repair through mechanisms that include angiogenesis, reduced apoptosis, stem cell proliferation, and immunomodulation (Lanfranconi *et al.*, 2009). In some cases, a rich clinical experience exists for growth factors outside of stroke indications, such as for patients with renal failure or infertility. The utility of growth factors to promote stroke recovery has been studied in preclinical stroke studies. In many cases, preclinical studies suggest that administration of exogenous growth factors 24 hours or longer after stroke onset significantly improves long-term behavioral outcome, e.g., brain-derived neurotrophic factor (BDNF) (Ren & Finklestein, 2005), epidermal growth factor plus erythropoietin (Kolb *et al.*, 2007), or human chorionic gonadotropin (hCG) plus erythropoietin (Belayev *et al.*, 2009).

Most human stroke studies to date have examined hematopoietic growth factors, some of which have an extensive record of safe application in humans. Granulocyte-colony stimulating factor (G-CSF) is one such growth factor. Favorable preclinical findings led to the AX200 for Ischemic Stroke (AXIS) study (Schabitz *et al.*, 2010), which found that G-CSF given within 12 hours of stroke was safe and well-tolerated in 44 patients. These good safety results were echoed by a separate study of 60 patients (England *et al.*, 2012). The follow-up study, AXIS-2 study (Ringelstein *et al.*, 2013), compared the middle G-CSF dose from the AXIS study (135 *u*g/kg) with placebo in 328 patients <9 hours post-stroke using a multicenter, randomized, placebo-controlled design. G-CSF was not different from placebo on the primary endpoint, modified Rankin Scale score at day 90.

Preclinical studies suggest that systemically administered erythropoietin may enter the brain and improve outcome when introduced with a delay, such as 24 hours post-stroke (Jerndal *et al.*, 2010). Other studies found favorable effects of sequential growth factor administration, giving epidermal growth factor (Kolb *et al.*, 2007) or beta-hCG (Belayev *et al.*, 2009) followed by erythropoietin, beginning 1-7 days after stroke, possibly by increasing neural stem cell levels. This approach was directly translated to humans in the beta-hCG+erythropoietin in acute stroke (BETAS) study, a single-dose, multisite, open-label, non-controlled safety trial that gave 3 hCG doses beginning 1–2 days post-stroke followed by 3 erythropoietin doses beginning 7–8 days after stroke. No safety concerns were identified (Cramer *et al.*, 2010). The safety of erythropoietin was also found in a randomized placebo-controlled study of 167 patients who received two doses of erythropoietin versus placebo beginning 48 post-stroke (Yip *et al.*, 2011). The BETAS study was followed by the REGENESIS study (Cramer & Hill, 2014). This was intended to be a randomized, placebo-controlled, double-blind proof of concept study of sequential hCG and erythropoietin using the BETAS study treatment schedule. This trial was put on hold by regulatory authorities due to concerns related to an acute stroke neuroprotective trial (Ehrenreich *et al.*, 2009) that initiated erythropoietin <6 hours of stroke onset, even though the REGENESIS trial initiated erythropoietin 7–8 days post-stroke. Subsequently, the REGENESIS trial was modified to be a dose-ranging safety study and, due to financial constraints, largely moved to India. Enrollment was terminated by the sponsor early after 96 enrollees. Treatment groups did not differ in safety or in the primary endpoint, NIHSS score change to day 90.

Growth factors are generally large proteins for which CNS ingress is limited. A number of strategies have been proposed to overcome this, such as helping growth factors to cross the blood–brain barrier via conjugation to a molecular Trojan horse (Zhang & Pardridge, 2006). Another strategy has been to transfect an exogenous stem cell with a gene encoding a growth factor, as has been studied for fibroblast growth factor-2 (Kim *et al.*, 2005), GDNF (Horita *et al.*, 2006), BDNF (Zhao *et al.*, 2004), VEGF (Iwase *et al.*, 2005), PIGF (Liu *et al.*, 2006), or HGF (Zhao *et al.*, 2006). Development of small ligands is another potential solution (Hermann, 2009; Yasuhara *et al.*, 2006; Zhang & Pardridge, 2001).

The post-stroke restorative effect of other large molecules, such as monoclonal antibodies, is also under study. A monoclonal antibody modulates activity within a targeted signaling pathway by binding to a specific target such as a receptor or cell surface marker. One approach in this context focuses on neutralizing factors that inhibit growth in the adult CNS, with the overall model being that neural repair after stroke after CNS injury is limited by the lack of a permissive growth environment. Axonal growth has long been known to occur in the peripheral nervous system (Rivers & Head, 1908); however, in the CNS three major inhibitors (myelin-associated glycoprotein [MAG], oligo-myelin glycoprotein, and Nogo-A) result in the lack of a permissive growth environment. Furthermore, levels of these molecules increase after stroke (Cheatwood *et al.*, 2008; Li & Carmichael, 2006). Blockade of these inhibitors, such as with a monocloncal antibody, promotes axonal growth (Buchli & Schwab, 2005; Domeniconi & Filbin, 2005). A monoclonal antibody can be used to neutralize these inhibitory CNS signals and promote axon growth after injury. A recent study randomized 42 patients with stroke to placebo versus one of three doses of intravenous GSK249320, a humanized IgG1 monoclonal antibody to MAG that has a disabled Fc region. Each patient received two infusions: the first administered 24–72 hours after stroke onset, and the second, 9 days later. No safety concerns were identified (Cramer *et al.*, 2013), and a phase II proof-of-concept study is ongoing.

Numerous small molecules have also been examined to improve outcome after stroke. Small molecules may have advantages in terms of transport through the blood–brain barrier (Pardridge, 2012) and thus increased access to healthy brain areas. In many cases, candidate small molecules represent repurposed drugs, i.e., those already approved for other indications. Many of the small molecules studied for neural repair target a specific brain neurotransmitter system.

3.3.2 *Monoaminergic drugs*

Monoaminergic drugs have been most frequently studied. An early focus was on amphetamine (Feeney *et al.*, 1982), which acts on multiple monoaminergic targets. Initial human experience in small studies was favorable (Crisostomo *et al.*, 1988; Walker-Batson *et al.*, 1995), but the Subacute Therapy with Amphetamine and Rehabilitation for Stroke (STARS) study was not. This randomized, double-blind, placebo-controlled trial did not demonstrate a benefit (Gladstone *et al.*, 2006). The authors examined 5 weeks of twice-weekly amphetamine in 71 patients enrolled 5–10 days post-stroke. The drug was safe but did not improve the primary outcome, motor recovery over 3 months using the arm/leg Fugl–Meyer motor score (Gladstone *et al.*, 2006). Strengths of this study include the use of a single therapist to administer all physiotherapy, and use of a single examiner to assess study outcomes, which reduces variance and increases study power. A weakness of the study is that the treatment protocol was not directly translated from preclinical findings, and so the optimal dose, timing, and frequency of amphetamine to promote stroke recovery remains uncertain. It is difficult to determine whether STARS showed that amphetamine is not useful overall, or simply that the one protocol examined is not useful.

3.3.2.1 *Dopamine*

Dopamine regulates many aspects of neural functioning, including excitability, synaptic transmission, plasticity, protein trafficking, and gene transcription (Tritsch & Sabatini, 2012). Not surprisingly, therefore, dopamine has a key role in wide-ranging brain processes such as movement, reward, learning, and plasticity (McAllister, 2009). The role of this neurotransmitter is movement is well established: dopaminergic terminals in motor cortex contribute to cortical plasticity and indeed are necessary for motor skill learning (Hosp *et al.*, 2011; Molina-Luna *et al.*, 2009).

Drugs increasing dopaminergic neurotransmission can improve learning and plasticity in healthy subjects (Floel & Cohen, 2010). A randomized, double-blind, placebo-controlled study in patients with stroke was consistent with these results, finding 100 mg L-Dopa/day (given as Sinemet and combined with physical therapy) to be significantly better than placebo on the primary endpoint, motor status by the Rivermead Motor Assessment after 3 weeks (Scheidtmann *et al.*, 2001). The authors randomized 53 patients within 6 months of stroke onset to 3 weeks of daily L-Dopa or placebo coupled with physiotherapy. Dopaminergic drugs have the potential advantage that measures of genetic variability may help predict inter-subject differences in treatment response (MacDonald *et al.*, 2016; Pearson-Fuhrhop *et al.*, 2013); this study awaits replication. Smaller studies using other dopamine agonists have been largely negative, e.g., a placebo-controlled, double-blind study of 33 patients 1–12 months post-stroke did not find a difference between a 9-week course of ropinirole+physiotherapy compared to placebo+physiotherapy on gait velocity (Cramer *et al.*, 2009). Small studies hint at the potential for noradrenergic drugs (Fasotti *et al.*, 2013; Small & Llano, 2009; Wang *et al.*, 2011).

3.3.2.2 *Serotonin*

Serotonin is another monoaminergic neurotransmitter potentially helpful in promoting neural repair and thereby improving stroke recovery. Serotonin normally plays a role in modulating multiple cognitive functions, particularly response inhibition and memory consolidation, and modulates the impact of punishment-related signals on learning and emotion (Cools *et al.*, 2008; Cowen & Sherwood, 2013; Logue & Gould, 2014). Early reports supported the potential clinical utility of selective serotonin reuptake inhibitor (SSRI) drugs for promoting improved motor outcome after stroke (Dam *et al.*, 1996; Fruehwald *et al.*, 2003; Miyai & Reding, 1998; Pariente *et al.*, 2001). Recent reports suggest that increasing serotonin neurotransmission improves stroke recovery. Robinson *et al.* (Robinson *et al.*, 2008) performed a multisite, randomized controlled trial for prevention of depression among 176 non-depressed patients enrolled within 3 months of stroke onset. Patients randomized to the placebo arm were significantly ($p < 0.001$) more likely to reach the primary outcome (development of major or minor depression), compared with patients in either of the two active study arms (SSRI escitalopram or problem-solving therapy). One analysis of a subgroup of these patients found that cognitive outcomes at 12 months were significantly better among those randomized to escitalopram, independent of depression, while a separate subgroup analysis found a lower incidence of generalized anxiety disorder with escitalopram or problem-solving therapy (Mikami *et al.*, 2014).

The strongest evidence in support of an SSRI to improve outcomes after stroke comes from The Fluoxetine for Motor Recovery After Acute Ischemic Stroke (FLAME) study (Chollet *et al.*, 2011). This was a double-blind, placebo-controlled trial that enrolled non-depressed hemiplegic/hemiparetic patients within 10 days of ischemic stroke onset. Patients were randomized to 3 months of oral fluoxetine (20 mg/day) or placebo. Those randomized to fluoxetine showed significantly greater gains on the primary endpoint, change in the arm/leg Fugl–Meyer motor score to day 90 ($p = 0.003$), a remarkable 9.7 point difference on this 100 point scale. Measures of genetic variability may inform likelihood of response to a serotonergic drugs (Kato & Serretti, 2010; Kohen *et al.*, 2011).

3.3.2.3 *Norepinephrine*

Drugs that modulate transmission in norepinephrine have received limited study in the context of stroke recovery. Noradrenergic neurotransmission broadly amplifies neuronal activity, increases the general level of excitability, and selectively amplifies activities evoked by unexpected inputs (Merzenich *et al.*, 2014). This effect of norepinephrine on regulating overall arousal levels has a modulatory effect on executive function. To date, there has been only a handful of studies on noradrenergic drugs to promote stroke recovery. These have been small in size but show promising results (Beltran *et al.*, 2010; Wang *et al.*, 2011; Zittel *et al.*, 2007).

3.3.2.4 *Other drugs*

Drugs that modulate neurotransmission in acetylcholinergic pathways have also received limited study in relation to neural repair. In the cortex, acetylcholine inputs positively enable plasticity by (1) selectively amplifying only anticipated ("selectively attended"), and (2) selectively weakening non-anticipated inputs (Merzenich *et al.*, 2014). Modulation of nicotinic cholinergic neurotransmission alters attention, while muscarinic receptors play a greater role in cognitive flexibility (Logue & Gould, 2014). Luria, long ago, advocated for cholinergic therapies to enhance recovery (Luria, 1963). Studies in rodents (Kilard & Merzenich, 1998; Wang *et al.*, 2016) and primates (Ramanathan *et al.*, 2009) with experimental stroke support this view, but data from human subjects with stroke remain sparse (Berthier *et al.*, 2006). Data with respect to non-motor aspects of stroke recovery are limited in quantity but potentially promising (Barrett *et al.*, 2011; Berthier *et al.*, 2006; Whyte *et al.*, 2008), and a recent study in 33 patients found donepezil to be safe when initiated within 24 hours of stroke onset (Barrett *et al.*, 2011).

Other small molecules have been studied. Sildenafil (Bednar, 2008; Zhang *et al.*, 2002) has a time window of at least 7 days and has been tested in human subjects recovering from stroke (Silver *et al.*, 2009). An extended-release formulation of the potassium channel blocker 4-aminopyridine has been approved by the U.S. Food and Drug Administration to improve walking ability in patients with multiple sclerosis (Goodman *et al.*, 2010), and is now being studied after stroke (Iaci *et al.*, 2013; Simpson *et al.*, 2015). A randomized, placebo-controlled trial of citicoline initiated within 24 hours of stroke onset was stopped for futility (Davalos *et al.*, 2012).

3.3.3 *Traditional and alternative medicines*

Traditional and alternative medicines have also been studied though few using formal research methods. One exception is the Chinese Medicine Neuroaid Efficacy on Stroke recovery (CHIMES) study (Chen *et al.*, 2013), a multicenter, double-blind, placebo-controlled clinical trial that randomized 1,100 patients to 3 months of Neuroaid versus placebo, provided in the form of three daily doses of 4 oral capsules each. Neuroaid is a traditional Chinese medicine containing extracts of 9 herbal and 5 animal components. Patients started therapy within

72 hours of stroke onset. No difference between treatment arms was seen in the primary outcome measure, shift in the modified Rankin Scale at 3 months.

3.3.4 *Cell-based therapies*

Cell-based therapies are receiving increased attention, with numerous cell types under consideration (Lindvall & Kokaia, 2011; Savitz *et al.*, 2014). Most often, the strategy is to directly administer exogenous stem cells, examples of which include transformed tumor cells, adult stem cells such as marrow stromal cells, stem cells with modified genes or a bioscaffold, umbilical cord cells, placental cells, embryonic stem cells, and fetal stem cells. Such exogenous cells can be autologous, allogeneic, or xenografts. Among the stem cell populations under study, considerable attention has been drawn to mesenchymal stromal cells (MSC; also known as mesenchymal stem cells or multipotent stromal cells). MSCs are adult non-hematopoietic pluripotent cells. Human MSCs have been defined by the International Society for Cellular Therapy (Dominici *et al.*, 2006) on the basis of three criteria, i.e., to be MSC the cells must: (1) adhere to plastic in standard culture conditions; (2) express surface antigens such as CD105 and CD90 and not express antigens such as CD45 or HLA-DR; and (3) be able to differentiate *in vitro* to osteoblasts, adipocytes, and chondroblasts. MSCs are the basis for the first clinically approved human stem cell therapy in North America (Cyranoski, 2012).

Preclinical evidence suggests that MSCs improve behavioral outcomes after experimental stroke via several different mechanisms in parallel (Eckert *et al.*, 2013), a potential advantage over pharmacological therapies that act via a single treatment mechanism (Honmou *et al.*, 2012; Li & Chopp, 2009; Phinney & Prockop, 2007). One set of mechanisms pertains to local changes in the brain, including neurogenesis, synaptogenesis, and angiogenesis (Dharmasaroja, 2009). MSCs orchestrate numerous cellular proliferative events such as subventricular and subgranular zone cell proliferation, consistent with endogenous neuronal precursors mobilization, and in addition reduce apoptosis (Chen *et al.*, 2003; Steffenhagen *et al.*, 2012). MSCs have been found to infiltrate ischemic brain regions and upregulate genes related to restorative events (Yilmaz *et al.*, 2010), resulting in increased local levels of numerous growth factors and chemokines. These paracrine events promote improved recovery after stroke, and when administered early post-stroke, MSCs might also be associated with a neuroprotective effect and thus decreased infarct volume. MSCs secrete extracellular matrix components that promote neuronal survival (Maltman *et al.*, 2011; Sakai *et al.*, 2001). Although MSCs can differentiate into several mesodermal lineages, these cells do not replace functional neurons or glia (Hilfiker *et al.*, 2011; Krabbe *et al.*, 2005; Miller *et al.*, 2010).

A second set of MSC mechanisms after stroke relates to immumomodulatory mechanisms, which is often remote in distribution (Ankrum & Karp, 2010; Li & Chopp, 2009). Numerous inflammatory events occur after stroke, both within the brain and systemically (Iadecola & Anrather, 2011; Siniscalchi *et al.*, 2014; Tobin *et al.*, 2014). MSCs affect numerous immune events, e.g., regulating T-cells via inhibiting T-cell proliferation, promoting T-cell regulatory effects, and exerting suppressive effects on CD4+ and CD8+ T-cells (Aggarwal & Pittenger, 2005; Di Nicola *et al.*, 2002; Nasef *et al.*, 2007; Yoo *et al.*, 2013). MSCs may also have effects on leukocyte proliferation (Bacigalupo *et al.*, 2005; Bartholomew *et al.*, 2002; Sato *et al.*, 2007) and B-cell and functions (Beyth *et al.*, 2005; Corcione *et al.*, 2006; Marigo & Dazzi, 2011; Ribeiro *et al.*, 2011). Other immunological effects include a shift in macrophage profiles (Marigo & Dazzi, 2011; Maggini *et al.*, 2010) as well as remote immunomodulation, e.g., from the lung (Ankrum & Karp, 2010; Li & Chopp, 2009; Oh *et al.*, 2010) or spleen (Ajmo *et al.*, 2008; Detante *et al.*, 2009; Gao *et al.*, 2001; Sahota *et al.*, 2013). Importantly, MSCs themselves are considered to be relatively immunoprivileged (Marigo & Dazzi, 2011; Nauta *et al.*, 2006; Cohen, 2013), which opens the door to administration of allogenic MSC, greatly increasing the translational potential of MSC by avoiding the need for immunosuppression or lengthy delays required to expand autologous cells.

MSCs can be administered intracerebrally (IC), intra-arterially (IA), or intravenously (IV). The fate and distribution pattern of MSCs vary according to the route by which they are introduced (Gao *et al.*, 2001; Walczak *et al.*, 2008). MSCs are relatively large cells, having a diameter of 10–30 μm (Lee *et al.*, 2009). Not surprising, therefore, most MSCs are passively entrapped in the lung vasculature following IV transfusion (Detante *et al.*, 2009). In healthy animals, IV MSCs are mainly found in lungs and liver at 3 hours post-injection; in lungs, liver, and spleen at 24 hours; in bone marrow, spleen, and lung over the next several days; and in bone marrow, spleen, lung, bone, muscle, cartilage, and liver months thereafter (Allers *et al.*, 2004; Detante *et al.*, 2009; Erices *et al.*, 2003) After stroke, MSCs also home to regions of brain injury (Chen *et al.*, 2015; Li & Chopp, 2009), with the vast majority found in the ischemic core and its peri-infarct boundary zone. Introduction of MSCs via the IA or the IC route is more invasive, but does avoid first-pass pulmonary effects and so reduces the amount of MSCs in the lungs (Grieve *et al.*, 2010; Willing *et al.*, 2003).

MSCs have a very strong track record of efficacy in preclinical studies of stroke. A recent meta-analysis examined 46 preclinical studies in which MSC was given after cerebral ischemia (Vu *et al.*, 2014). A total of 46 studies with 62 MSC treatment arms were identified. MSC improved outcomes in 44 of the 46 studies, and in 54 of the 62 treatment arms. The mean effect size for MSC administration was consistently very large, e.g., averaging 1.78 for the modified Neurological Severity Score across 28 studies; results were similar overall when analyses were restricted to studies that initiated MSC ≥24 hours after stroke onset. In general, effect sizes remained substantial after adjusting for potential publication bias and were robust across species, delivery route, time of administration in relation to stroke, MSC immunogenicity, and MSC dose.

MSCs have an overall excellent safety record in clinical trials of human subjects across numerous non-cerebrovascular diagnoses (Devine, 2002; Figueroa *et al.*, 2012; Hilfiker *et al.*, 2011; Lalu *et al.*, 2012; Williams & Hare, 2011). Results from the limited number of studies to date examining MSC in patients with stroke show a similarly good safety record. Initial human stroke trials focused on autologous MSCs. Bang *et al.* (2005) randomized patients with ischemic stroke into control and experimental groups, with the experimental groups receiving IV infusion of 10^8 autologous BM-MSC 4 to 9 weeks following onset of symptoms. MSC therapy was associated with significantly improved functional status (modified Rankin score and Barthel index) up to 6 months after transplantation, as compared to controls. Honmou *et al.* (2011) cultured MSCs in autologous human serum rather than fetal calf serum, reducing cell preparation time as well as any potential risk of transmissible disorders. Patients received IV autologous MSC 36–133 days post-stroke, using a study design that did not include a control group. MSC administration was not associated with adverse events or neurological deterioration. Bhasin *et al.* (2011) found that IV transplantation of autologous MSC in 6 patients with chronic (3 to 12 months following) stroke was safe as compared with 6 controls, and that MSC treatment was not associated with any significant change in neurological function.

Subsequent efforts have extended to allogeneic MSCs. Steinberg *et al.* (2016) studied patients who were 6–60 months after an ischemic stroke in the territory of the middle cerebral artery (MCA), with moderate-moderately severe disability, 18 of whom received one of three MSC (with transient Notch transfection) doses (2.5, 5, or 10 million) the primary endpoint was safety, with 2-year follow-up; no concerns were identified, although six patients experienced serious treatment-emergent adverse events, of which two were probably or definitely related to the surgical procedure. A number of secondary clinical and imaging endpoints were examined, with some behavioral measures showing significant improvement over time.

Yavagal *et al.*, building on preclinical studies (Yavagal *et al.*, 2014), reported the results of the RECOVER-Stroke trial (Yavagal, 2015), a phase 2A, double-blind, randomized, sham-controlled trial of 48 patients who were 13–19 days after an ischemic stroke in the territory of the MCA or anterior cerebral artery. The primary objective focused on the safety of intracarotid infusion of ALD-401 cells, a marrow-based cell selected on the basis of high aldehyde dehydrogenase expression, in patients with subacute anterior circulation ischemic stroke. Other key entry criteria included mRS score ≥3 at time of randomization and patent ipsilesional carotid artery. All patients received IC injection, the content of which was randomized 3:2 (cells:sham) and transfused at a mean dose of 3.1 million cells. Over 1 year of follow-up, no safety concerns were identified. Secondary analyses did not identify any differences between the two treatment groups regarding change in behavioral outcome measures.

Hess *et al.* (2015) reported the results of a double-blind, placebo-controlled, dose-escalation study of Multistem in patients with acute stroke. The primary objective of the study focused on two primary endpoints, safety and efficacy. MultiStem is a biologic product of adherent progenitor cells depleted of CD45(+) cells derived from adult bone marrow or other non-embryonic tissue sources, and transfused IV 24–48 hours after an ischemic stroke affecting cerebral cortex. Patients had baseline NIHSS score 8–20 that was stable and lacked significant medical comorbidities. Three doses were studied, up to 1.2 billion cells, a dose substantially higher than that typically encountered in most human studies in any medical context. Over 1 year of follow-up, there were no safety concerns, defined in terms of frequency of dose-limiting adverse events. However, there was also no difference between the two treatment arms in efficacy, defined as stroke recovery to day 90 based on a global test analysis that included mRS, NIHSS, and Barthel Index. Some secondary analyses suggested reduced short-term mortality in the Multistem arm, and also the possibility of greater Multistem effects on stroke recovery when administered earlier (<36 hours post-stroke).

Cellular therapies hold great promise but can introduce complexities not always encountered with other restorative therapies. For example, the treatment is not a lyophilized power but rather living cells, the potency and identity of which could change over time or during storage. As a reuslt, the stability of such measures must be confirmed prior to release (Food and Drug Administration Center, 2011). Also, some stem cells persist for months or years after administration, requiring extended assessment of their long-term fate — 10 years for the first embryonic stem cell trial approved by the U.S. FDA (Chapman & Scala, 2012). Note too that some patients and scientists have ethical concerns with use of certain cells (Hyun, 2010), a consideration that is rarely raised in relation to many other therapy classes such as small molecules.

3.3.5 *Other therapies*

A number of intensive activity-based therapy regimens have been studied. A key example is constraint-induced therapy, an approach that trains the affected limb while restraining the non-affected limb, in order to overcome learned disuse of the affected limb. In the EXCITE trial, constraint-induced therapy was associated with significant gains in motor outcome in 222 patients enrolled 3–9 months after stroke onset (Wolf *et al.*, 2006), with these effects enduring for years (Wolf *et al.*, 2008). The constraint-induced approach is also being studied in non-motor domains, such as aphasia (Berthier & Pulvermuller, 2011). The timing of such intensive therapies may be as important as the content. The VECTORS trial examined arm motor constraint-induced therapy early after stroke and found that in 52 patients enrolled within 1 month of stroke onset, higher intensity of therapy was associated with poorer behavioral outcome at day 90 (Dromerick *et al.*, 2007) — too much too soon can be harmful. The LEAPS trial compared two therapies focused on gait in 408 patients within 2 months of stroke and found that treadmill training with body-weight support did not differ from progressive exercise at home managed by a physical therapist in effects on walking ability 1 year after stroke (Duncan *et al.*, 2011). While effects of these therapeutic approaches did not differ, the LEAPS trial did find that a majority of patients with stroke can experience significant behavioral gains when therapy is initiated weeks after stroke onset, with 52% of treated patients showing improved gait velocity 1 year after stroke onset. An intervention started on the first day post-stroke that improved gait velocity in a majority of patients 12 months post-stroke would garner great praise; this should be no less true when the therapy is initiated weeks after stroke onset.

Important questions remain for activity-based regimens focused on intensive therapy, such as the optimal timing and amount of training, how to design the content of training in order to maximize generalizability of therapy effects across real-life challenges, adjusting content of training in relation to time after stroke, and the potential to maximize treatment effects by combining with pharmacological therapies. A major area requiring additional research is the dose of activity. Two recent studies (Lang *et al.*, 2016; Winstein *et al.*, 2016) randomized patients in the chronic phase of stroke to varying quality or quantity of activity-based therapy and found no difference between groups. One possible explanation is that the doses explored, up to 30 (Winstein *et al.*, 2016) and 32.8 (Lang *et al.*, 2016) hours, were far smaller than needed to most robustly test the study hypotheses.

The effect of therapy delivered by robotic devices has also been examined. Numerous robotic devices have been studied (Balasubramanian *et al.*, 2010;

Brewer *et al.*, 2007; Reinkensmeyer *et al.*, 2004; Veerbeek *et al.*, 2017; Volpe *et al.*, 2009). These devices offer potential advantages, such as consistent and long-lasting output, programmability, utility for virtual reality applications, safety, high precision, the potential for an improved therapist: patient ratio, and great potential for telerehabilitation and therefore ability to reach underserved regions (Cramer, 2010). However, challenges remain. Lo *et al.* found, in 127 patients who were in the chronic phase of stroke, that robot-assisted therapy did not significantly improve motor function after 12 weeks, as compared with usual care or intensive therapy; in secondary analyses, robot-assisted therapy improved outcomes over 36 weeks as compared with usual care but not with intensive therapy (Lo *et al.*, 2010). A recent review noted that: (1) effects on motor control are small and specific to the joints targeted by the robotic intervention; (2) limited data support generalization of robot-derived gains to broader functions; and (3) little data exist among patients in the initial weeks following stroke onset (Veerbeek *et al.*, 2017). Factors that might represent avenues for improving the impact of robotic therapy include more fully defining the relationship between robotic therapy and traditional physiotherapy, and matching the right patients with the right robotic devices and protocols.

3.3.6 *Brain stimulation*

The brain is an electrical organ, suggesting the potential for brain stimulation to promote neural repair and thereby improve outcomes after stroke. Many forms of brain stimulation have been examined, including repetitive TMS, theta burst stimulation, epidural cortical stimulation, transcranial direct current stimulation, transcranial alternating current stimulation, and stimulation via a laser-based device (Edwardson *et al.*, 2013). There is precedence for a focus on brain stimulation, as the gold standard therapy for major depression remains a form of brain stimulation — electroconvulsive therapy (American Psychiatric Association, 2010). Some results with brain stimulation to promote improved outcomes after stroke, mainly targeting motor outcomes, have been favorable (Ackerley *et al.*, 2010; Lindenberg *et al.*, 2011) while others have not (Levy *et al.*, 2008; Pomeroy *et al.*, 2007). A Phase III trial aiming to improve arm motor outcomes in patients with chronic hemiparetic stroke examined neurosurgically implanted epidural cortical stimulation + physical therapy but did not find this intervention to be significantly different from physical therapy alone (Levy *et al.*, 2016); *post hoc* analysis indicated response to brain stimulation was substantially greater among subjects with preservation of physiological integrity or with subtotal injury to key motor system anatomical structures, suggesting the ability to stratify patients to reduce trial variance and increase effect sizes (Nouri & Cramer, 2011).

Many questions remain to be answered to develop brain stimulation as a treatment to improve outcomes after stroke. The source of, and means to deal with, high intra- and inter-subject variability requires further study (Nicolo *et al.*, 2015). It remains unclear how to match patients with the optimal mode of stimulation: (1) some brain stimulation therapies can have bidirectional effects and so, for example, can be set to reduce activity, such as in brain areas that might be interfering with recovery; or (2) they may be set to increase activity, such as in dormant brain areas where increased activity is desired. The maximum impact of stimulation-based approaches to neural repair might be seen when protocols take fullest advantage of the potential of this approach to differentially stimulate various brain areas (Liew *et al.*, 2014), with high temporal resolution and possibly in association with behavioral reinforcement. Application to non-motor behaviors also requires further development (Elsner *et al.*, 2013; Shah *et al.*, 2013).

3.4 Principles of Neural Repair After Stroke

Neural repair after stroke is not a one-size-fits-all undertaking. Instead, the effectiveness of restorative therapies can be maximized with attention to certain principles (Cramer *et al.*, 2011). The evidence for several of these is considered below.

3.4.1 *Time-sensitive*

Neural repair is time-sensitive. Treating stroke recovery is a 4D issue, i.e., time is a factor as much as choice of 3D treatment targets. The cellular and biochemical underpinnings of recovery, many of which are potential therapeutic targets, evolve over time (Jones & Schallert, 1992; Li & Carmichael, 2006; Stroemer *et al.*, 1998). Some biological targets are only relevant to repair during a specific time period after stroke. Furthermore, some therapies can have different effects on stroke depending on timing. For example, long-term effects of a GABA agonist or *N*-methyl-D-aspartate receptor blocker can be favorable if administered in the early hours after stroke (Green *et al.*, 2000; Ovbiagele *et al.*, 2003) but deleterious if initiated days later (Kozlowski *et al.*, 1994; Wahlgren & Martinsson, 1998; Barth *et al.*, 1998), and the reverse may be true for matrix metalloproteinases (Zhao *et al.*, 2007). Similarly, therapies that modulate inflammation are likely time-sensitive given that inflammatory mediators can have detrimental effects acutely but beneficial effects later during repair (Allan & Rothwell, 2003; Lucas *et al.*, 2006).

3.4.2 *Experience-dependent*

Neural repair is experience-dependent. Since the classic study by Feeney *et al.* (1982), which showed that a stimulant promoted improved motor outcome after experimental stroke in rodents only when the drug was paired with training, increasing evidence suggests that a restorative therapy needs reenforcement and shaping through experience to produce best behavioral results. A number of subsequent studies support this conclusion (Adkins *et al.*, 2008; Fang *et al.*, 2010; Garcia-Alias *et al.*, 2009; Hovda & Feeney, 1984; Starkey & Schwab, 2012). As pharmacological and cellular therapies gain traction, increased attention will be needed to patient experiences that occur concomitant with the therapy of interest, as these experiences can influence therapeutic efficacy. Points of interest might include the timing, content, and intensity of such experiences. A related issue is that the psychosocial milieu in which patients experience post-stroke recovery is also a critically important experiential covariate (Johansson, 2000).

An extension of this point is that trials of restorative interventions after stroke must be concerned with any therapy that is provided to patients outside of study procedures, such as standard of care occupational, physical, or speech therapy, which could complicate interpretation of clinical trial results. These will therefore be provided as standard of care to most patients enrolled in a stroke recovery clinical trial, and in a variable manner across treatment sites (Freburger *et al.*, 2011), where functional walking ability was the primary endpoint, 81.9% of enrollees received physical therapy outside of study procedures (an average of 25 such sessions). Several options for addressing this issue have been suggested, including: (1) limiting outside therapy (Dromerick *et al.*, 2009), though patients might be non-compliant and this could slow recruitment; (2) measuring the amount of outside therapy, to be used as a covariate in data analysis (Cramer *et al.*, 2009); and (3) restricting enrollment to healthcare systems where such therapy is relatively standardized (Cramer, 2008).

3.4.3 *Patient stratification*

Patient stratification is critical to maximize the impact of post-stroke neural repair therapies. Patients vary substantially on a wide number of key covariates, each of which can significantly and substantially impact a treatment's effect size. Numerous variables have been found to be potential predictors of stroke outcome, including location and size of injury (Riley *et al.*, 2011; Stinear *et al.*, 2007), genotype (Cramer & Procaccio, 2012; Kato & Serretti, 2010; Kohen *et al.*, 2010; Pearson-Fuhrhop *et al.*, 2013; Siironen *et al.*, 2007), measures of brain function (Cramer *et al.*, 2007a; Wang *et al.*, 2011; Wu *et al.*, 2015), and affective disorders (Gillen *et al.*, 2001; Lai *et al.*, 2002). Such measures may be of pivotal value in definining the population most likely to benefit from a given therapy, as stroke is a very heterogeneous condition. Likely, a combination of measures will be needed to best select and stratify patients in trials and treatment of neural repair after stroke (Burke *et al.*, 2015; Nouri & Cramer, 2011; Stinear *et al.*, 2007, 2012).

3.4.4 *Modality-specific measures*

Modality-specific measures might be useful to capture treatment effects when the target is neural repair (Cramer *et al.*, 2007b). The reasons for this approach revolve around the fact that restorative therapies achieve their effect by improving the function of specific neural systems. Improvement is seen in neural systems where the surviving substrate is amenable to repair: a behavior for which the underlying brain regions are destroyed by stroke is less likely to improve than a behavior for which the underlying brain regions are accessible to a restorative therapy. For example, a patient whose stroke partially spares the motor system but utterly destroys the language system may show gains in strength but not aphasia in response to a restorative therapy. Furthermore, this patient may or may not show gains in global stroke outcome measures such as the modified Rankin Scale score. Such motor gains might be considered functionally worthwhile by many patients and so important to measure in clinical trials.

Modality-specific endpoints are aligned with the strategy of providing increased granularity for measuring treatment gains in individual neural systems and have the same value for understanding spontaneous stroke recovery, where the rate and degree of gains often vary widely across behavioral modalities (Hier *et al.*, 1983; Markgraf *et al.*, 1992; Marshall *et al.*, 2000). Improvement in global clinical status is of course a goal of paramount importance in neural repair, but a treatment that provides gains by promoting neuroplasticity might demonstrate maximum effect in brain networks that have subtotal injury, or that are receiving the highest level of concomitant therapy (such as physical therapy for gait), underscoring the complementary value of endpoints that measure these modality-specific treatment effects. A number of modality-specific endpoints have been successfully incorporated into clinical trials and indeed have been the basis for regulatory approval (Goodman *et al.*, 2010).

3.4.5 *Brain organization*

The nature of brain organization prior to stroke influences post-stroke brain plasticity. For example, in healthy subjects, some behaviors such as language or hand movement tend to be highly lateralized (i.e., generation of the behavior involves mainly one hemisphere), and other behaviors such as bulbar and facial movement tend to be less lateralized (i.e., generation of the behavior involves both hemispheres). These differences remain apparent after stroke. Cramer and Crafton (Cramer & Crafton, 2006) found that face movement is more bilaterally organized than shoulder or arm movement in healthy subjects (i.e., before any stroke), and that this remains true after stroke. Such a difference could have functional implications. Hamdy *et al.* (1996) found that the cortical representation for swallowing is normally bilaterally organized. Not surprisingly, therefore, dysphagic patients who recovered after stroke showed an increase in their cortical pharyngeal map size within the contralesional hemisphere, whereas patients who remained dysphagic did not show this change (Hamdy *et al.*, 1998). For reorganization of brain maps after stroke, the pattern of brain reorganization can influence behavioral status, and this pattern is at times constrained by features of normal brain organization.

Acknowledgments

This research is supported by a grant from the U.S. National Institutes of Health (K24 HD074722) to S.C.C.

References

Ackerley, S.J., Stinear, C.M., *et al.* (2010). Combining theta burst stimulation with training after subcortical stroke. *Stroke* **41**(7): 1568–1572.

Adeoye, O., Hornung, R., *et al.* (2011). Recombinant tissue-type plasminogen activator use for ischemic stroke in the United State: A doubling of treatment rates over the course of 5 years. *Stroke* **42**(7): 1952–1955.

Adkins, D., L, Hsu, J.E., *et al.* (2008). Motor cortical stimulation promotes synaptic plasticity and behavioral improvements following sensorimotor cortex lesions. *Exp Neurol* **212**(1): 14–28.

Aggarwal, S., Pittenger, M.F. (2005). Human mesenchymal stem cells modulate allogeneic immune cell responses. *Blood* **105**(4): 1815–1822.

Ajmo, C.T., Jr., Vernon, D.O., *et al.* (2008). The spleen contributes to stroke-induced neurodegeneration. *J Neurosci Res* **86**(10): 2227–2234.

Allan, S.M., Rothwell, N.J. (2003). Inflammation in central nervous system injury. *Philos Trans R Soc Lond B Biol Sci* **358**(1438): 1669–1677.

Allers, C., Sierralta, W.D., *et al.* (2004). Dynamic of distribution of human bone marrow-derived mesenchymal stem cells after transplantation into adult unconditioned mice. *Transplantation* **78**(4): 503–508.

American Psychiatric Association. (2010). Practice guideline for the treatment of patients with major depressive disorder. 3rd edn. American Psychiatric Association, Arlington, VA.

Ankrum, J., Karp, J.M. (2010). Mesenchymal stem cell therapy: Two steps forward, one step back. *Trends Mol Med* **16**(5): 203–209.

Arai, K., Lo, E.H. (2009). Experimental models for analysis of oligodendrocyte pathophysiology in stroke. *Exp Transl Stroke Med* 1: 6.

Bacigalupo, A., Valle, M., *et al.* (2005). T-cell suppression mediated by mesenchymal stem cells is deficient in patients with severe aplastic anemia. *Exp Hematol* **33**(7): 819–827.

Balasubramanian, S., Klein, J., *et al.* (2010). Robot-assisted rehabilitation of hand function. *Curr Opin Neurol* **23**(6): 661–670.

Bang, O.Y., Lee, J.S., *et al.* (2005). Autologous mesenchymal stem cell transplantation in stroke patients. *Ann Neurol* **57**(6): 874–882.

Barrett, K.M., Brott, T.G., *et al.* (2011). Enhancing recovery after acute ischemic stroke with donepezil as an adjuvant therapy to standard medical care: Results of a phase IIA clinical trial. *Stroke* **20**(3): 177–182.

Barth, T., Hoane, M., *et al.* (1998). Effects of glutamate antagonists on the recovery and maintenance of behavioral function after brain injury. In: Goldstein L., (ed.) *Restorative Neurolog: Advances in Pharmacotherapy for Recovery After Stroke*. Armonk, NY: Futura Publishing Co.

Bartholomew, A., Sturgeon, C., *et al.* (2002). Mesenchymal stem cells suppress lymphocyte proliferation in vitro and prolong skin graft survival *in vivo*. *Exp Hematol* **30**(1): 42–48.

Bednar, M.M. (2008). The role of sildenafil in the treatment of stroke. *Curr Opin Investig Drugs* **9**(7): 754–759.

Belayev, L., Khoutorova, L., *et al.* (2009). A novel neurotrophic therapeutic strategy for experimental stroke. *Brain Res* **128**: 117–123.

Beltran, E.J., Papadopoulos, C.M., *et al.* (2010). Long-term motor improvement after stroke is enhanced by short-term treatment with the alpha-2 antagonist, atipamezole. *Brain Res* **134**: 174–182.

Berthier, M.L., Green, C., *et al.* (2006). A randomized, placebo-controlled study of donepezil in poststroke aphasia. *Neurology* **67**(9): 1687–1689.

Berthier, M.L., Pulvermuller, F. (2011). Neuroscience insights improve neurorehabilitation of poststroke aphasia. *Nat Rev Neurol* **7**(2): 86–97.

Beyth, S., Borovsky, Z., *et al.* (2005). Human mesenchymal stem cells alter antigen-presenting cell maturation and induce T-cell unresponsiveness. *Blood* **105**(5): 2214–2219.

Bhasin, A., Srivastava, M.V., *et al.* (2011). Autologous mesenchymal stem cells in chronic stroke. *Cerebrovasc Dis Extra* **1**(1): 93–104.

Bonilha, L., Nesland, T., *et al.* (2014). Mapping remote subcortical ramifications of injury after ischemic strokes. *Behav Neurol* **2014**: 215380.

Bradnam, L.V., Stinear, C.M., *et al.* (2012). Contralesional hemisphere control of the proximal paretic upper limb following stroke. *Cereb Cortex* **22**(11): 2662–2671.

Brewer, B.R., McDowell, S.K., *et al.* (2007). Poststroke upper extremity rehabilitation: A review of robotic systems and clinical results. *Top Stroke Rehabil* **14**(6): 22–44.

Buchkremer-Ratzmann, I., August, M., *et al.* (1996). Electrophysiological transcortical diaschisis after cortical photothrombosis in rat brain. *Stroke* **27**(6): 1105–1109.

Buchli, A.D., Schwab, M.E. (2005). Inhibition of Nog: A key strategy to increase regeneration, plasticity and functional recovery of the lesioned central nervous system. *Ann Med* **37**(8): 556–567.

Burke Q.E., Dodakian, L., *et al.* (2015). Neural function, injury, and stroke subtype predict treatment gains after stroke. *Ann Neurol* **77**(1): 132–145.

Carmichael, S.T. (2006). Cellular and molecular mechanisms of neural repair after stroke: Making waves. *Ann Neurol* **59**(5): 735–742.

Carmichael, S.T., Tatsukawa, K., *et al.* (2004). Evolution of diaschisis in a focal stroke model. *Stroke* **35**(3): 758–763.

Cassidy, J.M., Cramer, S.C. (2016). Spontaneous and therapeutic-induced mechanisms of functional recovery after stroke. *Transl Stroke Res* **8**(1): 33–46.

Chapman, A.R., Scala, C.C. (2012). Evaluating the first-in-human clinical trial of a human embryonic stem cell-based therapy. *Kennedy Inst Ethics J* **22**(3): 243–261.

Cheatwood, J.L., Emerick, A.J., *et al.* (2008). Nogo—A expression after focal ischemic stroke in the adult rat. *Stroke* **39**(7): 2091–2098.

Chen, J., Li, Y., *et al.* (2003). Intravenous bone marrow stromal cell therapy reduces apoptosis and promotes endogenous cell proliferation after stroke in female rat. *J Neurosci Res* **73**(6): 778–786.

Chen, M., Li, X., *et al.* (2015). The inhibitory effect of mesenchymal stem cell on blood-brain barrier disruption following intracerebral hemorrhage in rats: Contribution of TSG-6. *J Neuroinflammation* **1:** 61.

Chen, C.L., Young, S.H., *et al.* (2013). Chinese medicine neuroaid efficacy on stroke recovery: A double-blind, placebo-controlled, randomized study. *Stroke* **44**(8): 2093–2100.

Chollet, F., Tardy, J., *et al.* (2011). Fluoxetine for motor recovery after acute ischaemic stroke (FLAME): A randomised placebo-controlled trial. *Lancet Neurol* **10**(2): 123–130.

Clarkson, A.N., Huang, B.S., *et al.* (2010). Reducing excessive GABA-mediated tonic inhibition promotes functional recovery after stroke. *Nature* **468:** 305–309.

Clarkson, A.N., Overman, J.J., *et al.* (2011). AMPA receptor-induced local brain-derived neurotrophic factor signaling mediates motor recovery after stroke. *J Neurosci* **31**(10): 3766–3775.

Cohen, J.A. (2013). Mesenchymal stem cell transplantation in multiple sclerosis. *J Neurol Sci* **333**(1–2): 43–49.

Cools, R., Roberts, A.C., *et al.* (2008). Serotoninergic regulation of emotional and behavioural control processes. *Trends Cogn Sci* **12**(1): 31–40.

Corcione, A., Benvenuto, F., *et al.* (2006). Human mesenchymal stem cells modulate B-cell functions. *Blood* **107**(1): 367–372.

Cowen, P., Sherwood, A.C. (2013). The role of serotonin in cognitive functio: Evidence from recent studies and implications for understanding depression. *J Psychopharmacol* **27**(7): 575–583.

Cramer, S.C. (2008a) Repairing the human brain after stroke: I. Mechanisms of spontaneous recovery. *Ann Neurol* **63**(3): 272–287.

Cramer, S.C. (2008b) Repairing the human brain after stroke. II. Restorative therapies. *Ann Neurol* **63**(5): 549–560.

Cramer, S.C. (2010). Brain repair after stroke. *N Engl J Med* **362**(19): 1827–1829.

Cramer, S.C., Abila, B., *et al.* (2013). Safety, pharmacokinetics, and pharmacodynamics of escalating repeat doses of GSK249320 in patients with stroke. *Stroke* **44**(5): 1337–1342.

Cramer, S.C., Chopp, M. (2000). Recovery recapitulates ontogeny. *Trends Neurosci* **23**(6): 265–271.

Cramer, S.C., Crafton K.R. (2006). Somatotopy and movement representation sites following cortical stroke. *Exp Brain Res* **168**(1–2): 25–32.

Cramer, S.C., Dobkin, B.H., *et al.* (2009). Randomized, placebo-controlled, double-blind study of ropinirole in chronic stroke. *Stroke* **40**(9): 3034–3038.

Cramer, S.C., Fitzpatrick. C., *et al.* (2010). The beta-hCG+erythropoietin in acute stroke (BETAS) study: A 3-center, single-dose, open-label, noncontrolled, phase IIa safety trial. *Stroke* **41**(5): 927–931.

Cramer, S.C., Hill, M.D. (2014). Human choriogonadotropin and epoetin alfa in acute ischemic stroke patients (REGENESIS-LED trial). *Stroke* **9**(3): 321–327.

Cramer, S.C., Koroshetz, W.J., *et al.* (2007a) The case for modality-specific outcome measures in clinical trials of stroke recovery-promoting agents. *Stroke* **38**(4): 1393–1395.

Cramer, S.C., Parrish, T.B., *et al.* (2007b). Predicting functional gains in a stroke trial. *Stroke* **38**(7): 2108–2114.

Cramer, S.C., Procaccio, V. (2012). Correlation between genetic polymorphisms and stroke recovery: Analysis of the GAIN Americas and GAIN International Studies. *Eur J Neurol* **19**(5): 718–724.

Cramer, S.C., Sur, M., *et al.* (2011). Harnessing neuroplasticity for clinical applications. *Brain* **134**(Pt 6): 1591–1609.

Crisostomo, E.A., Duncan, P.W., *et al.* (1988). Evidence that amphetamine with physical therapy promotes recovery of motor function in stroke patients. *Ann Neurol* **23**(1): 94–97.

Cumberland Consensus Working Group, Cheeran, B., *et al.* (2009). The future of restorative neurosciences in stroke: Driving the translational research pipeline from basic science to rehabilitation of people after stroke. *Neurorehabil Neural Repair* **23**(2): 97–107.

Cyranoski, D. (2012). Canada approves stem cell product. *Nat Biotechnol* **30**(7): 571.

Dam, M., Tonin, P., *et al.* (1996). Effects of fluoxetine and maprotiline on functional recovery in poststroke hemiplegic patients undergoing rehabilitation therapy. *Stroke* **27**(7): 1211–1214.

Dancause, N., Barbay, S., *et al.* (2005). Extensive cortical rewiring after brain injury. *J Neurosci* **25**(44): 10167–10179.

Davalos, A., Alvarez-Sabin, J., *et al.* (2012). Citicoline in the treatment of acute ischaemic stroke: An international, randomised, multicentre, placebo-controlled study (ICTUS trial). *Lancet* **380**(9839): 349–357.

Detante, O., Moisan, A., *et al.* (2009). Intravenous administration of 99mTc-HMPAO-labeled human mesenchymal stem cells after stroke: *In vivo* imaging and biodistribution. *Cell Transplant* 18(12): 1369–1379.

Devine, S. (2002). Mesenchymal stem cell: Will they have a role in the clinic? *J Cell Biochem Suppl* **38**: 73–79.

Dharmasaroja, P. (2009). Bone marrow-derived mesenchymal stem cells for the treatment of ischemic stroke. *J Clin Neurosci* **16**(1): 12–20.

Di Nicola, M., Carlo-Stella, C., *et al.* (2002). Human bone marrow stromal cells suppress T-lymphocyte proliferation induced by cellular or nonspecific mitogenic stimuli. *Blood* **99**(10): 3838–3843.

Dijkhuizen, R.M., van der Marel, K., *et al.* (2012). Functional MRI and diffusion tensor imaging of brain reorganization after experimental stroke. *Transl Stroke Res* 3(1): 36–43.

Ding, G., Jiang, Q., *et al.* (2008). Magnetic resonance imaging investigation of axonal remodeling and angiogenesis after embolic stroke in sildenafil-treated rats. *J Cereb Blood Flow Metab* **28**(8): 1440–1448.

Domeniconi, M., Filbin, M.T. (2005). Overcoming inhibitors in myelin to promote axonal regeneration. *J Neurol Sci* **233**(1–2): 43–47.

Dominici, M., Le Blanc, K., *et al.* (2006). Minimal criteria for defining multipotent mesenchymal stromal cells. The International Society for Cellular Therapy position statement. *Cytotherapy* 8(4): 315–317.

Dromerick, A.W., Lang, C.E., *et al.* (2007). Very Early Constraint-induced Movement Therapy (VECTORS: Phase II Trial Results (Abstract). *Stroke* 38(2): 465.

Dromerick, A.W., Lang, C.E., *et al.* (2009). Very Early Constraint-induced Movement during Stroke Rehabilitation (VECTORS): A single-center RCT. *Neurology* **73**(3): 195–201.

Duncan, P.W., Sullivan, K.J., *et al.* (2011). Body-weight-supported treadmill rehabilitation after stroke. *New Engl J Med* **364**(21): 2026–2036.

Eckert, M.A., Vu, Q., *et al.* (2013). Evidence for high translational potential of mesenchymal stromal cell therapy to improve recovery from ischemic stroke. *J Cereb Blood Flow Metab* **33**(9): 1322–1334.

Edwardson, M.A., Lucas, T.H., *et al.* (2013). New modalities of brain stimulation for stroke rehabilitation. *Exp Brain Res* **224**(3): 335–358.

Ehrenreich, H., Weissenborn, K., *et al.* (2009). Recombinant human erythropoietin in the treatment of acute ischemic stroke. *Stroke* **40**(12): e647–e656.

Elsner, B., Kugler, J., *et al.* (2013). Transcranial direct current stimulation (tDCS) for improving function and activities of daily living in patients after stroke. *Cochrane Database Syst Rev* **15**(11): CD009645.

England, T.J., Abaei, M., *et al.* (2012). Granulocyte-colony stimulating factor for mobilizing bone marrow stem cells in subacute strok: The stem cell trial of recovery enhancement after stroke 2 randomized controlled trial. *Stroke* **43**(2): 405–411.

Erices, A.A., Allers, C.I., *et al.* (2003). Human cord blood-derived mesenchymal stem cells home and survive in the marrow of immunodeficient mice after systemic infusion. *Cell Transplant* **12**(6): 555–561.

Fang, P.C., Barbay, S., *et al.* (2010). Combination of NEP 1–40 treatment and motor training enhances behavioral recovery after a focal cortical infarct in rats. *Stroke* **41**(3): 544–549.

Fasotti, L., van Kessel, M. (2013). Novel insights in the rehabilitation of neglect. *Front Hum Neurosci* **7**: 780.

Feeney, D., Baron, J. (1986). Diaschisis. *Stroke* **17**(5): 817–830.

Feeney, D., Gonzalez, A., *et al.* (1982). Amphetamine, halperidol, and experience interact to affect the rate of recovery after motor cortex injury. *Science* **217**(4562): 855–857.

Figueroa, F.E., Carrion, F., *et al.* (2012). Mesenchymal stem cell treatment for autoimmune diseases: A critical review. *Biol Res* **45**(3): 269–277.

Finklestein, S.P., Caday, C.G., *et al.* (1990). Growth factor expression after stroke. *Stroke* **21**(11 Suppl): III122-III124.

Floel, A., Cohen, L.G. (2010). Recovery of function in humans: Cortical stimulation and pharmacological treatments after stroke. *Neurobiol Dis* **37**(2): 243–251.

Food and Drug Administration Center for Biologics Evaluation and Research. (2011). Guidance for industr: Potency tests for cellular and gene therapy products. http://www.fda.gov/downloads/BiologicsBloodVaccines/GuidanceComplianceRegulatoryInformation/Guidances/CellularandGeneTherapy/UCM243392.pdf. Last accessed 3 August 2017.

Freburger, J.K., Holmes, G.M., *et al.* (2011). Disparities in postacute rehabilitation care for stroke: An analysis of the state inpatient databases. *Arch Phys Med Rehabil* **92**(8): 1220–1229.

Fruehwald, S., Gatterbauer, E., *et al.* (2003). Early fluoxetine treatment of post-stroke depression — a three-month double-blind placebo-controlled study with an open-label long-term follow up. *J Neurol* **250**(3): 347–351.

Gao, J., Dennis, J.E., *et al.* (2001). The dynamic *in vivo* distribution of bone marrow-derived mesenchymal stem cells after infusion. *Cells Tissues Organs* **169**(1): 12–20.

Garcia-Alias, G., Barkhuysen, S., *et al.* (2009). Chondroitinase ABC treatment opens a window of opportunity for task-specific rehabilitation. *Nat Neurosci* **12**(9): 1145–1151.

Gillen, R., Tennen, H., *et al.* (2001). Depressive symptoms and history of depression predict rehabilitation efficiency in stroke patients. *Arch Phys Med Rehabil* **82**(12): 1645–1649.

Gladstone, D.J., Danells, C.J., *et al.* (2006). Physiotherapy coupled with dextroamphetamine for rehabilitation after hemiparetic stroke: A randomized, double-blind, placebo-controlled trial. *Stroke* **37**(1): 179–185.

Goodman, A.D., Brown, T.R., *et al.* (2010). A phase 3 trial of extended release oral dalfampridine in multiple sclerosis. *Ann Neurol* **68**(4): 494–502.

Goyal, M., Menon, B.K., *et al.* (2016). Endovascular thrombectomy after large-vessel ischaemic stroke: A meta-analysis of individual patient data from five randomised trials. *Lancet* **387**(10029): 1723–1731.

Green, A.R., Hainsworth, A.H., *et al.* (2000). GABA potentiation: A logical pharmacological approach for the treatment of acute ischaemic stroke. *Neuropharmacology* **39**(9): 1483–1494.

Grefkes, C., Ward, NS. (2014). Cortical reorganization after stroke: How much and how functional? *Neuroscientist* **20**(1): 56–70.

Grieve, S.M., Bhindi, R., *et al.* (2010). Microvascular obstruction by intracoronary delivery of mesenchymal stem cells and quantification of resulting myocardial infarction by cardiac magnetic resonance. *Circ Heart Fail* 3(3): e5–e6.

Hacke, W., Kaste, M., *et al.* (2008). Thrombolysis with alteplase 3 to 4.5 hours after acute ischemic stroke. *N Engl J Med* **359**(13): 1317–1329.

Hamdy, S., Aziz, Q., *et al.* (1996). The cortical topography of human swallowing musculature in health and disease. *Nat Med* **2**(11): 1217–1224.

Hamdy, S., Aziz, Q., *et al.* (1998). Recovery of swallowing after dysphagic stroke relates to functional reorganization in the intact motor cortex. *Gastroenterology* **115**(5): 1104–1112.

Heiss, W.D., Thiel, A. (2006). A proposed regional hierarchy in recovery of post-stroke aphasia. *Brain Lang* 98:118–123.

Hermann, D.M. (2009). Enhancing the delivery of erythropoietin and its variants into the ischemic brain. *ScientificWorldJournal* **9**: 967–969.

Hermann, D.M., Chopp, M. (2012). Promoting brain remodelling and plasticity for stroke recover: Therapeutic promise and potential pitfalls of clinical translation. *Lancet Neurol* **11**: 369–380.

Hess, D., Clark, W., *et al.* (2015). Results of the double-blind, randomized, placebo controlled, Phase 2 safety and efficacy trial of multistem in adults with ischemic stroke (Abstract). Presented a: European Stroke Organisation Conference, Apr 17–19, Glasgow, UK.

Hier, D., Mondlock, J., *et al.* (1983). Recovery of behavioral abnormalities after right hemisphere stroke. *Neurology* **33**(3): 345–350.

Hilfiker, A., Kasper, C., *et al.* (2011). Mesenchymal stem cells and progenitor cells in connective tissue engineering and regenerative medicine: Is there a future for transplantation? *Langenbecks Arch Surg* **396**(4): 489–497.

Honmou, O., Houkin, K., *et al.* (2011). Intravenous administration of auto serum-expanded autologous mesenchymal stem cells in stroke. *Brain* **134**(Pt 6):1790–1807.

Honmou, O., Onodera, R., *et al.* (2012). Mesenchymal stem cells: Therapeutic outlook for stroke. *Trends Mol Med* **18**(5): 292–297.

Horita, Y., Honmou, O., *et al.* (2006). Intravenous administration of glial cell line-derived neurotrophic factor gene-modified human mesenchymal stem cells protects against injury in a cerebral ischemia model in the adult rat. *J Neurosci Res* **84**(7): 1495–1504.

Hosp, J.A., Pekanovic, A., *et al.* (2011). Dopaminergic projections from midbrain to primary motor cortex mediate motor skill learning. *J Neurosci* **31**(7): 2481–2787.

Hovda, D.A., Feeney, D.M. (1984). Amphetamine with experience promotes recovery of locomotor function after unilateral frontal cortex injury in the cat. *Brain Res* **298**(2): 358–361.

Hyun I. (2010). The bioethics of stem cell research and therapy. *J Clin Invest* **120**(1): 71–75.

Iaci, J.F., Parry, T.J., *et al.* (2013). Dalfampridine improves sensorimotor function in rats with chronic deficits after middle cerebral artery occlusion. *Stroke* **44**(7): 1942–1950.

Iadecola, C., Anrather, J. (2011). The immunology of stroke: From mechanisms to translation. *Nat Med* **17**(7): 796–808.

Iwase, T., Nagaya, N., *et al.* (2005). Comparison of angiogenic potency between mesenchymal stem cells and mononuclear cells in a rat model of hindlimb ischemia. *Cardiovasc Res* **66**(3): 543–551.

Jerndal, M., Forsberg, K., *et al.* (2010). A systematic review and meta-analysis of erythropoietin in experimental stroke. *J Cereb Blood Flow Metab* **30**(5): 961–968.

Johansen-Berg, H., Dawes, H., *et al.* (2002). Correlation between motor improvements and altered fMRI activity after rehabilitative therapy. *Brain* **125**(Pt 12): 2731–2742.

Johansson, B. (2000). Brain plasticity and stroke rehabilitation. The Willis Lecture. *Stroke* **31**(1): 223–230.

Johnston, S.C., Hauser, S.L. (2008). Neurological disease on the global agenda. *Ann Neurol* **64**(1): A11-A12.

Jones, T., Kleim, J., *et al.* (1996). Synaptogenesis and dendritic growth in the cortex opposite unilateral sensorimotor cortex damage in adult rats: A quantitative electron microscopic examination. *Brain Res* **733**(1): 142–148.

Jones, T., Schallert, T. (1992). Overgrowth and pruning of dendrites in adult rats recovering from neocortical damage. *Brain Res* **581**(1): 156–160.

Kato, M., Serretti, A. (2010). Review and meta-analysis of antidepressant pharmacogenetic findings in major depressive disorder. *Mol Psychiatry* **15**(5): 473–500.

Kawamata, T., Dietrich, W., *et al.* (1997b) Intracisternal basic fibroblast growth factor (bFGF) enhances functional recovery and upregulates the expression of a molecular marker of neuronal sprouting following focal cerebral infarction. *Proc Natl Acad Sci USA* **94**: 8179–8184.

Kawamata, T., Speliotes, E., *et al.* (1997a). The role of polypeptide growth factors in recovery from stroke. I: Freund, H.J., Sabel, B.A., *et al.*, (eds.). *Brain Plasticity*. Philadelphia: Lippincott-Raven.

Kilgard, M.P., Merzenich, M.M. (1998). Cortical map reorganization enabled by nucleus basalis activity. *Science* **279**(5367): 1714–1718.

Kim, B.O., Tian, H., *et al.* (2005). Cell transplantation improves ventricular function after a myocardial infarction: A preclinical study of human unrestricted somatic stem cells in a porcine model. *Circulation* **112**(9 Suppl): I96–I104.

Knecht, S., Hesse, S., *et al.* (2011). Rehabilitation after stroke. *Dtsch Arztebl Int* **108**(36): 600–606.

Kohen, R., Cain, K.C., *et al.* (2011). Response to psychosocial treatment in poststroke depression is associated with serotonin transporter polymorphisms. *Stroke* **42**(7): 2068–2070.

Kolb, B., Morshead, C., *et al.* (2007). Growth factor-stimulated generation of new cortical tissue and functional recovery after stroke damage to the motor cortex of rats. *J Cereb Blood Flow Metab* **27**(5): 983–997.

Kozlowski, D., Jones, T., *et al.* (1994). Pruning of dendrites and restoration of function after brain damag: Role of the NMDA receptor. *Restor Neurol Neurosci* **7**(2): 119–126.

Krabbe, C., Zimmer, J., *et al.* (2005). Neural transdifferentiation of mesenchymal stem cells — a critical review. *APMIS* **113**(11–12): 831–844.

Lai, S.M., Duncan, P.W., *et al.* (2002). Depressive symptoms and independence in BADL and IADL. *J Rehabil Res Dev* **39**(5): 589–596.

Lalu, M.M., McIntyre, L., *et al.* (2012). Safety of cell therapy with mesenchymal stromal cells (SafeCell): A systematic review and meta-analysis of clinical trials. *PloS One* **7**(10): e47559.

Lanfranconi, S., Locatelli, F., *et al.* (2009). Growth factors in ischemic stroke. *J Cell Mol Med* **15**(8): 1645–1687.

Lang, C.E., Strube, M.J., *et al.* (2016). Dose response of task-specific upper limb training in people at least 6 months poststroke: A phase II, single-blind, randomized, controlled trial. *Ann Neurol* **80**(3): 342–354.

Lee, R.H., Pulin, A.A., *et al.* (2009). Intravenous hMSCs improve myocardial infarction in mice because cells embolized in lung are activated to secrete the anti-inflammatory protein TSG-6. *Cell Stem Cell* **5**(1): 54–63.

Levy, R.M., Benson, R.R., *et al.* (2008). Cortical stimulation for upper-extremity hemiparesis from ischemic stroke: Everest Study primary endpoint results (Abstract). *Stroke* **39**(2): 568.

Levy, R.M., Harvey R.L, *et al.* (2016). Epidural electrical stimulation for stroke rehabilitation: Results of the prospective, multicenter, randomized, single-blinded Everest Trial. *Neurorehabil Neural Repair* **30**(2): 107–119.

Li, S., Carmichael, S.T. (2006). Growth-associated gene and protein expression in the region of axonal sprouting in the aged brain after stroke. *Neurobiol Dis* **23**(2): 362–373.

Li, Y., Chopp, M. (2009). Marrow stromal cell transplantation in stroke and traumatic brain injury. *Neurosci Lett* **456**(3): 120–123.

Liew, S.L., Santarnecchi, E., *et al.* (2014). Non-invasive brain stimulation in neurorehabilitation: Local and distant effects for motor recovery. *Front Hum Neurosci* **8**: 378.

Lindenberg, R., Zhu, L.L., *et al.* (2011). Predicting functional motor potential in chronic stroke patients using diffusion tensor imaging. *Hum Brain Mapp* **33**(5): 1040–1051.

Lindvall, O., Kokaia, Z. (2011). Stem cell research in stroke: How far from the clinic? *Stroke* **42**(8): 2369–2375.

Liu, H., Honmou, O., *et al.* (2006). Neuroprotection by PlGF gene-modified human mesenchymal stem cells after cerebral ischaemia. *Brain* **129**(Pt 10): 2734–2745.

Lloyd-Jones, D., Adams, R.J., *et al.* (2010). Heart disease and stroke statistics — 2010 update: A report from the American Heart Association. *Circulation* **121**(7): e46-e215.

Lo, A.C., Guarino, P.D., *et al.* (2010). Robot-assisted therapy for long-term upper-limb impairment after stroke. *N Engl J Med* **362**(19): 1772–1783.

Logue, S.F., Gould, T.J. (2014). The neural and genetic basis of executive function: Attention, cognitive flexibility, and response inhibition. *Pharmacol Biochem Behav* **123**: 45–54.

Lotze, M., Markert, J., *et al.* (2006). The role of multiple contralesional motor areas for complex hand movements after internal capsular lesion. *J Neurosci* **26**(22): 6096–6102.

Lucas, S.M., Rothwell, N.J., *et al.* (2006). The role of inflammation in CNS injury and disease. *Br J Pharmacol* **147** (Suppl 1): S232–S240.

Luria, A. (1963). *Restoration of Function After Brain Injury*. New York: Macmillan Company.

MacDonald, H.J., Stinear, C.M., *et al.* Dopamine gene profiling to predict impulse control and effects of dopamine agonist ropinirole. *J Cogn Neurosci* **28**(7): 909–919.

Maggini, J., Mirkin, G., *et al.* (2010). Mouse bone marrow-derived mesenchymal stromal cells turn activated macrophages into a regulatory-like profile. *PloS One* **5**(2): e9252.

Maltman, D.J., Hardy, S.A., *et al.* (2011). Role of mesenchymal stem cells in neurogenesis and nervous system repair. *Neurochem Int* **59**(3): 347–356.

Marigo, I., Dazzi, F. (2011). The immunomodulatory properties of mesenchymal stem cells. *Semin Immunopathol* **33**(6): 593–602.

Markgraf, C., Green, E., *et al.* (1992). Sensorimotor and cognitive consequences of middle cerebral artery occlusion in rats. *Brain Res* **575**(2): 238–246.

Marshall, R., Perera, G., *et al.* (2000). Evolution of cortical activation during recovery from corticospinal tract infarction. *Stroke* **31**(3): 656–661.

McAllister, T.W. (2009). Polymorphisms in genes modulating the dopamine system: Do they influence outcome and response to medication after traumatic brain injury? *J Head Trauma Rehabil* **24**(1): 65–68.

Merzenich, M.M., Van Vleet, T.M., *et al.* (2014). Brain plasticity-based therapeutics. *Front Human Neurosci* **8**: 385.

Mikami, K., Jorge, R.E., *et al.* (2014). Prevention of post-stroke generalized anxiety disorder, using escitalopram or problem-solving therapy. *J Neuropsychiatry Clin Neurosci* **26**(4): 323–328.

Miller, R.H., Bai, L., *et al.* (2010). The potential of mesenchymal stem cells for neural repair. *Discov Med* **9**(46): 236–242.

Miyai, I, Reding, M.J. (1998). Effects of antidepressants on functional recovery following stroke. *J Neuro Rehab* **12**(1): 5–13.

Molina-Luna, K., Pekanovic, A., *et al.* (2009). Dopamine in motor cortex is necessary for skill learning and synaptic plasticity. *PLoS One* **4**(9): e7082.

Murphy, T.H., Corbett, D. (2009). Plasticity during stroke recovery: From synapse to behaviour. *Nat Rev Neurosci* **10**(12): 861–872.

Nasef, A., Mathieu, N., *et al.* (2007). Immunosuppressive effects of mesenchymal stem cells: Involvement of HLA-G. *Transplantation* **84**(2): 231–237.

National Institute of Neurological Disorders (NINDS) and Stroke rt-PA Stroke Study Group. (1995). Tissue plasminogen activator for acute ischemic stroke. *N Engl J Med* **333**(24): 1581–1587.

Nauta, A.J., Westerhuis, G., *et al.* (2006). Donor-derived mesenchymal stem cells are immunogenic in an allogeneic host and stimulate donor graft rejection in a nonmyeloablative setting. *Blood* **108**(6): 2114–2120.

Netz, J., Lammers, T., *et al.* (1997). Reorganization of motor output in the non-affected hemisphere after stroke. *Brain* **120**(Pt 9): 1579–1586.

Nhan, H., Barquist, K., *et al.* (2004). Brain function early after stroke in relation to subsequent recovery. *J Cereb Blood Flow Metab* **24**(7): 756–763.

Nicolo, P., Ptak, R., *et al.* (2015). Variability of behavioural responses to transcranial magnetic stimulatio: Origins and predictors. *Neuropsychologia* **74**: 137–144.

Nouri, S., Cramer, S.C. (2011). Anatomy and physiology predict response to motor cortex stimulation after stroke. *Neurology* **77**(11): 1076–1083.

Nudo, R.J. (2011). Neural bases of recovery after brain injury. *J Commun Disord* **44**(5): 515–520.

Oh, J.Y., Roddy, G.W., *et al.* (2010). Anti-inflammatory protein TSG-6 reduces inflammatory damage to the cornea following chemical and mechanical injury. *Proc Natl Acad Sci USA* **107**(39): 16875–16880.

Ovbiagele, B., Kidwell, C.S., *et al.* (2003). Neuroprotective agents for the treatment of acute ischemic stroke. *Curr Neurol Neurosci Rep* **3**(1): 9–20.

Overman, J.J., Carmichael, S.T. (2014). Plasticity in the injured brain: More than molecules matter. *Neuroscientist* **20**(1): 15–28.

Pappata, S., Levasseur, M., *et al.* (2000). Thalamic microglial activation in ischemic stroke detected *in vivo* by PET and [11C] PK11195. *Neurology* **55**(7): 1052–1054.

Pardridge, W.M. (2012). Drug transport across the blood-brain barrier. *J Cereb Blood Flow Metab* **32**(11): 1959–1972.

Pariente, J., Loubinoux, I., *et al.* (2001). Fluoxetine modulates motor performance and cerebral activation of patients recovering from stroke. *Ann Neurol* **50**(6): 718–729.

Pearson-Fuhrhop, K.M., Minton, B., *et al.* (2013). Genetic variation in the human brain dopamine system influences motor learning and its modulation by L-dopa. *PloS One* **8**(4): e61197.

Phinney, D.G., Prockop, D.J. (2007). Concise review: mesenchymal stem/multipotent stromal cell: The state of transdifferentiation and modes of tissue repair — current views. *Stem Cells* **25**(11): 2896–2902.

Pomeroy, V.M., Cloud, G., *et al.* (2007). Transcranial magnetic stimulation and muscle contraction to enhance stroke recovery: A randomized proof-of-principle and feasibility investigation. *Neurorehabil Neural Repair* **21**(6): 509–517.

Que, M., Schiene, K., *et al.* (1999). Widespread up-regulation of N-methyl-D-aspartate receptors after focal photothrombotic lesion in rat brain. *Neurosci Lett* **273**(2): 77–80.

Ramanathan, D., Tuszynski, M.H., *et al.* (2009). The basal forebrain cholinergic system is required specifically for behaviorally mediated cortical map plasticity. *J Neurosci* **29**(18): 5992–6000.

Redecker, C., Wang, W., *et al.* (2002). Widespread and long-lasting alterations in GABA(A)-receptor subtypes after focal cortical infarcts in rats: Mediation by NMDA-dependent processes. *J Cereb Blood Flow Metab* **22**(12): 1463–1475.

Reed, S., Cramer, S.C., *et al.* (2001). Treatment with tissue plasminogen activator and inpatient mortality rates for patients with ischemic stroke treated in community hospitals. *Stroke* **32**(8): 1832–1840.

Reinkensmeyer, D., Emken, J., *et al.* (2004). Robotics, motor learning, and neurologic recovery. *Annu Rev Biomed Eng* **6**: 497–525.

Ren, J.M., Finklestein, S.P. (2005). Growth factor treatment of stroke. *Curr Drug Targets CNS Neurol Disord* **4**(2): 121–125.

Ribeiro, C.A., Salgado, A.J., *et al.* (2011). The secretome of bone marrow mesenchymal stem cells-conditioned media varies with time and drives a distinct effect on mature neurons and glial cells (primary cultures). *J Tissue Eng Regen Med* **5**(8): 668–672.

Riley, J.D., Le, V., *et al.* (2011). Anatomy of stroke injury predicts gains from therapy. *Stroke* **42**(2): 421–426.

Ringelstein, E.B., Thijs, V., *et al.* (2013). Granulocyte colony-stimulating factor in patients with acute ischemic strok: Results of the AX200 for Ischemic Stroke Trial. *Stroke* **44**(10): 2681–2687.

Rivers, W.H., Head, H. (1908). A human experiment in nerve division. *Brain* **3**: 323–450.

Robinson, R.G., Jorge, R.E., *et al.* (2008). Escitalopram and problem-solving therapy for prevention of poststroke depression: A randomized controlled trial. *JAMA* **299**(20): 2391–2400.

Sahota, P., Vahidy, F., *et al.* (2013). Changes in spleen size in patients with acute ischemic stroke: A pilot observational study. *Stroke* **8**(2): 60–67.

Sakai, T., Johnson, K.J., *et al.* (2001). Plasma fibronectin supports neuronal survival and reduces brain injury following transient focal cerebral ischemia but is not essential for skin-wound healing and hemostasis. *Nat Med* **7**(3): 324–330.

Sato, K., Ozaki, K., *et al.* (2007). Nitric oxide plays a critical role in suppression of T-cell proliferation by mesenchymal stem cells. *Blood* **109**(1): 228–234.

Savitz, S.I., Cramer, S.C., *et al.* (2014). Stem cells as an emerging paradigm in stroke: Enhancing the development of clinical trials. *Stroke* **45**(2): 634–639.

Schabitz, W.R., Laage, R., *et al.* (2010). AXIS: A trial of intravenous granulocyte colony-stimulating factor in acute ischemic stroke. *Stroke* **41**(11): 2545–2551.

Scheidtmann, K., Fries, W., *et al.* (2001). Effect of levodopa in combination with physiotherapy on functional motor recovery after stroke: A prospective, randomised, double-blind study. *Lancet* **358**(9284): 787–790.

Seevinck, P.R., Deddens, L.H., *et al.* (2010). Magnetic resonance imaging of brain angiogenesis after stroke. *Angiogenesis* **13**(2): 101–111.

Shah, P.P., Szaflarski, J.P., *et al.* (2013). Induction of neuroplasticity and recovery in post-stroke aphasia by non-invasive brain stimulation. *Front Hum Neurosci* **7**: 888.

Siironen, J., Juvela, S., *et al.* (2007). The Met allele of the BDNF Val66Met polymorphism predicts poor outcome among survivors of aneurysmal subarachnoid hemorrhage. *Stroke* **38**(10): 2858–2860.

Silver, B., McCarthy, S., *et al.* (2009). Sildenafil treatment of subacute ischemic strok: A safety study at 25-mg daily for 2 weeks. *J Stroke Cerebrovasc Dis* **18**(5): 381–383.

Simpson, D.M., Goldenberg, J., *et al.* (2015). Dalfampridine in chronic sensorimotor deficits after ischemic strok: A proof of concept study. *J Rehabil Med* **47**(10): 924–931.

Siniscalchi, A., Gallelli, L., *et al.* (2014). Cerebral stroke injury: The role of cytokines and brain inflammation. *J Basic Clin Physiol Pharmacol* **25**(2): 131–137.

Sist, B., Fouad, K., *et al.* (2014). Plasticity beyond peri-infarct cortex: Spinal up regulation of structural plasticity, neurotrophins, and inflammatory cytokines during recovery from cortical stroke. *Exp Neurol* **252**: 47–56.

Small, S.L., Llano, D.A. (2009). Biological approaches to aphasia treatment. *Curr Neurol Neurosci Rep* **9**(6): 443–450.

Sobesky, J., Thiel, A., *et al.* (2005). Crossed cerebellar diaschisis in acute human stroke: A PET study of serial changes and response to supratentorial reperfusion. *J Cereb Blood Flow Metab* **25**(12): 1685–1691.

Starkey, M.L., Schwab, M.E. (2012). Anti-Nogo-A and training: Can one plus one equal three? *Exp Neurol* **235**(1): 53–61.

Steffenhagen, C., Dechant, F.X., *et al.* (2012). Mesenchymal stem cells prime proliferating adult neural progenitors toward an oligodendrocyte fate. *Stem Cells Dev* **21**(11): 1838–1851.

Steinberg, G.K., Kondziolka, D., *et al.* (2016). Clinical outcomes of transplanted modified bone marrow-derived mesenchymal stem cells in stroke: A Phase 1/2a study. *Stroke* **47**(7): 1817–1824.

Stinear, C.M., Barber, P.A., *et al.* (2007). Functional potential in chronic stroke patients depends on corticospinal tract integrity. *Brain* **130**(Pt 1): 170–180.

Stinear, C.M., Barber, P.A., *et al.* (2012). The PREP algorithm predicts potential for upper limb recovery after stroke. *Brain* **135**(Pt 8): 2527–2535.

Stinear, C.M., Byblow, W.D., *et al.* (2014). An update on predicting motor recovery after stroke. *Ann Phys Rehabil Med* **57**(8): 489–498.

Stroemer, R., Kent, T., *et al.* (1998). Enhanced neocortical neural sprouting, synaptogenesis, and behavioral recovery with D-amphetamine therapy after neocortical infarction in rats. *Stroke* **29**(11): 2381–2395.

Teng, H., Zhang, Z.G., *et al.* (2008). Coupling of angiogenesis and neurogenesis in cultured endothelial cells and neural progenitor cells after stroke. *J Cereb Blood Flow Metab* **28**(4): 764–771.

Tobin, M.K., Bonds, J.A., *et al.* (2014). Neurogenesis and inflammation after ischemic stroke: What is known and where we go from here. *J Cereb Blood Flow Metab* **34**(10): 1573–1584.

Tritsch, N.X., Sabatini, B.L. (2012). Dopaminergic modulation of synaptic transmission in cortex and striatum. *Neuron* **76**(1): 33–50.

Turton, A., Wroe, S., *et al.* (1996). Contralateral and ipsilateral EMG responses to transcranial magnetic stimulation during recovery of arm and hand function after stroke. *Electroencephalogr Clin Neurophysiol* **101**(4): 316–328.

Veerbeek, J.M., Langbroek-Amersfoort, A.C., *et al.* (2017). Effects of robot-assisted therapy for the upper limb after stroke. *Neurorehabil Neural Repair* **31**(2): 107–121.

Volpe, B.T., Huerta, P.T., *et al.* (2009). Robotic devices as therapeutic and diagnostic tools for stroke recovery. *Arch Neurol* **66**(9): 1086–1090.

Volz, L.J., Rehme, A.K., *et al.* (2016). Shaping early reorganization of neural networks promotes motor function after stroke. *Cereb Cortex* **26**(6): 2882–2894.

von Monakow, C. (1969). Diaschisis, 1914. In: Pribram, K., (ed.). *Brain and Behavior 1. Mood, States and Mind*. Baltimore: Penguin Books.

Vu, Q., Xie, K., *et al.* (2014). Meta-analysis of preclinical studies of mesenchymal stromal cells for ischemic stroke. *Neurology* **82**(14): 1277–1286.

Wahlgren, N., Martinsson, L. (1998). New concepts for drug therapy after stroke. Can we enhance recovery? *Cerebrovasc Dis* **8** (Suppl 5): 33–38.

Walczak, P., Zhang, J., *et al.* (2008). Dual-modality monitoring of targeted intraarterial delivery of mesenchymal stem cells after transient ischemia. *Stroke* **39**(5): 1569–1574.

Walker-Batson, Dd, Smith, P., *et al.* (1995). Amphetamine paired with physical therapy accelerates motor recovery after stroke. Further evidence. *Stroke* **26**(12): 2254–2259.

Wang, L., Conner, J.M., *et al.* (2016). Rehabilitation drives enhancement of neuronal structure in functionally relevant neuronal subsets. *Proc Natl Acad Sci USA* **113**(10): 2750–2755.

Wang, L.E., Fink, G.R., *et al.* (2011). Noradrenergic enhancement improves motor network connectivity in stroke patients. *Ann Neurol* **69**(2): 375–388.

Ward, N., Brown, M., *et al.* (2003). Neural correlates of motor recovery after stroke: A longitudinal fMRI study. *Brain* **126**(Pt 11): 2476–2496.

Whyte, E.M., Lenze, E.J., *et al.* (2008). An open-label pilot study of acetylcholinesterase inhibitors to promote functional recovery in elderly cognitively impaired stroke patients. *Cerebrovasc Dis* **26**(3): 317–321.

Wieloch, T., Nikolich, K. (2006). Mechanisms of neural plasticity following brain injury. *Curr Opin Neurobiol* **16**(3): 258–264.

Williams, A.R., Hare, J.M. (2011). Mesenchymal stem cell: Biology, pathophysiology, translational findings, and therapeutic implications for cardiac disease. *Circ Res* **109**(8): 923–940.

Willing, A., Lixian, J., *et al.* (2003). Intravenous versus intrastriatal cord blood administration in a rodent model of stroke. *J Neurosci Res* **73**(3): 296–307.

Winstein, C.J., Wolf, S.L., *et al.* (2016). Effect of a task-oriented rehabilitation program on upper extremity recovery following motor stroke: The ICARE Randomized Clinical Trial. *JAMA* **315**(6): 571–581.

Wolf, S.L., Winstein, C.J., *et al.* (2006). Effect of constraint-induced movement therapy on upper extremity function 3 to 9 months after stroke: The EXCITE randomized clinical trial. *JAMA* **296**(17): 2095–2104.

Wolf, S.L., Winstein, C.J., *et al.* (2008). Retention of upper limb function in stroke survivors who have received constraint-induced movement therapy: The EXCITE randomised trial. *Lancet Neurol* **7**(1): 33–40.

Writing Group Members, Mozaffarian, D., *et al.* (2016). Heart Disease and Stroke Statistics — 2016 Update: A report from the American Heart Association. *Circulation* **133**(4): e38–e360.

Wu, J., Quinlan, E.B., *et al.* (2015). Connectivity measures are robust biomarkers of cortical function and plasticity after stroke. *Brain* **138**(Pt 8): 2359–2369.

Yasuhara, T., Borlongan, C., *et al.* (2006). *Ex vivo* gene therapy: Transplantation of neurotrophic factor-secreting cells for cerebral ischemia. *Front Biosci* **11**: 760–775.

Yavagal, D., Huang, D., *et al.* (2015). Intra-arterial delivery of autologous ALDHbr cells in ischemic strok: Final 1-year results of the RECOVER-Stroke Trial (Abstract). Presented at European Stroke Organisation Conference, Apr 17–19, Glasgow, UK.

Yavagal, D.R., Lin, B., *et al.* (2014). Efficacy and dose-dependent safety of intra-arterial delivery of mesenchymal stem cells in a rodent stroke model. *PloS One* **9**(5): e93735.

Yilmaz, G., Alexander, J.S., *et al.* (2010). Induction of neuro-protective/regenerative genes in stem cells infiltrating post-ischemic brain tissue. *Exp Transl Stroke Med* **2**(1): 11.

Yip, H.K., Tsai, T.H., *et al.* (2011). Effect of erythropoietin on level of circulating endothelial progenitor cells and outcome in patients after acute ischemic stroke. *Crit Care* **15**(1): R40.

Yoo, S.W., Chang, D.Y., *et al.* (2013). Immune following suppression mesenchymal stem cell transplantation in the ischemic brain is mediated by TGF-beta. *Neurobiol Dis* **58**: 249–257.

Zhang, Y., Pardridge, W.M. (2001). Conjugation of brain-derived neurotrophic factor to a blood-brain barrier drug targeting system enables neuroprotection in regional brain ischemia following intravenous injection of the neurotrophin. *Brain Res* **889**(1–2): 49–56.

Zhang, R., Wang, Y., *et al.* (2002). Sildenafil (Viagra) induces neurogenesis and promotes functional recovery after stroke in rats. *Stroke* **33**(11): 2675–2680.

Zhang, Y., Pardridge, W.M. (2006). Blood-brain barrier targeting of BDNF improves motor function in rats with middle cerebral artery occlusion. *Brain Res* **1111**(1): 227–229.

Zhang, Z.G., Chopp, M. (2009). Neurorestorative therapies for stroke: Underlying mechanisms and translation to the clinic. *Lancet Neurol* **8**(5): 491–500.

Zhao, L.X., Zhang, J., *et al.* (2004). Modification of the brain-derived neurotrophic factor gene: A portal to transform mesenchymal stem cells into advantageous engineering cells for neuroregeneration and neuroprotection. *Exp Neurol* **190**(2): 396–406.

Zhao, M.Z., Nonoguchi, N., *et al.* (2006). Novel therapeutic strategy for stroke in rats by bone marrow stromal cells and ex vivo HGF gene transfer with HSV-1 vector. *J Cereb Blood Flow Metab* **26**(9): 1176–1188.

Zhao, B.Q., Tejima, E., *et al.* (2007). Neurovascular proteases in brain injury, hemorrhage and remodeling after stroke. *Stroke* **38**(2 Suppl): 748–752.

Zittel, S., Weiller, C., *et al.* (2007). Reboxetine improves motor function in chronic stroke. A pilot study. *J Neurol* **254**(2): 197–201.

Brain AVM: Current Treatments and Challenges

4

W. Caleb Rutledge and Michael T. Lawton

Department of Neurological Surgery,
University of California, San Francisco, San Francisco, California, USA

An arteriovenous malformation (AVM) is a tangle of dysplastic vessels that forms a high-flow, low-resistance arteriovenous shunt. In the absence of an intervening capillary bed, dilated arteries shunt blood directly into arterialized veins. High flow through the feeding arteries, AVM nidus, and draining veins may cause AVM rupture and intracerebral hemorrhage.

AVM's have an estimated prevalence of about 50 cases per 100,000 (95% CI 10–100) (Morris *et al.*, 2009). Improved access and utilization of non-invasive brain imaging, such as magnetic resonance imaging/angiography (MRI/MRA) and computed tomography/angiography (CT/CTA), have led to increased, incidental detection of unruptured AVMs. Patients commonly present with intracerebral hemorrhage from AVM rupture, seizures, and focal or progressive neurologic deficits from functional and ischemic steal or venous congestion. For unruptured AVMs, the hemorrhage rate is about 1–2% per year. Once ruptured, the risk of future hemorrhage increases up to five-fold. Risk factors for additional hemorrhage include increasing patient age, Hispanic ethnicity, deep venous drainage, and deep AVM location (Kim *et al.*, 2007, 2014; Stapf *et al.*, 2006).

4.1 Management

Current management of brain AVMs is multimodal and includes observation with follow-up or treatment with microsurgical resection, endovascular embolization, stereotactic radiosurgery, or some combination of the above. The risks of intervention must be balanced against the risk of spontaneous rupture and hemorrhage in the AVM's natural history when selecting patients for treatment. In a recent systematic review and meta-analysis of treatment of brain AVMs, including over 13,000 patients and 46,000 years of patient follow-up, successful AVM obliteration was achieved in 96% of patients after microsurgery, 38% after radiosurgery, and 13% after endovascular embolization (van Beijnum *et al.*, 2011). Regardless of the treatment selected, once the decision has been made to treat a brain AVM, the goal of treatment is complete obliteration of the AVM to eliminate the risk of future hemorrhage, as partially treated AVMs remain at risk for rupture.

4.1.1 *Observation*

Currently, there is little consensus on how to manage patients with unruptured AVMs (Lawton *et al.*, 2015; Rutledge *et al.*, 2014). In recent trials comparing conservative medical management to intervention, treated patients had significantly higher rates of stroke and death than those managed conservatively (Al-Shahi Salman *et al.*, 2014; Mohr *et al.*, 2014; Wedderburn *et al.*, 2008). For unruptured AVMs, the risk of hemorrhage is about 1–2% per year, but once ruptured, the risk of subsequent hemorrhage increases and treatment is recommended (Halim *et al.*, 2004; Hernesniemi *et al.*, 2008; Kim *et al.*, 2014). While outcomes after AVM hemorrhage are better than after spontaneous intracerebral hemorrhage, brain AVMs are associated with excess long-term morbidity and mortality, especially once ruptured (Laakso *et al.*, 2008; van Beijnum *et al.*, 2009). Unfortunately, there is no primary medical therapy to prevent AVM formation, stabilize AVM vessels, or induce obliteration and prevent rupture. Patients with associated seizures are treated with anticonvulsants, and headaches are treated symptomatically.

4.1.2 *Microsurgical resection*

Microsurgical resection is the mainstay of AVM treatment. With careful patient selection, outcomes after surgery are excellent with obliteration rates of about 96% and little morbidity and mortality (Davidson & Morgan, 2010; Potts *et al.*, 2015; Rutledge *et al.*, 2014; van Beijnum *et al.*, 2011). The Spetzler–Martin (SM) and Lawton–Young supplemental grading systems predict surgical risks and guide patient selection for surgery. The SM system includes AVM size, eloquence of surrounding brain parenchyma, and venous drainage (Spetzler & Martin, 1986). Based on these factors, AVMs are graded I to V. SM grade I, II, and III lesions are good candidates for microsurgical resection with low rates of surgical morbidity and mortality, while observation is recommended for grade IV and V AVMs, except in cases of recurrent hemorrhage or progressive neurological deficits (Spetzler & Ponce, 2011). The supplemental grading system incorporates additional factors important to surgical selection and outcomes, including patient age, hemorrhagic presentation, and nidal diffuseness (Lawton *et al.*, 2010). Patients with supplemented scores of 6 or less are good candidates for surgery, while those with scores greater than 6 have high surgical morbidity and mortality rates (Tables 1 and 2).

In addition to these grading systems, AVMs are categorized into different types and subtypes. There are seven types of AVMs in the frontal, temporal, and parieto-occipital lobes, ventricles, deep central core, brainstem, and cerebellum (Lawton, 2014). Additionally, there are 4–6 subtypes within each type based on the brain surface on which the AVM is located (e.g., lateral, medial, basal) and other specific anatomy (e.g., midbrain, pons, and medulla) (Table 3). Each subtype is characterized by its own arterial supply, draining veins, eloquent structures, surgical approach, and management strategy.

For example, lateral temporal AVMs are fed by branches descending from the sylvian fissure from the inferior trunk of the MCA. Also, the temporopolar and anterior temporal arteries may course around the temporal pole anteriorly, while branches from the PCA occasionally feed posteriorly located lateral temporal AVMs. Drainage is usually superficial to the vein of Labbé and transverse sinus

Table 1. Comparison of SM and Supplementary grading systems.

SM Grading		Points	Supplementary grading		Points
Size	<3 cm	1	Age	<20 years	1
	3–6 cm	2		20–40 years	2
	>6 cm	3		>40 years	3
Venous Drainage	Superficial	0	Bleeding	Yes	0
	Deep	1		No	1
Eloquence	No	0	Compactness	Yes	0
	Yes	1		No	1
Total		5			5

Table 2. Application of SM, supplementary, and supplemented SM grading systems.

		Total	Improved/ Unchanged		Worse/ Dead	
		N	*N*	%	*N*	%
SM grade						
	I	56	51	91	5	9
	II	123	93	76	30	24
	III	90	63	70	27	30
	IV	29	20	69	9	31
	V	2	0	0	2	100
Supplementary grade						
	I	27	26	96	1	4
	II	67	59	88	8	12
	III	113	88	78	25	22
	IV	79	47	59	32	41
	V	14	7	50	7	50
Supplemented SM grade						
	1	0	0	0	0	0
	2	7	7	100	0	0
	3	21	21	100	0	0
	4	55	50	91	5	9
	5	90	71	79	19	21
	6	70	51	73	19	27
	7	44	20	45	24	55
	8	8	4	50	4	50
	9	5	3	60	2	40
	10	0	0	0	0	0

(Figure 1). Lateral temporal AVMs may involve Wernicke's area when they are posteriorly located in the dominant hemisphere.

AVM surgery follows sequential steps that include exposure, subarachnoid dissection, defining the draining vein, defining the feeding arteries, pial dissection,

Table 3. Summary of AVM types and subtypes in the author's experience.

Type	Subtype	*N*	%
Frontal AVMs		135	22.5
	Lateral frontal	48	8.0
	Medial frontal	21	3.5
	Paramedian frontal	24	4.0
	Basal frontal	21	3.5
	Sylvian frontal	21	3.5
Temporal AVMs		112	18.7
	Lateral temporal	76	12.7
	Basal temporal	12	2.0
	Medial temporal	17	2.8
	Sylvian temporal	7	1.2
Parieto-occipital AVMs		108	18.0
	Lateral parieto-occipital	52	8.7
	Medial parieto-occipital	17	2.8
	Paramedian parieto-occipital	29	4.8
	Basal occipital	10	1.7
Ventricular/Periventricular AVMs		42	7.0
	Callosal	23	3.8
	Ventricular body	5	0.8
	Atrial	9	1.5
	Temporal horn	5	0.8
Deep AVMs		56	9.3
	Pure sylvian	8	1.3
	Insular	25	4.2
	Basal ganglial	10	1.7
	Thalamic	13	2.2
Brainstem AVMs		26	4.3
	Anterior midbrain	1	0.2
	Posterior midbrain	6	1.0
	Anterior pontine	6	1.0
	Lateral pontine	7	1.2

(*Continued*)

Table 3. (*Continued*)

Type	Subtype	*N*	%
	Anterior medullary	1	0.2
	Lateral medullary	5	0.8
Cerebellar AVMs		82	13.7
	Suboccipital cerebellar	18	3.0
	Tentorial cerebellar	16	2.7
	Vermian cerebellar	25	4.2
	Tonsillar cerebellar	7	1.2
	Petrosal cerebellar	16	2.7
Mixed AVMs		39	6.5
Total AVMs		600	

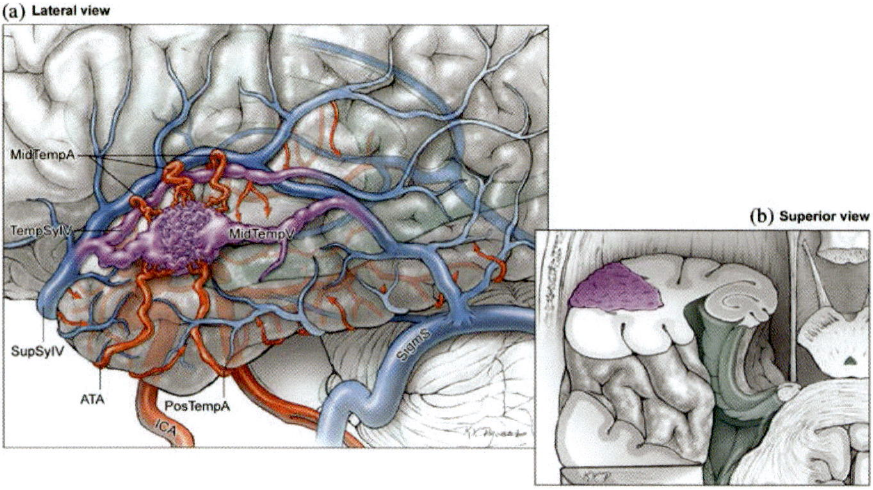

Figure 1. The lateral temporal AVM subtype. (a) Lateral and (b) superior cross-sectional views.

parenchymal dissection, ependymal or deep dissection, and finally resection. Preservation of the draining vein until the end of the resection is critical to prevent intraoperative rupture. Occluding venous outflow prematurely causes increased intranidal pressure and distension of the malformation, resulting in rupture and bleeding in the surgical field. Tamponade and suction are often not effective to control the bleeding and clear the field. Feeding arteries should be disconnected as close as possible to the AVM to prevent sacrifice of normal branches and infarcts in adjacent normal brain. As the AVM is circumferentially dissected and feeding arteries disconnected, the draining vein changes color from red to blue, indicating complete de-arterialization of nidus and resectability.

For lateral temporal AVMs, a pterional craniotomy is used to expose the AVM if it is located in front of the external auditory canal, while a temporal craniotomy is used for AVMs behind the external auditory canal. The dural opening is based on the pterion or inferiorly at the middle fossa floor depending on the craniotomy performed. Usually, only sulcal dissection is needed to define the margins of the nidus, feeding arteries, and draining veins. Drainage to the vein of Labbé is at the AVM's posterior margin, while sylvian veins are at the superior margin. Feeding arteries are identified superiorly descending from inferior MCA trunk branches, anteriorly from the M1 MCA branches coursing around the temporal lobe, or inferiorly from the P2 PCA branches. These arterial feeders are interrupted during parenchymal circumdissection, which may extend to the temporal horn with anterior choroidal artery supply. Once the draining vein is disconnected, the AVM is removed from the temporal lobe (Figure 2).

Outcomes after microsurgical resection of low-grade AVMs are excellent. In two recent large series, morbidity and mortality after microsurgery ranged from only 0.7 to 2.4% and 0 to 0.5%, respectively, with cure rates of 97–98% (Davidson & Morgan, 2010; Potts *et al.*, 2015).

4.1.3 *Stereotactic radiosurgery*

Stereotactic radiosurgery is a viable alternative to microsurgical resection. Delivery of a sufficient dose of radiation to the AVM induces closure of the lumen of AVM vessels and obliteration of the AVM over a 2–3 year latency period. Outcomes are excellent with low morbidity, particularly for small AVMs (Kano *et al.*, 2012, 2014; Lunsford *et al.*, 1991; Steiner *et al.*, 1992). While less invasive compared to microsurgical resection, there are disadvantages to radiosurgery. There is a risk of hemorrhage during the latency period and adverse radiation effects (AREs) from damage to surrounding tissue (Barker *et al.*, 2003; Cohen-Inbar *et al.*, 2015; Kano *et al.*, 2017). Successful AVM obliteration depends directly on the amount of radiation delivered, but the rate of AREs also increases with the amount of radiation delivered.

4.1.4 *Endovascular embolization*

AVMs selected for microsurgical resection are now routinely evaluated for preoperative embolization to obliterate feeding artery aneurysms and deep, surgically

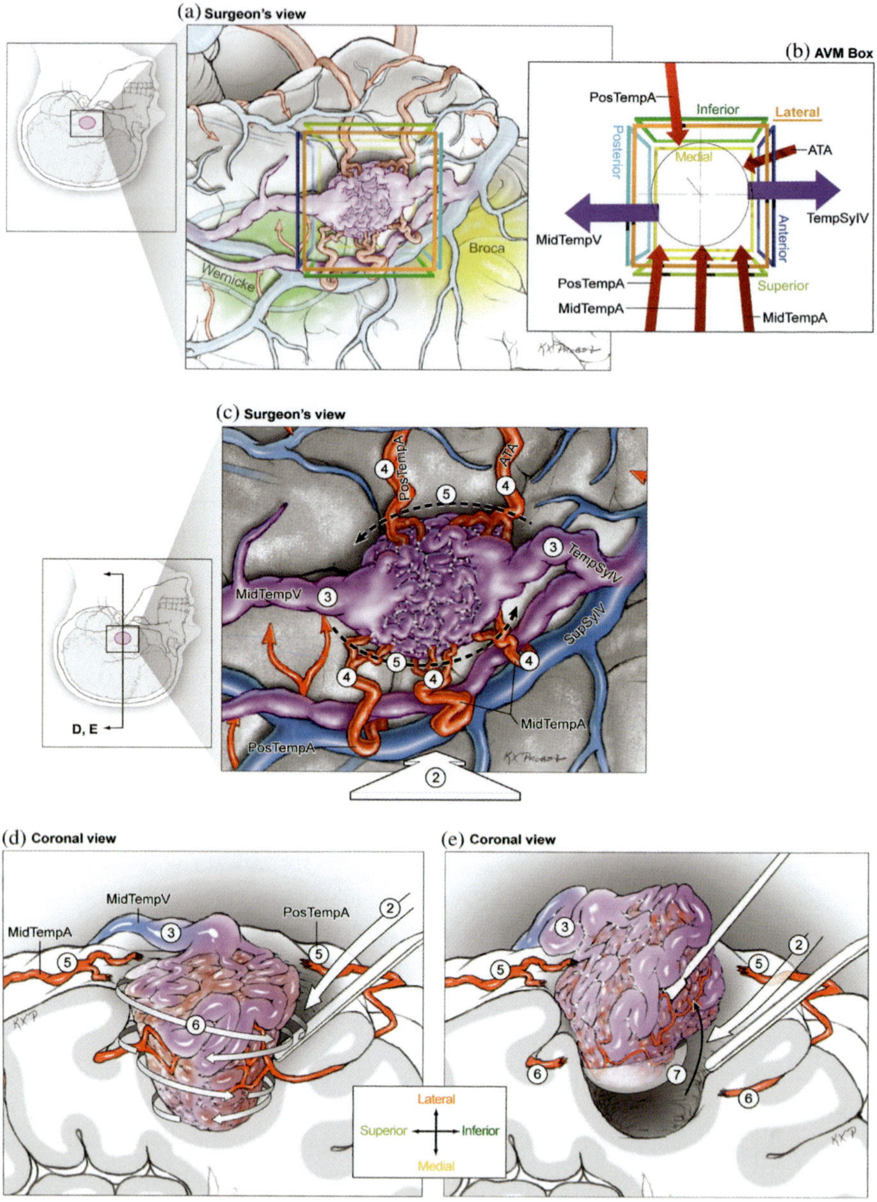

Figure 2. Resection strategy for lateral temporal AVMs. (a) Step 1, exposure with temporal craniotomy. (b) AVM box showing arterial inputs. (c) Step 2, sulcal dissection; Step 3, identifying cortical draining vein; Step 4, identifying cortical feeders; Step 5, interrupting arterial feeders along the superior and inferior margins. (d) Step 6, circumdissection into the temporal lobe. (e) Step 7, delivering the nidus from the temporal lobe.

inaccessible feeding arteries to reduce blood flow to the AVM, minimize blood loss during surgery, and reduce operative times (Vinuela *et al.*, 1991). Despite advances in endovascular techniques, including microcatheter systems and liquid embolic agents, there remains significant risk of complications with long-term morbidity and death from AVM embolization (Haw *et al.*, 2006; Nakstad & Nornes, 1994; Taylor *et al.*, 2004). A variety of techniques are used in AVM embolization, including NBCA glue, Onyx, and coils in both transarterial and transvenous approaches. Obliteration rates are higher with the use of Onyx compared to NBCA (Saatci *et al.*, 2011; Sahlein *et al.*, 2012; van Rooij *et al.*, 2012).

Curative embolization is possible, usually small AVMs with a single direct feeding artery (feeding only the AVM nidus) (van Rooij *et al.*, 2012). *En passage* feeders supplying the AVM nidus and normal brain distal to it must be preserved. Similar to surgical grading systems, endovascular grading systems have been developed for risk assessment and outcome determination in brain AVM patients treated by endovascular techniques. A recent classification includes the number of feeding arteries, eloquence, and presence of an arteriovenous fistula component (Feliciano *et al.*, 2010). Fewer feeding arteries, non-eloquence, and absence of an arteriovenous fistula are associated with endovascular cure and fewer complications (Bell *et al.*, 2015). Like the surgical grading systems, the Buffalo score is a 5-point system. Points are assigned for the number of arterial pedicles (1–3), the arterial pedicle diameter (0 or 1), and eloquence (0 or 1). In a retrospective review of 50 patients, higher scores were associated with higher complication rates (Dumont *et al.*, 2015).

Follow-up imaging after treatment is recommended to confirm complete AVM obliteration and is essential after embolization and radiosurgery. Angiography is necessary to confirm complete obliteration. Delayed angiogram is recommended in pediatric patients sometime after obliteration because of the recurrence risk in pediatric patients.

4.2 Challenges in the Treatment of Brain AVMs

Management of brain AVMs continues to evolve with advances in microsurgical, radiosurgical, and endovascular techniques, including irrigating bipolar forceps and AVM microclips, and refinement of surgical approaches and strategies based on the development of AVM subtypes and grading systems, volume-staged radiosurgery, microcatheter systems, and Onyx and other liquid embolysates. For low-grade AVMs, cure rates with both microsurgery and stereotactic radiosurgery are excellent with little risk to the patient. Newer embolic agents and delivery systems have improved obliteration rates with fewer complications after endovascular treatment. Future advances will likely further decrease its complication rate and increase the rate of complete obliteration and cure. However, significant challenges remain in the management and treatment of AVMs, particularly unruptured and high-grade AVMs.

4.2.1 *Unruptured AVMs*

While treatment is recommended for ruptured AVMs to prevent future hemorrhage, management of unruptured AVMs is controversial. A recent randomized, prospective, multicenter study (Randomized Trial of Unruptured Brain Arteriovenous Malformations [ARUBA]) showed that patients with brain AVMs treated with interventional therapy had a much higher rate of stroke and death than patients managed medically; however, the study's design was flawed by selection bias, participating site characteristics and practice patterns, and the short length of follow-up (Mohr *et al.*, 2014). Patients were treated with embolization alone (32%) and radiosurgery alone (33%), yet few patients (18%) were treated surgically despite superior cure rates observed with surgery with few complications (van Beijnum *et al.*, 2011). More studies are needed to establish indications for treatment of unruptured AVMs. Neurosurgeons have advocated for another randomized trial comparing surgery and observation for low-grade AVMs and the establishment of national, multicenter, adjudicated registries of brain AVMs with detailed outcome analyses (Bambakidis *et al.*, 2014; Lawton *et al.*, 2015).

4.2.2 *High-grade AVMs*

Low-grade SM AVMs with supplemented scores less than or equal to 6 are excellent candidates for microsurgical resection or SRS. These AVMs are generally small, superficial, and located in non-eloquent cortex. Outcomes with microsurgical resection are excellent with little surgical morbidity (Davidson & Morgan, 2010; Potts *et al.*, 2015). Stereotactic radiosurgery also has high obliteration rates with little morbidity, making either modality acceptable. However, high-grade SM AVMs with supplemented scores greater than 6 have unacceptably high rates of surgical morbidity and mortality, approaching 50% (Table 2). Similarly, SRS is associated with low obliteration rates for high-grade AVMs, as dose reductions are required to prevent complications from damage to surrounding tissue for AVMs with larger nidal volumes. While observation is preferred for many high-grade AVMs, hemorrhage, pre-existing deficits, failure of radiosurgery, and young age are compelling indications to intervene. Volume-staged radiosurgery followed by surgical resection once the AVM is downgraded is one alternative to observation for these patients (Abla *et al.*, 2015).

Defining anatomical subtypes of AVMs allows for development of established approaches, refinement of surgical techniques, and identification of specific subtypes amenable to a surgical cure with acceptable mortality and morbidity rates (Lawton, 2014). While many neurosurgeons consider high-grade deep or brainstem AVMs inoperable, lateral pontine AVMs are resectable AVMs. Unlike other brainstem AVMs, their superior, inferior, and lateral borders are made up of non-eloquent cerebellum, allowing for parenchymal circumdissection. Favorable outcomes are possible in carefully selected patients (Batjer & Samson, 1986; Han *et al.*, 2015; Solomon & Stein, 1986).

References

Abla, A.A., Rutledge, W.C., *et al.* (2015). A treatment paradigm for high-grade brain arteriovenous malformations: Volume-staged radiosurgical downgrading followed by microsurgical resection. *J Neurosurg* **122**(2): 419–432.

Al-Shahi Salman, R., White, P.M., *et al.* (2014). Outcome after conservative management or intervention for unruptured brain arteriovenous malformations. *JAMA* **311**(16): 1661–1669.

Bambakidis, N.C., Cockroft, K.M., *et al.* (2014). The case against a randomized trial of unruptured brain arteriovenous malformations: Misinterpretation of a flawed study. *Stroke* **45**(9): 2808–2810.

Barker, F.G., 2nd, Butler, W.E., *et al.* (2003). Dose-volume prediction of radiation-related complications after proton beam radiosurgery for cerebral arteriovenous malformations. *J Neurosurg* **99**(2): 254–263.

Batjer, H., Samson, D. (1986). Arteriovenous malformations of the posterior fossa. Clinical presentation, diagnostic evaluation, and surgical treatment. *J Neurosurg* **64**(6): 849–856.

Bell, D.L., Leslie-Mazwi, T.M., *et al.* (2015). Application of a novel brain arteriovenous malformation endovascular grading scale for transarterial embolization. *AJNR Am J Neuroradiol* **36**(7): 1303–1309.

Cohen-Inbar, O., Lee, C.C., *et al.* (2015). A quantitative analysis of adverse radiation effects following Gamma Knife radiosurgery for arteriovenous malformations. *J Neurosurg* **123**(4): 945–953.

Davidson, A.S.M., Morgan, M.K. (2010). How safe is arteriovenous malformation surgery? A prospective, observational study of surgery as first-line treatment for brain arteriovenous malformations. *Neurosurgery* **66**(3): 498–505.

Dumont, T.M., Kan, P., *et al.* (2015). A proposed grading system for endovascular treatment of cerebral arteriovenous malformations: Buffalo score. *Surg Neurol Int* **6**: 3.

Feliciano, C.E., de Leon-Berra, R., *et al.* (2010). A proposal for a new arteriovenous malformation grading scale for neuroendovascular procedures and literature review. *P R Health Sci J* **29**(2): 117–120.

Halim, A.X., Johnston, S.C., *et al.* (2004). Longitudinal risk of intracranial hemorrhage in patients with arteriovenous malformation of the brain within a defined population. *Stroke* **35**(7): 1697–1702.

Han, S.J., Englot, D.J., *et al.* (2015). Brainstem arteriovenous malformations: Anatomical subtypes, assessment of "occlusion *in situ*" technique, and microsurgical results. *J Neurosurg* **122**(1): 107–117.

Haw, C.S., terBrugge, K., *et al.* (2006). Complications of embolization of arteriovenous malformations of the brain. *J Neurosurg* **104**(2): 226–232.

Hernesniemi, J.A., Dashti, R., *et al.* (2008). Natural history of brain arteriovenous malformations: A long-term follow-up study of risk of hemorrhage in 238 patients. *Neurosurgery* **63**(5): 823–831.

Kano, H., Flickinger, J.C., *et al.* (2017). Estimating the risks of adverse radiation effects after gamma knife radiosurgery for arteriovenous malformations. *Stroke* **48**(1): 84–90.

Kano, H., Flickinger, J.C., *et al.* (2014). Stereotactic radiosurgery for Spetzler-Martin Grade III arteriovenous malformations. *J Neurosurg* **120**(4): 973–981.

Kano, H., Lunsford, L.D., *et al.* (2012). Stereotactic radiosurgery for arteriovenous malformations, Part 1: Management of Spetzler–Martin Grade I and II arteriovenous malformations. *J Neurosurg* **116**(1): 11–20.

Kim, H., Al-Shahi Salman, R., *et al.* (2014). Untreated brain arteriovenous malformation: Patient level meta-analysis of hemorrhage predictors. *Neurology* **83**(7): 590–597.

Kim, H., Sidney, S., *et al.* (2007). Racial/ethnic differences in longitudinal risk of intracranial hemorrhage in brain arteriovenous malformation patients. *Stroke* **38**(9): 2430–2437.

Laakso, A., Dashti, R., *et al.* (2008). Long-term excess mortality in 623 patients with brain arteriovenous malformations. *Neurosurgery* **63**(2): 244–253; discussion 253–245.

Lawton, M.T. (2014) *Seven AVMs: Tenets and Techniques for Resection*. New York: Thieme.

Lawton, M.T., Kim, H., *et al.* (2010). A supplementary grading scale for selecting patients with brain arteriovenous malformations for surgery. *Neurosurgery* **66**(4): 702–713; discussion 713.

Lawton, M.T., Rutledge, W.C., *et al.* (2015). Brain arteriovenous malformations. *Nat Rev Dis Prim* **1**: 15008.

Lunsford, L.D., Kondziolka, D., *et al.* (1991). Stereotactic radiosurgery for arteriovenous malformations of the brain. *J Neurosurg* **75**(4): 512–524.

Mohr, J.P., Parides, M.K., *et al.* (2014). Medical management with or without interventional therapy for unruptured brain arteriovenous malformations (ARUBA): A multicentre, non-blinded, randomised trial. *Lancet* **383**(9917): 614–621.

Morris, Z., Whiteley, W.N., *et al.* (2009). Incidental findings on brain magnetic resonance imaging: Systematic review and meta-analysis. *BMJ* **339**: b3016.

Nakstad, P.H., Nornes, H. (1994). Superselective angiography, embolisation and surgery in treatment of arteriovenous malformations of the brain. *Neuroradiology* **36**(5): 410–413.

Potts, M.B., Lau, D., *et al.* (2015). Current surgical results with low-grade brain arteriovenous malformations. *J Neurosurg* **122**(4): 912–920.

Rutledge, W.C., Abla, A.A., *et al.* (2014). Treatment and outcomes of ARUBA eligible patients with unruptured brain arteriovenous malformations at a single institution. *Neurosurg Focus* **37**(3): E8.

Saatci, I., Geyik, S., *et al.* (2011). Endovascular treatment of brain arteriovenous malformations with prolonged intranidal Onyx injection technique: Long-term results in 350 consecutive patients with completed endovascular treatment course. *J Neurosurg* **115**(1): 78–88.

Sahlein, D.H., Mora, P., *et al.* (2012). Nidal embolization of brain arteriovenous malformations: Rates of cure, partial embolization, and clinical outcome. *J Neurosurg* **117**(1): 65–77.

Solomon, R.A., Stein, B.M. (1986). Management of arteriovenous malformations of the brain stem. *J Neurosurg* **64**(6): 857–864.

Spetzler, R.F., Martin, N.A. (1986). A proposed grading system for arteriovenous malformations. *J Neurosurg* **65**(4): 476–483.

Spetzler, R.F., Ponce, F.A. (2011). A 3-tier classification of cerebral arteriovenous malformations. Clinical article. *J Neurosurg* **114**(3): 842–849.

Stapf, C., Mast, H., *et al.* (2006). Predictors of hemorrhage in patients with untreated brain arteriovenous malformation. *Neurology* **66**(9): 1350–1355.

Steiner, L., Lindquist, C., *et al.* (1992). Clinical outcome of radiosurgery for cerebral arteriovenous malformations. *J Neurosurg* **77**(1): 1–8.

Taylor, C.L., Dutton, K., *et al.* (2004). Complications of preoperative embolization of cerebral arteriovenous malformations. *J Neurosurg* **100**(5): 810–812.

van Beijnum, J., Lovelock, C.E., *et al.* (2009). Outcome after spontaneous and arteriovenous malformation-related intracerebral haemorrhage: Population-based studies. *Brain* **132**(Pt 2): 537–543.

van Beijnum, J., van der Worp, H.B., *et al.* (2011). Treatment of brain arteriovenous malformations: A systematic review and meta-analysis. *JAMA* **306**(18): 2011–2019.

van Rooij, W.J., Jacobs, S., *et al.* (2012). Curative embolization of brain arteriovenous malformations with onyx: Patient selection, embolization technique, and results. *AJNR Am J Neuroradiol* **33**(7): 1299–1304.

Vinuela, F., Dion, J.E., *et al.* (1991). Combined endovascular embolization and surgery in the management of cerebral arteriovenous malformations: Experience with 101 cases. *J Neurosurg* **75**(6): 856–864.

Wedderburn, C.J., van Beijnum, J., *et al.* (2008). Outcome after interventional or conservative management of unruptured brain arteriovenous malformations: A prospective, population-based cohort study. *Lancet Neurol* **7**(3): 223–230.

Animal Models and Prospective Therapeutic Targets for Brain Arteriovenous Malformation

5

Wan Zhu, Rui Zhang, Li Ma, and Hua Su

Center for Cerebrovascular Research,
Department of Anesthesia and Perioperative Care,
University of California, San Francisco, San Francisco, California, USA

Brain arteriovenous malformations (AVMs) are abnormal connections of arteries and veins in the brain shunting blood directly from artery to vein. Due to high pressure in the vein and abnormal vessel wall structure, AVMs tend to rupture, causing life-threatening intracranial hemorrhage and long-term disability. Unfortunately, current therapies are mostly invasive and associated with treatment-related high morbidities. Development of new, non-invasive, and safe therapies is urgent. Animal models are important tools for identifying therapeutic targets and test therapies. In this chapter, we first review current bAVM animal models and the knowledge gained from developing and studying these models. We then summarize putative therapeutic targets that have been identified based on animal and human studies.

5.1 The Development of bAVM Models

Various animal models for bAVM have been established over the years. These models can be divided into two major categories: hemodynamic and genetic. In this section, we discuss the pros and cons of current bAVM models in pathogenesis studies and therapeutic developments.

5.1.1 *Hemodynamic models*

AVM changes hemodynamics and perfusion pressure in the vascular beds in the brain and other organs, which leads to an abnormal physiological change in the endothelium. Hence, many bAVM models were established by creating arteriovenous shunts through surgical anastomosis of extracranial arteries and veins to mimic hemodynamic changes in the brain of bAVM patients. Known as hemodynamic models, they are of incredible research and clinical value, particularly in the early days, and included testing and evaluating embolization and radiosurgery therapy. Some of these models are still being used. The following hemodynamic models are commonly discussed in the literature.

5.1.1.1 *Carotid-jugular fistulae models*

Carotid-jugular fistulae (CJF) models were created based on the normal perfusion pressure breakthrough (NPPB) theory proposed by Robert F. Spetzler (Rangel-Castilla *et al.*, 2015; Spetzler *et al.*, 1978). Spetzler *et al.* first established a cat model through anastomosis between the rostral end of the common carotid artery (CCA) and the caudal end of the external jugular vein (EJV), which caused a retrograde blood flow from the cephalic carotid system to the fistula, causing fistula dilation (Spetzler *et al.*, 1987). However, the change in cerebral blood flow (CBF) in this model was minimal and insignificant (Miyasaka *et al.*, 1994; Tokiwa *et al.*, 1995). Later, Morgan and colleagues modified this model in rats by generating an end-to-end anastomosis of both rostral ends of CCA and EJV on the right side of rats, and ligated the caudal ends of both vessels and the ipsilateral external carotid artery (ECA) (Morgan *et al.*, 1989b). In their rat model, they were able to create significant difference in regional CBF, reduced CBF in the open fistula group, and increased CBF in the closed fistula group compared to controls (Morgan *et al.*, 1989a, 1989b). Later, several groups made some modifications based on Morgan's rat model (Bederson *et al.*, 1991; Hai *et al.*, 2002, 2004; Kojima *et al.*, 2007). All models resembled some clinical features of bAVM, such as alteration of blood pressure in feeding arteries, draining veins and cerebral perfusion, elevation of VEGF level, and occurrence of brain edema, hemorrhage, and leakiness of the blood-brain barrier (BBB) (Bederson *et al.*, 1991; Hai *et al.*, 2004; Kojima *et al.*, 2007). However, because the fistulas in these models were not in the brain parenchyma, the models are considered surrogate models for bAVMs. CJF models represent the first attempt to create animal models for bAVM.

5.1.1.2 *Rete mirabile models*

Carotid rete mirabile (RM) is a vascular structure with a tangle of microarteries and arterioles situated at the termination of each ascending pharyngeal artery (APA) as it perforates the cranial base. It is present in animals with an even number of toes, such as swine and sheep. Because this unique structure looks like an AVM nidus, several groups used RM in swine or sheep as a surrogate (Chaloupka *et al.*, 1994; Massoud *et al.*, 1994; Qian *et al.*, 1999). The body size of swine makes these models better tools to study therapeutic procedures, such as embolization (Amiridze *et al.*, 2008) and radiosurgery (De Salles *et al.*, 1996; Jahan *et al.*, 2006). However, it is very important to be aware that unlike bAVMs, which have abnormal vessel structure and a higher pressure gradient between feeding arteries and draining veins, vessel structure and blood pressure in RM are normal. In addition, RMs are located outside the brain, which does not make them true bAVM models.

In addition to these major classes of surrogate models mentioned above, there are some other types of bAVM animal models that are generated thorough surgical interventions, including a dog intracranial arteriovenous fistula model created through a femoral venous graft with end-to-side anastomosis to the cortical branch of the middle cerebral artery (MCA) and the superior sagittal sinus (SSS) (Numazawa *et al.*, 2005); a rat model using the extracranial venous plexus as a surrogate for AVM nidus (Yassari *et al.*, 2004); a dog model generated through implantation of a superficial temporal artery segment to create an AVM-like lesion (Pietila *et al.*, 2000); and a xenograft arteriovenous fistula model in rats (Lawton *et al.*, 2004). Most of these models have been used to study the disease mechanism and evaluate radiosurgery. Again, none of these models has intra-brain parenchymal nidus and therefore cannot be used to study nidus growth and hemorrhage.

5.1.2 *Genetic models*

5.1.2.1 *Mutation of HHT causative genes*

In humans, more than 95% of bAVM are sporadic cases. Some somatic mutations in sporadic bAVM have recently been identified. There are a few familiar bAVM cases. Familiar cases that we studied the most are found in patients with an autosomal dominant disease, hereditary hemorrhagic telangiectasia (HHT). In addition to bAVM, HHT patients also develop AVM in multiple organs, including lung, intestine, liver and skin. The two main subtypes of HHT are caused by haploinsufficiency of endoglin (*Eng*) gene (type 1) and activin receptor-like kinase 1 (*ALK1* or

ACVRL1) gene (type 2). In addition, mutations in the *SMAD4* (or *MADH4*) gene are associated with juvenile polyposis syndrome (JPS) and type 3 HHT (Gallione *et al.*, 2004). Further, mutations in the *BMP9* (bone morphogenetic protein 9 or *GDF2*) gene causes vascular phenotypes, including epistaxis and telangiectasia that overlap with HHT phenotypes (Wooderchak-Donahue *et al.*, 2013). Although some additional loci were mapped in some affected HHT families (Davidson *et al.*, 2009; Cole *et al.*, 2005), no precise gene was found.

All gene mutations in HHTs are in the TGFβ superfamily. TGFβ signaling is important for vascular development. Both *ALK1* and *ENG* genes are involved in the TGFβ/BMP9 signaling pathway (Shovlin, 2010). While ALK1 is classified as type I TGFβ/BMP9 receptor, ENG is classified as an associate TGFβ type II TGFβ receptor. SMAD4 is a downstream signaling effector in the TGFβ-ALK1/ENG pathway. BMP9 is a ligand of ALK1 and ENG (Tillet & Bailly, 2015). So far, most genetic bAVM models are established on *Alk1* or *Eng* mutant mice.

5.1.2.2 *Zebrafish model*

The zebrafish is commonly used as a model organism in genetic studies and biomedical research. The major advantage of zebrafish in studying vascular diseases is that their embryos are nearly transparent and live blood vessel development can be easily studied under an optical microscope (Gore *et al.*, 2012). It is also easier to introduce multiple gene mutations into zebrafish than in rodents and other animals. Zebrafish *alk1*$^{-/-}$ embryos show dramatically enlarged and high-flow cranial AVMs that shunt blood from arteries to veins, bypassing the central arteries (Corti *et al.*, 2011; Roman *et al.*, 2002). Interestingly, zebrafish studies also suggested that *alk1* gene expression influences blood flow (Corti *et al.*, 2011) and loss of *alk1* gene affects arterial endothelial cell migration but not proliferation (Rochon *et al.*, 2016).

5.1.2.3 *Mouse models*

5.1.2.3.1 Alk1 or Eng knockout

Two lines of *Alk1* knockout mice have been generated through disruption of its transcriptional initiation or the kinase domain (Oh *et al.*, 2000; Urness *et al.*, 2000), both of which cause loss-of-function Alk1. Homozygous *Alk1*$^{-/-}$ mice in both lines die in mid-gestation (E10.5) with massive vascular defects, including dilation and fusion of arteries and veins (Oh *et al.*, 2000; Urness *et al.*, 2000), indicating the importance of Alk1 function in vascular development. *Alk1*$^{+/-}$ mice survive into adult stage with vascular malformations in different organs, including liver, intestine, skin, and lung (Srinivasan *et al.*, 2003) and thus were used as an HHT2 model. However, brain vascular malformation is rare: a dilated vessel in the cerebellum was found in 1 of 47 *Alk1*$^{+/-}$ mice (Srinivasan *et al.*, 2003). Therefore, neither homozygous nor heterozygous *Alk1* mutant mice are ideal models for studying bAVM in adult mice.

Similarly, the homozygous loss-of-function *Eng* gene generated by adding a premature stop codon in mice causes embryonic lethality (E10.5) (Arthur *et al.*, 2000). Mice with heterozygous mutation of *Eng* gene exhibited some phenotypes resembling the phenotypes of HHT1 patients, including mucocutaneous telangiectases, external bleeding, and vascular malformation of liver, lungs, brain, and gastrointestines (Bourdeau *et al.*, 2000b). However, the occurrence of cerebral AVMs is also low (7%) (Bourdeau *et al.*, 2000b).

Although *Alk1*$^{+/-}$ and *Eng*$^{+/-}$ mice have provided an important resource to study gene expression, gene function in the endothelium and pathophysiological abnormalities of AVM in affected organs (Ardelean *et al.*, 2014; Bourdeau *et al.*, 2001; Jerkic *et al.*, 2012; Mahmoud *et al.*, 2010; Toporsian *et al.*, 2010; Torsney *et al.*, 2003), due to the extremely low penetrance of bAVM, *Alk1*$^{+/-}$ and *Eng*$^{+/-}$ mice are not ideal for studying bAVM.

5.1.2.3.2 Alk1 or Eng conditional knockout

In order the get living adult *Alk1*$^{-/-}$ or *Eng*$^{-/-}$, conditional knockout using Cre–lox recombination system is employed. This system requires two elements: expression of the Cre recombinase, and existence of two lox sites (loci of recombination, or loxP site, since the sequence was originated from P1 bacteriophage). When both elements are present, depending on the orientation of loxP sequences, DNA recombination occurs to generate sequence deletion or inversion.

Park *et al* and Allinson *et al* have generated *Alk1* floxed (*Alk1*$^{2f/2f}$) and *Eng* floxed (*Eng*$^{2f/2f}$) mouse lines that have *Alk1* gene exons 4–6 and *Eng* gene exons 5–6 flanked by loxP sites, respectively (Allinson *et al.*, 2007; Park *et al.*, 2008). Gene mutation can be induced in adult mice through conditional activation of Cre gene to avoid embryonic lethality found in homozygous mutation of *Alk1* or *Eng*. Also, inducible cell type-specific expression of Cre recombinase allows the mutation of the genes in specific cell types.

5.1.2.3.3 Angiogenic stimulation

In HHT patients, the penetration of bAVM is relatively low (about 5–20%) compared to AVMs in the lung, liver, or intestine (Dupuis-Girod *et al.*, 2010; Govani

& Shovlin, 2009). *Alk1*$^{+/-}$ or *Eng*$^{+/-}$ mice rarely have bAVM (Bourdeau *et al.*, 2000b; Srinivasan *et al.*, 2003). Homozygous mutation of *Alk1* in adult mice resulted in robust AVM formation in many organs, but not in the brain (Park *et al.*, 2009). Skin AVM formed in adult *Alk1*$^{-/-}$ and *Eng*$^{-/-}$ mice only around the skin wound (Park *et al.*, 2009), suggesting that in some organs, genetic defects alone are not enough to trigger bAVM formation. Other factors, e.g., wound or angiogenesis, are also needed.

Clinical studies show elevated vascular endothelial growth factor (VEGF) in surgically resected bAVM lesions (Hashimoto *et al.*, 2005b; Koizumi *et al.*, 2002) and in the plasma of HHT patients (Sadick *et al.*, 2005b). Our group at the Center for Cerebrovascular Research at the University of California, San Francisco, tested whether angiogenic stimulation can enhance bAVM formation in *Alk1*$^{+/-}$ or *Eng*$^{+/-}$ mice. Using intra-brain injection of an adeno-associated virus expressing VEGF (AAV-VEGF), vascular dysplasia (abnormal vessels) was induced at the injection sites in almost all of the mice (Hao *et al.*, 2010). Therefore, we proposed that bAVM formation is an abnormal response-to-injury (Hao *et al.*, 2010; Kim *et al.*, 2011b). However, the vascular dysplasia in the brain of *Alk1*$^{+/-}$ or *Eng*$^{+/-}$ was at the capillary level (Hao *et al.*, 2010) and could not fully recapitulate the human bAVM phenotype.

5.1.2.3.4 Adult onset bAVM mouse models

5.1.2.3.4.1 Brain focal Alk1 or Eng homozygous deletion plus angiogenic stimulation

Brain focal *Alk1* homozygous deletion and angiogenic stimulation were induced in *Alk1*$^{2f/2f}$ mice that have two loxP sites franking exon 4 and 6 of mouse *Alk1* gene (Park *et al.*, 2008) through intra-brain injection of an adeno virus expressing a Cre recombinase (Ad-Cre) and AAV-VEGF (Walker *et al.*, 2011). Around 8 weeks later, bAVM phenotypes, including tortured vasculature, AV shunting, and reduction of vascular smooth muscles, were detected at the injection site (Figure 1) (Walker *et al.*, 2011). This is the first adult-onset bAVM animal model that had a true AVM nidus resembling human lesion in the brain parenchyma.

A similar strategy of co-injection of Ad-Cre and AAV-VEGF into the brain of the *Eng*$^{2f/2f}$ mice, which has two loxP sites franking exon 5 and 6 of *Eng* gene (Allinson *et al.*, 2007; Choi *et al.*, 2012), was also employed. Only capillary level of abnormal vessels was found (Choi *et al.*, 2012). This was most likely due to the low gene deletion efficiency of *Eng* compared to *Alk1* (1% versus 16%) (Choi *et al.*, 2012).

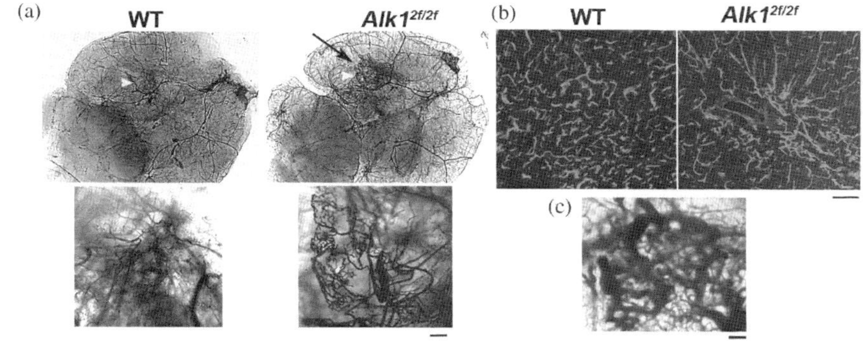

Figure 1. Adult-onset bAVM in the brain *Alk1*-deleted angiogenic region. (a) Large tangled vessels resembling human bAVM only detected in $Alk1^{2f/2f}$ brain co-injected with Ad-Cre and AAV-VEGF, but not in WT mice that received the same injection. (b) Representative images show lectin-perfused vessels. Enlarged irregular vessels presented in $Alk1^{2f/2f}$ mice but not in WT mice that received co-injection of Ad-Cre and AAV-VEGF. Scale bar: 50 μm. (c) Latex cast shows angled AVM vessels in $Alk1^{2f/2f}$ bAVM brain. Scale bar: 100 μm (Walker *et al.*, 2011).

5.1.2.3.4.2 Conditional deletion of Alk1 or Eng globally plus brain focal angiogenic stimulation

Cre recombinase expression can be regulated by inducible systems, such as the estrogen receptor (ER) ligand-binding domain system and the tetracycline-regulated systems (Saunders, 2011). In the former, an ER ligand-binding domain is fused with cre recombinase. When tamoxifen is present, creER-fused protein will enter the nucleus to trigger recombination of loxP sites causing gene deletion. In the latter, depending on whether a tet-on or tet-off system is used, gene expression can be activated or shut down in the presence of tetracycline/doxycycline. These inducible systems allow deletion of targeted genes in specific cell types and/or at a specific developmental stage.

To globally delete *Alk1* or *Eng* in adult mice, the R26CreER mouse line, in which the R26 promoter drives ubiquitous cre expression induced by tamoxifen (Badea *et al.*, 2003; Friedrich & Soriano, 1991), was bred with $Alk1^{2f/2f}$ or $Eng^{2f/2f}$ mice. The offsprings with genotypes of R26CreER;$Alk1^{2f/2f}$ or R26CreER;$Eng^{2f/2f}$ had *Alk1* or *Eng* deleted globally upon tamoxifen treatment. High mortality was found in R26CreER;$Alk1^{2f/2f}$ mice after cre activation through intra-peritoneal injection of tamoxifen (one dosage of 2.5mg/25 g of body weight) (Park *et al.*, 2009).

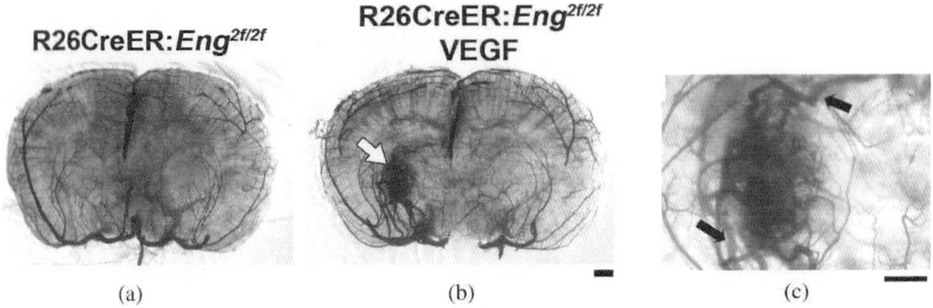

Figure 2. Brain AVM in R26CreER;*Eng*$^{2f/2f}$ mice. Representative latex-perfused images of R26CreER;*Eng*$^{2f/2f}$ mice without AAV-VEGF intra-brain injection (a) and with AAV-VEGF intra-brain injection (b). Scale bar: 1 mm. (c) Arrows indicate enlarged image of bAVM in (b) Scale bar: 500 μm (Choi *et al.*, 2014).

All mice were found dead between 9 and 21 days after tamoxifen treatment due to lung and gastrointestinal tract hemorrhage (Park *et al.*, 2009). No bAVM was detected in these mice (Park *et al.*, 2009).

Similarly, global deletion of *Eng* in adult R26CreER; *Eng*$^{2f/2f}$ mice through intraperitoneal injection of tamoxifen (2.5mg/25 g of body weight) once per day for 3 consecutive days did not cause bAVM formation (Choi *et al.*, 2014). Brain AVM developed only when brain angiogenesis was induced (Figure 2) (Choi *et al.*, 2014). Again, these data suggest that *Alk1* or *Eng* gene- deletion alone is insufficient to cause bAVM formation in adult mice. Angiogenic stimulation is required.

5.1.2.3.4.3 Endothelial Alk1 gene deletion plus brain angiogenic stimulation

The *Alk1* gene is predominantly expressed in the endothelium of developing arteries (Park *et al.*, 2008; Seki *et al.*, 2003). *Alk1* expression has also been found in chondrocytes (Davidson *et al.*, 2009) and mural cells (Milton *et al.*, 2012). The *Eng* gene is expressed by endothelial cells (Gougos & Letarte, 1988; Li *et al.*, 2000) and other cell types, such as activated monocytes/macrophages (Lastres *et al.*, 1992), mesenchymal cells, fibroblasts (St-Jacques *et al.*, 1994), and vascular smooth muscle cells (CSMCs) (Adam *et al.*, 1998; Conley *et al.*, 2000). It would be interesting to determine which cell type is crucial for bAVM development after mutation of the *Alk1* or *Eng* gene. By using different transgenes to induce cell-specific *Alk1* or *Eng* deletion, it was found that mutation of *Alk1* or *Eng* in endothelial cells is essential for AVM formation (Chen *et al.*, 2014; Choi *et al.*, 2014; Garrido-Martin *et al.*, 2014). Using the Pdgfb-iCreER transgene containing a platelet-derived growth factor b (*Pdgfb*) promoter, which drives tamoxifin-inducible codon optimized cre expression to mutant *Alk1* or *Eng* predominantly in endothelial cells, we were able to induce bAVM formation in the brain angiogenic region (Chen *et al.*, 2014; Claxton *et al.*, 2008). In addition to the bAVM phenotype, like R26CreER; *Alk1*$^{2f/2f}$ mice, Pdgfb-iCreER; *Alk1*$^{2f/2f}$ mice also exhibited profound AVM phenotypes in the gastrointestines, lung, and around the skin wound (Chen *et al.*, 2014), and died 6–13 days after tamoxifen administration (Chen *et al.*, 2014). Pdgfb-iCreER; *Eng*$^{2f/2f}$ mice did not die after tamoxifen treatment.

Deletion of *Alk1* in pericytes in adult mice using a pericyte-specific gene (NG2) promoter and an estrogen-inducible cre expression (NG2-iCreER; *Alk1*$^{2f/2f}$) did not cause AVM formation in the brain angiogenic foci (Chen *et al.*, 2014). Deletion of *Eng* in macrophages induced by a macrophage-specific gene (LysM) promoter driving cre expression (LysM-Cre; *Eng*$^{2f/2f}$), plus angiogenic stimulation in the brain did not cause bAVM either (Choi *et al.*, 2014). These studies demonstrate that endothelial deletion of *Alk1* or *Eng* is critical for bAVM initiation.

5.1.2.3.5 Developmental models

L1Cre(+) transgene mediates constitutive expression of cre in endothelial cells. L1Cre(+);*Alk1*$^{2f/2f}$ mice (Park *et al.*, 2008, 2009) have AV fistula in the brain and die by postnatal day 5 (P5) due to hemorrhage in the brain, lung, and gastrointestinal tract (Park *et al.*, 2009), indicating a critical role for Alk1 in endothelial function during neonatal and postnatal development. Because the mice do not survive into adulthood, this is not an ideal model for testing new therapies.

SM22α (smooth muscle actin or *ACTA2*) is naturally expressed in vascular SMCs (Shen *et al.*, 2012). However, in the SM22αCre transgenic mice, cre expresses not only in vascular smooth cells, but also in other cell types, such as endothelial cells (Milton *et al.*, 2012; Shen *et al.*, 2012). SM22αCre; *Alk1*$^{2f/2f}$ mice, and SM22αCre; *Alk1*$^{2f/-}$ mice, which have one *Alk1* allele deleted in all cell types, have AVMs in the brain and spinal cord, and have a high lethality before 2 weeks of age (Milton *et al.*, 2012). Malformed intestinal vessels have also been detected in these mice (Milton *et al.*, 2012). About 15% of SM22αCre; *Alk1*$^{2f/-}$ mice and 47% of SM22αCre; *Alk1*$^{2f/2f}$ survive beyond 2 weeks of age (Milton *et al.*, 2012). The bAVM vessels in these mice have reduced SMC coverage and changes of gene expression in AVM vessel walls, such as increased expression of angiopoietin 2 (*Angpt 2*), interleukin 1β (*Il1b*), and Cyclooxygenase 2 (*Cox2*) (Milton *et al.*, 2012).

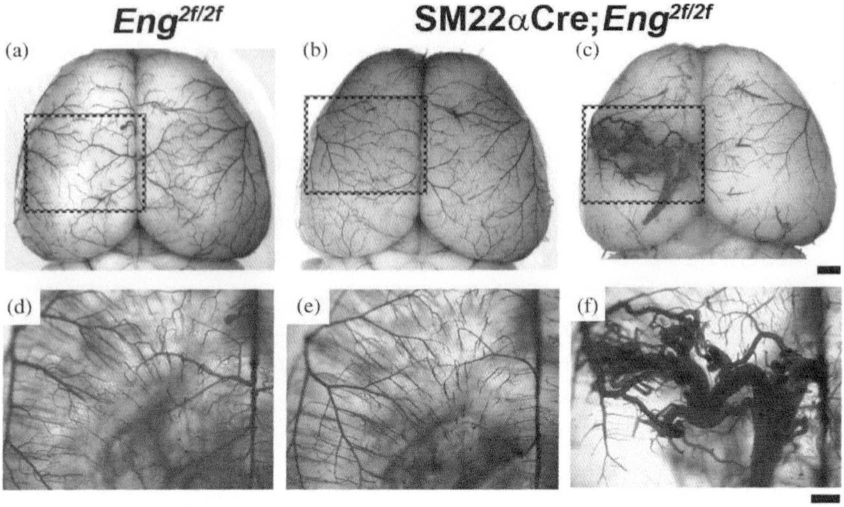

Figure 3. Spontaneously developed bAVM in SM22αCre; $Eng^{2f/2f}$ mice (Choi *et al.*, 2014). Representative latex-perfused brain of Eng2f/2f (a–b), SM22αCre;$Eng^{2f/+}$ (c–d), and SM22αCre; $Eng^{2f/2f}$ (e–f) mice. Scale bar: 1 mm in a, c, and e, and 500 μm in b, d, and f.

The mortality of SM22αCre; $Eng^{2f/2f}$ mice is lower than that of SM22αCre; $Alk1^{2f/2f}$ mice (Choi *et al.*, 2014). Approximately 90% of the SM22αCre; $Eng^{2f/2f}$ mice spontaneously develop bAVMs by 5 weeks of age and with a mortality of 50% by 6 weeks (Choi *et al.*, 2014). These mice also have spinal and intestinal AVMs (Figure 3) (Choi *et al.*, 2014).

Since some SM22αCre; $Alk1^{2f/2f}$ and SM22αCre;$Eng^{2f/2f}$ mice survive into adulthood with bAVMs (Choi *et al.*, 2014; Milton *et al.*, 2012) developing spontaneously without focal angiogenic stimulation, these models may be more suitable for dissecting bAVM formation and testing new therapies.

5.1.2.4.6 SMAD4 mutation

HHT3 patients have *SMAD4* mutations (Gallione *et al.*, 2004). However, *Smad4* heterozygous mice are phenotypically normal (Yang *et al.*, 1998). *Smad4* null mice die between embryonic day 6.5 and 8.5 with severe abnormalities, including decreased epiblast proliferation, abnormal egg cylinder formation, and lack of mesoderm induction (Yang *et al.*, 1998, 2002). Interestingly, brain focal deletion of *Smad4* induced by Ad-Cre plus angiogenic stimulation in $Smad4^{2f/2f}$ mice result only in capillary vascular dysplasia (Su *et al.*, 2010), and therefore, the *SMAD* gene family may regulate an overlay network of downstream genes that contribute to the AVM phenotype. In addition, clinical studies and animal research indicate that modifier genes may play roles in AVM development (Bourdeau *et al.*, 2000b).

5.1.2.3.7 Other genetic bAVM mouse models

In addition to mutation of genes involved in HHT, mutations of several other genes also cause abnormal cerebrovascular structure resembling bAVM structure in mice.

5.1.2.3.7.1 Mutation of matrix Gla protein (Mgp)

In vitro studies using bovine aortic endothelial cell show that activation of ALK1 by BMP9 induces matrix Gla protein (or *MGP*) expression (Yao *et al.*, 2006), and MGP in turn provides negative feedback, inhibits BMP4 and BMP7, and stimulates ALK1 expression (Yao *et al.*, 2006). However, the factors involved in this negative feedback have yet to be identified (Yao *et al.*, 2006). $Mgp^{-/-}$ mice express excessively high levels of Alk1 and its downstream effectors, including pSmad1/2/3. However, $Alk1^{+/-}$ mice have a decreased level of MGP (Yao *et al.*, 2011). Because of these contradictory data, it is too early to conclude if MGP and ALK1 are directly associated in a signaling pathway that regulates vascular formation/homeostasis.

Interestingly, severe AVMs were found in the lung, kidney (Yao *et al.*, 2011), and brain of $Mgp^{-/-}$ mice (Yao *et al.*, 2013). Clinically, *MGP* mutations are associated with Keutal syndrome, an autosomal recessive disorder. Patients have multifocal stenosis of pulmonary arteries in the peripheral regions of their lungs, as well as other abnormalities (Munroe *et al.*, 1999). Yet *MGP* mutations have not been identified in bAVM patients.

5.1.2.3.7.2 Notch gain or loss of function

Notch signaling plays a critical role in vascular development and maintenance of vascular homeostasis (Zhang *et al.*, 2016b). Both gain and loss of notch function in embryos cause AVM formation and embryonic lethality (Zhang *et al.*, 2016b). Upregulation of Notch signaling has been detected in human bAVM specimens (Murphy *et al.*, 2009; ZhuGe *et al.*, 2009). Expression of a constitutively endothelial-specific Notch4 intra-cellular domain, int3, in Tie2-tTA;TRE-int3 mice (or Tie2-tTA;TRE-Notch4*) during gestation resulted in the formation of dilated blood vessels in the brain. The mice died between postnatal 2 and 5 weeks due to hemorrhage in the cerebellum (Murphy *et al.*, 2008). Inhibition of int3 overexpression by tetracycline treatment starting from postnatal day 21 rescued mortality, indicating that normalization of Notch signaling reverses phenotypic abnormality in mutant mice (Murphy *et al.*, 2008). Int3 overexpression activated at birth also resulted in dilated vessels in the cerebral cortex (Murphy *et al.*, 2014). Increasing Notch signaling in the endothelium during the embryonic or neonatal stage results in the formation of dilated vessels in the brain. However, it is important to be aware that whether gain of Notch signaling during the adult period can result in bAVM is *still* unclear.

5.1.2.3.7.3 Rbpj mutation

The *Rbpj* gene encodes a transcription factor functioning in the canonical Notch signaling pathway. Endothelial deletion of the *Rbpj* gene at birth in Cdh5(PAC)-CreER; *Rbpj*$^{2f/2f}$ mice leads to AV shunt formation and hemorrhage in the brain of postnatal day 14 mice, as well as vascular abnormalities in the intestine and heart (Nielsen *et al.*, 2014). Hypoxic regions were noticed in the cerebellum of these mutant mice (Nielsen *et al.*, 2014). Interestingly, when *Rbpj* gene deletion was induced in adult mice (6 weeks of age), the cerebrovasculature was not significantly different from the control mice, although the intestine and uterus had some AVM phenotype (Nielsen *et al.*, 2014).

5.1.2.3.8 Integrin β8 (Itgb8) mutant

The *ITGB8* gene encodes the β8 subunit of integrin that forms a complex with integrin αv subunit. Integrin αvβ8 plays an important role in angiogenesis and vascular differentiation. β8 expression is reduced in human bAVM tissue (Su *et al.*, 2010). Focal *Itgb8* deletion plus angiogenic stimulation induce capillary level of cerebrovascular dysplasia in the brain of *Itgb8*$^{2f/2f}$ mice (Su *et al.*, 2010). No macroscopic level of bAVM was detected in these mice (Su *et al.*, 2010), indicating that loss of *Itgb8* alone is not sufficient to initiate bAVM formation. Deletion of *Itgb8* in the brain of *Alk1*$^{+/-}$ mice enhances the formation of dysplastic vessels and hemorrhage in the angiogenic foci of *Alk1*$^{+/-}$ brain (Ma *et al.*, 2016).

Although mutations of these genes in mice cause some cerebrovascular dysplasia and bAVM phenotypes, none of these mutations thus far has been found in bAVM patients.

5.1.3 *Future models*

Brain AVM phenotypes in genetic mouse models highly resemble human diseases; therefore, bAVM mouse models are very useful in studying the disease mechanism and testing new therapies. However, due to the mouse's body size, it is nearly impossible to monitor lesion progression using *in vivo* imaging and study bAVM hemodynamic changes. With the development of new gene editing techniques, genetic animal models may be generated in larger animals (e.g., rabbit or pig) in the near future.

5.2 Knowledge Gained from Animal Models Regarding AVM Pathogenesis

To date, the exact causative mechanism of bAVM pathogenesis is not clear, but many mechanisms have been proposed. Under normal conditions, an injury upregulates the expression of angiogenic factors, such as VEGF, which induces endothelial cell mitosis. Endothelial cell proliferation leads to formation of new vessels, which then develop into a stable neovasculature through remodeling. In the presence of certain abnormal genetic backgrounds, this otherwise normal injury repair process can lead to a vascular dysplastic response (Kim *et al.*, 2011b).

By studying mouse and zebrafish bAVM models, the following signaling pathways and mechanisms involved in bAVM pathogenesis have been identified.

5.2.1 *Molecular pathways*

5.2.1.1 *TGFβ*

TGFβ functions in many biological processes including regulation of cellular proliferation, differentiation, migration, and extracellular matrix formation. The TGFβ family signals into the cell via two distinct transmembrane proteins, known as type I (the activin-like receptors, ALK1–7) and type II receptors. After binding to their ligands, the type I receptors associate with and are phosphorylated by type II receptors. Phosphorylation activates the kinase domain of the type I receptor, and the activated type I receptor then signals to the SMAD transcription factor family.

In most cell types, TGFβ signals through ALK5; in endothelial cells, it also signals through ALK1 (Goumans *et al.*, 2003; Oh *et al.*, 2000). ALK1 is expressed predominantly in endothelial cells and is also found in the mesenchyme of the lung, submucosal layer of the stomach and intestines, and at epithelial–mesenchymal cell interaction sites (Roelen *et al.*, 1997). ENG is an accessory protein that interacts with multiple heteromeric receptor complexes containing TGFβs, activins, and BMPs (Barbara *et al.*, 1999). It is expressed at high levels in vascular endothelial cells and syncytiotrophoblast of full-term human placenta, as well as transiently in extravillous cytotrophoblasts and in the cardiac endocardium during development (Gougos & Letarte, 1990; Qu *et al.*, 1998; St-Jacques *et al.*, 1994). In endothelial cells, ENG can also interact with ALK1 and ALK5 in the absence of TGFβ (Abdalla *et al.*, 2000; Guerrero-Esteo *et al.*, 2002).

About 85% of HHT patients have ALK1 or ENG haploinsufficiency. As described above, mutation of *Alk1* or *Eng* results in AVM in mice and zebrafish. A minority of HHT patients have mutations in the *SMAD4* gene (Gallione *et al.*, 2004). These patients exhibit clinical features overlapping with juvenile polyposis. Mice with endothelial-specific *Smad4* mutation die at embryonic day 10.5 due to cardiovascular defects, including attenuated vascular sprouting and remodeling, collapsed dorsal aortas, enlarged hearts with reduced trabeculae, and failed endocardial cushion formation, suggesting that Smad4 activity in the developing endothelial cells is essential for blood vessel remodeling, maturation, and integrity (Lan *et al.*, 2007). Deletion of *Smad4* in the brain angiogenic region causes a capillary level vascular malformation (Su *et al.*, 2010).

5.2.1.2 *Notch*

The Notch pathway is an intercellular signaling pathway. Both the receptor and the ligand are transmembrane proteins located on adjacent cells. Upon ligand–receptor interaction, Notch is cleaved and the Notch Intercellular Domain (NICD) is released into the cytoplasm. Subsequently, NICD translocates into the nucleus where it binds to the DNA-binding protein RBPJ to target a specific DNA sequence (or transcriptional binding site) and initiates transcription of its downstream targets (Jarriault *et al.*, 1995). Notch gain-of-function and loss-of-function both cause abnormal arterial and venous specification. Ablation of Notch in mice (Krebs *et al.*, 2004) or zebrafish (Lawson *et al.*, 2001) results in fusion of the dorsal aorta and cardinal vein. Ablation of *Rbpj* also causes defects in arterial specification and AVM development (Krebs *et al.*, 2004). Deletion of the endothelial *Rbpj* gene in mouse neonates causes AVMs in the brain, intestine, and heart by day 14 (Nielsen *et al.*, 2014). Partial loss of the Notch ligand Dll4 also results in embryonic lethality with severe vascular defects reminiscent of those seen in mouse embryos lacking Notch receptors (Gale *et al.*, 2004). In 7- to 10-week-old adult mice, expression of a constitutively active Notch4 (Tie2-tTA;TRE-int3) causes AVMs in the liver, uterus, skin, and brain (Carlson *et al.*, 2005; Miniati *et al.*, 2010; Murphy *et al.*, 2008, 2014). Similar to Notch4, endothelial cell-specific, constitutively active Notch1 results in vascular defects and AVM formation (Krebs *et al.*, 2010). The sizes of the AVMs formed in these gain-of-function mice are reduced when Notch4 overexpression is turned off (Carlson *et al.*, 2005; Murphy *et al.*, 2012).

Besides Notch signaling components, AVMs have also been reported in mutation of transcription factors upstream of Notch. Deletion of the *Foxc* (Forkhead) gene, a group of transcription factors (*Foxc1*$^{+/-}$ or *Foxc2*$^{-/-}$), in mice result in AVMs (Kume *et al.*, 2001; Seo *et al.*, 2006). The endothelial cells of *Foxc1*$^{+/-}$ or *Foxc2*$^{-/-}$ mice fail to express Dll4, a ligand of Notch. These mice also show reduced expression of other arterial markers such as Notch1, Notch4, Jag1, Hey2, and EphrinB2.

5.2.1.3 *RASA1*

RASA1, encoding p120-RasGAP, is involved in signaling for various growth factor receptors that control proliferation, migration, and survival of several cell types, including vascular endothelial cells. Heterozygous *RASA1* mutation leads to capillary malformation–arteriovenous malformation (CM-AVM). In addition to AVM, patients present with atypical CM symptoms (Eerola *et al.*, 2003).

In summary, despite this genetic correlation, most bAVMs are non-familial and idiopathic (Young & Yang, 2004). In addition to the genes related to HHT, the other genes discussed above have not been identified in human bAVM. The variable presentation of HHT-related AVMs, along with the higher incidence of idiopathic AVMs, suggests that interactions with downstream factors must also be involved in AVM pathogenesis.

5.2.2 *Loss of function of the normal allele (2nd hit)*

Patients carrying AVM causative gene mutation do not exhibit a complete phenotype at birth. The nature of bAVM lesions suggests that lesion formation may require not only a mutation in one copy of a given AVM causative gene within a particular group of cells, but also a "second hit" that causes loss or reduction of function of the normal allele. This second hit could be genetic, in the form of a somatic mutation to the second copy of that gene, or a mutation in another gene acting in the same cellular pathway (Leblanc *et al.*, 2009).

For HHT, the prevailing view is that haploinsufficiency of one of its causative genes in somatic endothelial cells causes vascular malformations. However, analysis of human brain and lung AVMs in HHT1 indicates that haploinsufficiency of ENG is not sufficient to cause lesion development (Bourdeau *et al.*, 2000a), supported by mouse studies that loss of a single allele of one of the HHT causative genes is ineffective for bAVM formation (Bourdeau *et al.*, 2000b; Srinivasan *et al.*, 2003). Both $Eng^{+/-}$ and $Alk1^{+/-}$ mice are viable. However, loss of both alleles of any HHT-causative gene is embryonically lethal (Sorensen *et al.*, 2003; Urness *et al.*, 2000). Conditional homozygous deletion of *Alk1* in the mouse brain results in robust AVM formation in the angiogenic region (Choi *et al.*, 2012; Walker *et al.*, 2011). Moreover, bAVMs developed only in $Mgp^{-/-}$ mice, not in $Mgp^{+/-}$ mice (Yao *et al.*, 2013). The loss of function of the wild-type allele has also been demonstrated in two vascular malformations: somatic mucocutaneous venous malformations (Limaye *et al.*, 2009) and cerebral cavernous malformations (Akers *et al.*, 2009). Therefore, inactivation of the remaining wild-type allele, regardless of the mechanism by which it is inactivated (e.g., loss of heterozygosity or loss of protein during inflammation), is an important factor in bAVM development (Mahmoud *et al.*, 2010).

5.2.2.1 *Modifier genes*

The fact that some mouse strains appear to be more susceptible to AVM formation than others suggests that modifier genes and even epigenetic factors may also influence bAVM formation (Bourdeau *et al.*, 2001).

A genetic modifier loci (*Tgfbm*'s) that suppresses the lethal vascular phenotype of $Tgfb1^{-/-}$ has been identified in mice (Bonyadi *et al.*, 1997; Tang *et al.*, 2003, 2005). The human non-receptor tyrosine phosphatase 14 (*PTPN14*) gene, located in a region syntonic to mouse *Tgfbm2* (Tang *et al.*, 2003), has been associated with the presence of pulmonary AVM (PAVM) in HHT patients (Benzinou *et al.*, 2012). In addition, *PTPN14* may interact with *ALK1* and *EphrinB2*, directly or indirectly, to regulate angiogenesis and/or arteriovenous specification, both processes being implicated as abnormal in HHT pathology (Benzinou *et al.*, 2012). siRNA knockdown of endogenous *PTPN14* modulates angiogenesis in a 3D human primary endothelial culture system (Benzinou *et al.*, 2012). PTPN14 also supports endothelial expression of EphrinB2, a key molecule required for angiogenesis (Herbert *et al.*, 2009), establishment of endothelial arteriovenous identity (Urness *et al.*, 2000), and recruitment of pericytes/SMCs during blood vessel wall assembly (Foo *et al.*, 2006).

Similarly, single-nucleotide polymorphisms (SNPs) in *EMILIN2* gene, which encodes an extracellular matrix protein, elastin microfibril interfacer 2, have been associated with PAVMs in two independent cohorts of HHT1 patients from the Netherlands and France, and have been reported to interact with transcriptional co-activator with PDZ-binding motif (TAZ) in angiogenesis (Letteboer *et al.*, 2015).

5.2.2.2 *Angiogenesis*

Angiogenesis, the process of new vessel formation from pre-existing blood vessels, consists of activation and resolution/maturation phases (Pepper, 1997). During the activation phase, endothelial cells degrade the perivascular membrane and invade and migrate into the extracellular space, where they proliferate and form a lumen. In the resolution or maturation phase, cell migration and proliferation stop. The basement membrane is reconstituted, and new vessels are formed. The sprouting of new blood vessels or angiogenesis is a normal physiological process during embryonic development. This process also happens in adults during wound healing, exercise when muscle develops, and formation of collateral vessels. During malignant tumor development, angiogenesis plays a role in producing extra blood vessels to increase the tumor blood supply (Folkman & Klagsbrun, 1987). This angiogenetic process is induced by angiogenic factors, including VEGF.

Abnormally high level of VEGF has been found in surgically removed bAVM specimens (Cirulli *et al.*, 2003; Hao *et al.*, 2010; Kim *et al.*, 2011b; Sadick *et al.*, 2005a, 2005b; Walker *et al.*, 2011) and plasma of HHT patients (Cirulli *et al.*, 2003; Hashimoto *et al.*, 2000, 2005b; Hatva *et al.*, 1996; Kilic *et al.*, 2000; Kim *et al.*, 2011a; Koizumi *et al.*, 2002; Rothbart *et al.*, 1996; Sadick *et al.*, 2005a, 2005b; Sonstein *et al.*, 1996; Sure *et al.*, 2001a, 2001b). In animal models, spontaneous bAVM develop only when *Alk1* or *Eng* is deleted during the embryonic or perinatal stage (Choi *et al.*, 2014; Milton *et al.*, 2012). However, gene deficiency alone is not sufficient to induce AVM in the adult brain (Park *et al.*, 2009). More pronounced cerebral microvascular dysplasia forms in the VEGF-stimulated brain angiogenic foci in $Eng^{+/-}$ or $Alk1^{+/-}$ mice (Hao *et al.*, 2008b, 2010; Xu *et al.*, 2004). A macroscopic level of vascular dysplasia that mimics many phenotypes of human bAVMs has been induced by injecting AAV-VEGF into the *Eng*- or *Alk1*-deleted brain (Chen *et al.*, 2014; Choi *et al.*, 2014; Han *et al.*, 2014; Walker *et al.*, 2011), indicating that angiogenesis is required for bAVM development.

5.2.2.3 *Endothelial gene mutation*

To analyze gene deletion in which cell types are necessary for bAVM formation, different cre transgenic mouse lines have been employed to mediate gene deletion in specific cell lineages. *Pdgfb*-icreER has been used to mediate *Alk1* deletion predominantly in endothelial cells in adult mice. De novo AVM formation developed in the brain angiogenic region in addition to the intestine, lung, and ear wound (Chen *et al.*, 2014). Similarly, Scl-CreER-mediated *Eng* deletion in endothelial cells result in AVM formation in the skin around the ear and back wound (Garrido-Martin *et al.*, 2014). Scl-CreER-mediated *Alk1* deletion in endothelial cells also causes AVM formation in many internal organs, such as the cecum, ileum, stomach, and uterus (Garrido-Martin *et al.*, 2014). SM22α-Cre-mediated *Eng* and *Alk1* deletion in SMC and other cell types, including endothelial cells in the embryo, also leads to AVM formation in the brain, spinal cord, and intestines (Choi *et al.*, 2014; Milton & Seki, 2011). However, Myh11-CreER-mediated *Eng* or *Alk1* deletion in SMC in adult mice do not cause AVM formation (Garrido-Martin *et al.*, 2014), and neither does *NG*2-icreER-mediated *Alk1* deletion in pericytes (Chen *et al.*, 2014). In addition, *LysM*-Cre-mediated *Eng* deletion in lysozyme M-positive macrophage in the embryo does not cause AVM in the postnatal brain even after angiogenic stimulation (Choi *et al.*, 2014). The endothelial cell, therefore, is the primary cell type pertinent to AVM development.

5.2.2.4 *Vascular integrity*

Structural imperfection and immaturity of the vascular wall suggest that AVM vessels are maldeveloped. The following might be the result of abnormal vascular remodeling in response to hemodynamic changes caused by arteriovenous shunts and could contribute to bAVM microhemorrhage and rupture: (1) prominent caliber dilation, (2) hypertrophy of muscular layer, (3) hyalinization, (4) abnormal increase of elastic fibers (Isoda *et al.*, 1981), (5) changes in collagen content (Guo *et al.*, 2013; Niu *et al.*, 2012), (6) abnormal differentiation of smooth muscles cells, (7) reduction of vascular pericytes, and (8) alterations of other peri-endothelial support structures.

Abnormal vessel wall structure has also been noticed in the bAVM vessels in mouse models (Chen *et al.*, 2013; Walker *et al.*, 2011). Compared with normal brain angiogenic foci, the lesion in bAVM mouse models have more vessels with diameters larger than 15 μm that have no α-SMA positive cells.

Pericytes are contractile cells that wrap around the endothelial cells of capillaries and venules and have a crucial role in vascular stability. Reduction of vascular pericytes impairs vascular integrity (Armulik *et al.*, 2010; Bell *et al.*, 2010). Pdgfb and Pdgfr-β both have an essential part in pericyte recruitment during angiogenesis. Knockout Pdgfb or Pdgfr-β in mice results in loss of pericytes from the microvessels (Lindahl *et al.*, 1997). The absence of pericytes also leads to endothelial hyperplasia (associated with abnormal endothelial junctions) and excessive endothelial luminal membrane folds (Hellstrom *et al.*, 2001). *Pdgfr-*β or *Pdgfb* null mice have cerebral hemorrhage with an absence of microvascular pericytes in the brain vessels and endothelial hyperplasia (Hellstrom *et al.*, 2001). AVM vessels in mouse models also have fewer pericytes and SMCs (Chen *et al.*, 2013; Walker *et al.*, 2011). Reduced SMC and pericyte coverage are associated with increased vascular permeability and microhemorrhage. Pdgfr-β expression is reduced in the bAVM lesions of *Alk1*-deficient mice (Chen *et al.*, 2013), which is associated with reduction of SMC and pericyte-coverage, suggesting a possible link between Alk1 and Pdfgr-β/Pdgfb signaling pathways. However, it is not clear whether the reduced expression of Pdfgr-β in bAVM is the cause or the result of of pericyte reduction.

Angiopoietin/TIE2 signaling also plays a role in the recruitment of periendothelial support structures. Alternations of angiopoietin/TIE2 expression have been found in human bAVM specimens (Hashimoto *et al.*, 2001). Angiopoietin-2 (ANG-2), which allows loosening of cell-to-cell contacts, is overexpressed in the perivascular region in AVM vascular channels (Hashimoto *et al.*, 2001). Therefore, imbalance of angiopoietin/TIE2 signaling could be another cause of vessel wall defects in bAVM; this pathway has not been studied in bAVM animal models.

5.2.2.5 *Inflammation and other environmental factors*

An abnormally high number of macrophages is present in and around vascular walls in human bAVM specimens, with or without hemorrhage, suggesting that macrophage accumulation is not simply a response to hemorrhage (Chen *et al.*, 2006a, 2006b; Chen *et al.*, 2008; Guo *et al.*, 2014; Ma *et al.*, 2015). Polymorphisms in inflammatory cytokines and elevated expressions of inflammation-related genes in AVM patients further suggest their active roles in AVM pathogenesis (Achrol *et al.*, 2006a; Hashimoto *et al.*, 2004; Pawlikowska *et al.*, 2004). There is a significant increase in the messenger RNA (mRNA) levels of IL-6, IL-1β, TNF-α, and IL-8 in human bAVM tissues (Chen *et al.*, 2006b). The cellular adhesion molecules (CAMs), including E-selectin, intercellular CAM-1 (ICAM-1), and vascular CAM-1 (VCAM-1), also increase in human bAVM tissues (Storer *et al.*, 2008).

The upregulation of inflammatory cytokines and the overexpression of CAMs in endothelial cells result in the enhanced recruitment of leukocytes to the AVM tissue. Inflammatory cells found in AVM tissue include neutrophils, macrophages, and microglia, T-lymphocytes, and B-lymphocytes (Chen *et al.*, 2008; Guo *et al.*, 2014). Neutrophils are mainly found in the inner vessel walls and in AVM parenchymal tissues, whereas macrophages are located primarily in the outer vessel walls.

Animal studies have shown that bone marrow (BM)-derived cells infiltrate into the brain angiogenic region. Macrophages are the major BM-derived cells recruited to the brain angiogenic foci (Hao *et al.*, 2008a). Since the number of BM-derived macrophages in VEGF-induced brain angiogenic regions peaks earlier than vessel density increase, macrophages likely play a role in the promotion of angiogenesis.

Increased macrophage burden has been found in bAVM in mouse models generated through conditional deletion of HHT causative genes, *Eng* or *Alk1*, in combination with focal angiogenic stimulation (Chen *et al.*, 2013, 2014; Choi *et al.*, 2014; Walker *et al.*, 2011). Although *Eng*-deficiency has been shown to impair monocyte migration into injured tissue (Dingenouts *et al.*, 2015; Post *et al.*, 2010; Shen *et al.*, 2014), a high number of BM-derived macrophages and activated microglia has been found clustering around AVM lesions in the brain. Compared with normal BM-derived cells, *Eng*-deficient cells show slower and persistent infiltration into the brain angiogenic region as well as delayed clearance, causing unresolved inflammation, which, in turn, enhances abnormal vascular remodeling and the severity of the bAVM phenotype (Zhang *et al.*, 2016a).

5.2.2.6 *BM-derived cells*

Through transplantation of BM from GFP donor mice to WT mice, Hao *et al.* discovered that BM-derived cells are recruited into the VEGF-induced angiogenic focus and are involved in angiogenesis by activating MMP-9 (Hao *et al.*, 2008a). Most BM-derived cells are macrophages (71%), and only a small fraction are endothelial cells. Transplantation of WT BM to MMP-9 knockout mice restores their brain angiogenic response (Hao *et al.*, 2011). Further, *Eng*-deficient BM-derived cells are sufficient to induce vascular dysplasia in the brain angiogenic region of adult mice (Choi *et al.*, 2013). Similarly, transplantation of BM with endothelial specific *Alk1* deletion to healthy WT mouse also causes AVM formation in the brain angiogeneic foci (unpublished data). These data suggest that gene mutation in BM is sufficient to cause bAVM formation.

5.2.3 *Therapeutic development*

Patients with bAVMs are at risk for intra-cranial hemorrhage that can cause long-term disability and death. Therefore, the main goal of treatment for bAVM is to reduce the risk of hemorrhage. Current treatment options for bAVMs include invasive interventions, such as surgical removal, embolization and radiotherapy, and conservative observation. There is no specific medical therapy available. The result of a recently completed clinical trial (ARUBA, a randomized trial of unruptured bAVM) showed that invasive intervention has a higher risk of unfavorable outcomes (i.e., stroke and death) than conservative management procedures (Mohr *et al.*, 2014). With current imaging technology, more and more asymptomatic bAVM cases are being diagnosed. Managing these patients becomes a challenge because of the potential risk of life-threatening hemorrhage and the mortality and morbidity associated with high-risk treatment. Therefore, developing specific and low-risk medical therapies is crucial. Identifying potential targets that can be utilized to develop new therapies can be achieved by analyzing animal models and surgically resected human bAVM specimens.

5.2.3.1 *Anti-angiogenesis*

Compelling evidence that present in humans and animal models of bAVM indicate that excessive VEGF expression appears to be a fundamental part of bAVM pathophysiology for both sporadic bAVM (Hao *et al.*, 2010; Kim *et al.*, 2011b; Walker *et al.*, 2011) and bAVM in HHT patients (Cirulli *et al.*, 2003; Sadick *et al.*, 2005a, 2005b). In addition, increased VEGF expression has been found in patients with *de novo* bAVM growth and is associated with AVM hemorrhage. A *VEGF* gene polymorphism-brain AVM hemorrhage correlation has been found in several cases (Gong *et al.*, 2011). Therefore, interruption of VEGF signaling might be a viable therapeutic strategy.

Bevacizumab (Avastin) is a recombinant humanized monoclonal antibody that specifically recognizes and binds to human VEGF-A, thereby inhibiting VEGF-induced angiogenesis. It was the first clinically available angiogenic inhibitor approved by the U.S. Food and Drug Administration in 2004. The initial target of bevacizumab was inhibiting angiogenesis in cancer, and later it was expanded to include several off-label usage to inhibit pathological angiogenesis, including in retina degeneration and HHT (Andreoli & Miller, 2007; Dupuis-Girod *et al.*, 2012; Flieger *et al.*, 2006; Lupu *et al.*, 2013; Mitchell *et al.*, 2008). Submucosal or topical intranasal delivery of bevacizumab has been shown to be effective in preventing HHT-associated epistaxis (Simonds *et al.*, 2009). Bevacizumab treatment also reduced edema and improved the symptoms of a brain AVM patient who developed headache and hemiparesis after stereotactic radiosurgery (Riss *et al.*, 2015; Williams *et al.*, 2012).

In a bAVM mouse model, systemic delivery of bevacizumab penetrated the BBB, entered the bAVM lesion, inhibited angiogenesis, and reduced the number of dysplastic vessels (Walker *et al.*, 2011, 2012). This study suggests that VEGF blockade might be a better alternative than resection and radiosurgery. A clinical trial using bevacizumab to treat high grade bAVM has been registered (NCT02314377), designed to evaluate the efficacy of intravenous (IV) bevacizumab treatment in reducing the nidus size in patients.

However, the antibody therapy has some drawbacks. Rare side effects include AVM hemorrhage and systemic thrombosis (Maestraggi *et al.*, 2015; Tanvetyanon *et al.*, 2009). Brain AVM is a chronic disease and intermittent IV infusions might be required.

A promising alternative is the use of AAV-mediated expression of soluble FMS-related tyrosine kinase 1 (sFLT1). FLT1, (VEGFR1), first identified as a decoy receptor for VEGF but later found to have more complicated functions (Chung & Ferrara, 2011; Shibuya, 2006). sFLT1 is an alternative transcript of FLT1, containing only the extracellular domain of VEGFR1. It binds to VEGF with high affinity, and thus may reduce its downstream signaling through membrane-bound receptors (Kendall & Thomas, 1993). Injection of an AAV2 vector expressing sFLT1 into the vitreous humor of non-human primates effectively inhibited laser-induced choroidal neovascularization (Lukason *et al.*, 2011). AAV2-sFLT1 is well tolerated and capable of mediating long-term sFLT1 expression (Maclachlan *et al.*, 2011). One study has shown that co-injection of AAV2-sFLT02 containing sFLT1 domain 2 with modifications of C-terminal structure (Pechan *et al.*, 2009) along with AAV1-VEGF into the brain completely blocked VEGF-induced brain angiogenesis (Shen *et al.*, 2015). IV injection of AAV9-sFLT1 that contains a full-length sFLT 4 weeks after induction of brain angiogenesis also reduced vessel density in the brain angiogenic foci (Shen *et al.*, 2015). These data suggest that AAV-mediated sFLT1 expression might be a promising anti-angiogenesis therapy for treating bAVM patients. The genome of AAV remains in non-dividing cells for an extended period, most likely for several years (Kaplitt *et al.*, 1994; Koo *et al.*, 2011; Rivera *et al.*, 1999, 2005, 2006), making it an ideal vector for treating AVM since repeated dosing is unlikely to be needed. Another advantage of AAV gene therapy is that it allows adding a regulatory system to restrict therapeutic gene expression in bAVM when therapy is needed, which can reduce sFLT1 side effects caused by overexpression.

Another agent that can be used in gene therapy is microRNA. These are small non-coding RNAs that regulate gene expression by directing their target mRNA for cleavage or translational repression. An *in vitro* study has shown that overexpression of microRNA-18a in endothelial cells isolated from bAVM reduced their aberrant angiogenic features and decreased their VEGF secretion (Ferreira *et al.*, 2014). The flourishing advances in gene regulation have opened new therapeutic perspectives for bAVM medical treatment.

5.2.3.2 *Anti-inflammation*

Inflammation plays a fundamental role in bAVM progression and rupture. Human bAVM lesions have increased levels of inflammatory cytokines, including myeloperoxidase (MPO), interleukin-6, interleukin-1β, tumor necrosis factor-α, and matrix metalloproteinase-9 (MMP-9) (Braverman *et al.*, 1990; Chen *et al.*, 2013; Guo *et al.*, 2012). These cytokines promote the recruitment of neutrophils and macrophages to the site of injury, releasing metalloproteinase that damages the vessel walls of AVMs. Inhibiting the pro-inflammatory cytokines is therefore one of the most important therapeutic targets to stabilize AVM and prevent hemorrhage.

The degradative activity of MMP-9 has been proven to be one of the primary mechanisms through which inflammation damages the AVM vasculature and leads to hemorrhage (Mouchtouris *et al.*, 2015). The expression of total MMP-9, pro-MMP-9, and active MMP-9 significantly increased in AVMs compared with control brain tissue (Hashimoto *et al.*, 2003). The levels of MMP inhibitors, tissue inhibitor of metalloproteinase 1 (TIMP-1), and TIMP-3 also increased in AVMs, although to a lesser extent than MMP-9. The proportionally lower level of TIMPs in AVMs explains the increased susceptibility to vascular breakdown and rupture. Tetracycline class drugs (minocycline, doxycycline, etc.) and pyrrolidine dithiocarbamate are clinically applicable non-specific MMP-9 inhibitors that have the potential to enhance vascular stability, thus reducing the risk of bAVM spontaneous hemorrhage. They also possess neuroprotective properties. To date, this class of drugs has been tested in cerebral ischemia (Yrjanheikki *et al.*, 1999), ICH (Power *et al.*, 2003; Sumii & Lo, 2002), neurodegenerative disorders (Bonelli *et al.*, 2003; Brundula *et al.*, 2002), traumatic brain injury (Sanchez-Mejia *et al.*, 2001), and atherosclerotic disease (Curci *et al.*, 2000). Furthermore, the tetracycline class of drugs generally displays anti-inflammatory properties in addition to anti-MMP effects. Previous studies in a mouse model show that MMP-9 inhibitors can reduce subsequent hemorrhage of abnormal angiogenesis (Frenzel *et al.*, 2008). Minocycline completely inhibits cerebral MMP-9 activity and effectively suppresses VEGF-induced hemorrhage (Lee *et al.*, 2007). Doxycycline reduces MMP-9 activity in

VEGF-stimulated brain angiogenic foci and reduces parenchymal angiogenesis (Lee *et al.*, 2004).

A pilot study has been conducted in 14 bAVM patients to determine the effects of doxycycline on MMP-9 level in bAVMs. A total of 10 patients received doxycycline 100 mg twice per day, and four patients received placebo for 1 week before AVM resection. Decreased levels of total and active MMP-9 in resected tissue suggest that doxycycline might be effective in inhibiting MMP-9 in human brain AVM (Hashimoto *et al.*, 2005a). A Phase I study assessing the feasibility of long-term use of minocycline and doxycycline has also been conducted. A total of bAVM patients were treated with minocycline or doxycycline for 2 years. Around 70% of the patients tolerated the 200-mg daily dose. One patient had bAVM hemorrhage. The most reported side effect was photosensitivity (23%), followed by vertigo (12%), yeast infection (8%), gastrointestinal symptoms (8%), hyperpigmentation (4%), and allergic reaction (4%). Most of the adverse events were mild. Also, minocycline treatment tends to have more frequent side effects and intolerance of increased dose than doxycycline. These studies indicate that it is feasible to propose a long-term trial to assess the potential benefit of tetracycline therapy to decrease hemorrhagic risk in bAVM (Frenzel *et al.*, 2008). However, previous reports mainly focus on a few MMPs (MMP9, etc.) in bAVM, while the other MMPs (MMP12, etc.) might have a protective role during angiogenesis and inflammation regulation (Chan *et al.*, 2013). Doxycycline is a strong inhibitor of MMP9, but it might also affect the expression of MMP1, MMP7, and MMP12 (Li *et al.*, 2013). The non-selective inhibition of various MMPs may not be ideal for bAVM treatment. Therefore, future studies should investigate the roles of different MMPs in bAVM pathogenesis to develop specific inhibitors for each MMP.

Aspirin, a classic non-steroidal anti-inflammatory drug, is a cyclooxygenase (COX) inhibitor widely used in preventing thrombosis and reducing acute inflammation and pain. Recent studies indicate that aspirin also has inhibitory effects on several inflammatory mediators that have a role in aneurysm and bAVM pathogenesis (Starke *et al.*, 2015). Aspirin inhibits MMP-2 and MMP-9 expression (Yiqin *et al.*, 2009), releases TNF-α *in vitro* (Sanchez de Miguel *et al.*, 1999), and inhibits inflammatory cell adhesion on endothelial cells by reducing nuclear factor κB activity (NFκB) (Weber *et al.*, 1995; Yotsui *et al.*, 2007). In mice, aspirin and COX-2 inhibitor do not affect cerebral aneurysm formation but significantly decrease the incidence of rupture (Chalouhi *et al.*, 2016). Recent human studies also suggest that aspirin reduces infiltration of inflammatory cells and the rupture rate of cerebral aneurysm (Hasan *et al.*, 2011). In a pilot study, 3 months of aspirin therapy reduced COX-2 expression and the number of macrophages in the wall of human aneurysm (Hasan *et al.*, 2013). In a spontaneous bAVM mouse model, upregulation of COX-2 has been identified in the hemorrhage-prone AVM vessel walls (Milton *et al.*, 2012). Thus, further study of aspirin or COX-2 inhibitor could be designed for AVM patients.

An important issue regarding anti-inflammation therapy is the appropriate tool for evaluating treatment efficacy. Anti-inflammation therapies will more likely aim to stabilize the lesion, thereby reducing the risk of hemorrhage, rather than to decrease the nidus size dramatically. In addition to long-term follow-up to the hemorrhagic event after medical treatment, new advances in inflammatory cell imaging *in vivo* may facilitate the monitoring of outcomes of anti-inflammation therapies. Ferumoxytol is a nano-iron particle which could be used as an MRI contrast reagent to assess inflammation burden in bAVM. In the vascular walls and intervening stroma of resected bAVM specimens, macrophage burden shown by ferumoxytol-enhanced MRI correlates with those identified by histological examination (Hasan *et al.*, 2012). Therefore, ferumoxytol-enhanced MRI can be a promising tool to evaluate bAVM inflammation before and after medical treatment.

5.2.3.3 *Improvement of vascular integrity*

Unlike cancer-related chemotherapy that aims to eliminate abnormal tumor tissue, the concept for the treatment of bAVMs is to stabilize vascular structure, thereby decreasing the risk of spontaneous rupture. Microhemorrhage presents in unruptured human bAVMs (Guo *et al.*, 2012) and in bAVM mouse model lesions (Chen *et al.*, 2013). The vessels in these models have less mural cell coverage, which could be due to decreased expression of PDGFB (Chen *et al.*, 2013; Walker *et al.*, 2011). Lebrin *et al* have demonstrated that thalidomide treatment increases PDGFB expression in endothelial cells and stimulated mural cell proliferation (Lebrin *et al.*, 2010). In a mouse model with retina AVM, thalidomide increases the mural cell coverage and rescues the vessel wall defects. Thalidomide stimulates pericyte (mural cells) recruitment with a low dose and inhibits angiogenesis with a high dose (Lebrin *et al.*, 2010).

Thalidomide belongs to a class of immunomodulatory drugs (IMIDs). Because of its well-known adverse effects that limit patient tolerance, e.g., peripheral neuropathy and drowsiness (Franks *et al.*, 2004), the search for related molecules has yielded a second generation of IMIDs. Lenalidomide, the most widely used in the group, is effective in treating multiple myeloma and myelodysplastic syndrome (Quach *et al.*, 2012). IMIDs possess a number of immunomodulatory, anti-inflammatory, and antiangiogenic properties that are pertinent to bAVM therapy (Aragon-Ching *et al.*, 2007; D'Amato *et al.*, 1994; Franks *et al.*, 2004). IMIDs inhibit endothelial migration, but do not inhibit endothelial cell proliferation (Dredge *et al.*, 2002, 2005). There is a remarkable overlap in the pathways targeted by IMIDs in neoplastic disorders (Franks *et al.*, 2004) with those diyregulated in bAVM, e.g., NF-κB activation (Aziz *et al.*, 2010), VEGF overexpression (Hashimoto *et al.*, 2005b), HIF-1α (Sure *et al.*, 2004), $\alpha V\beta 3$ (Hashimoto *et al.*, 2004; Lim *et al.*, 2005), and cytokine elaboration (Chen *et al.*, 2006b), including TNF-α (Achrol *et al.*, 2006b) and IL-1β (Kim *et al.*, 2009). In early-phase human studies, thalidomide decreased hemorrhage from gastrointestinal or nasal telangiectasias (small AVMs) (Alam *et al.*, 2011; Chen *et al.*, 2012; Lebrin *et al.*, 2010).

We recently tested thalidomide and lenalidomide treatments in an *Alk1*-deficient bAVM mouse model and found that thalidomide treatment attenuated dysplastic bAVM severity, decreased hemorrhage and inflammation, and improved vascular SMC and pericyte coverage in the bAVM lesion (unpublished data). We also found that lenalidomide has a comparative effect to thalidomide in treating bAVM. These results suggest that the new generation of IMIDs might offer a therapy with fewer side effects. Further studies of IMIDs' underlying mechanism will shed light on more specific approaches.

5.2.3.4 *Bone marrow or bone marrow-derived monocyte transfusion*

BM-derived cells participate in VEGF-stimulated brain angiogenesis (Hao *et al.*, 2008a, 2011). In bAVM animal models, increased BM-derived monocyte infiltration has been found on the AVM vessel wall (Zhang *et al.*, 2016a). Compared with normal people, HHT patients have more peripheral endothelial progenitor cells (Zucco *et al.*, 2014), which are more likely to differentiate into macrophages in the angiogenic niche (Zhang *et al.*, 2016a). Therefore, transfusion of healthy BM-derived monocytes to AVM patients might be a therapeutic option.

Transplantation of $Eng^{+/-}$ BM to WT mice causes formation of vascular dysplasia in the brain angiogenic region, which is similar to that in the brain angiogenic region of $Eng^{+/-}$ mice (Hao *et al.*, 2010). Transplantation of WT BM to $Eng^{+/-}$ mice reduces the severity of vascular dysplasia in the brain angiogenic foci of $Eng^{+/-}$ (Hao *et al.*, 2010). Deletion of *Alk1* in BM-derived endothelial cells also leads to AVM formation in the brain angiogenic region (unpublished data). Moreover, transfusion of normal human monocytes rescues the impairment of $Eng^{+/-}$ mice in repairing myocardial injury (Post *et al.*, 2010; van Laake *et al.*, 2006). These data suggest that correction of gene mutations in patients' BM monocytes or BM-derived endothelial cells through BM transplantation and monocyte or endothelial progenitor transfusion could provide additional therapeutic options.

5.2.3.5 *Promising drugs not tested for bAVM*

Statin, competitive inhibitors of 3-hydroxy-3-methylglutaryl-CoA reductase (HMGCR), is a drug that lowers 3-hydroxy-3-methylglutaryl-CoA (HMGCoA)-derived lipids, including cholesterol and isoprenoid. Its cholesterol-lowering effect is now widely used in the treatment of hyperlipidemia. Since the lipid mediates oxidative stress and inflammation, statin is also emerging as an anti-oxidative and anti-inflammation drug (Albert *et al.*, 2001; Liao & Laufs, 2005). Recent studies and clinical trials indicate the protective role of statin in acute coronary syndrome (Schwartz *et al.*, 2001), arteriosclerosis (Amarenco *et al.*, 2004), aortic aneurysm (Takagi *et al.*, 2010), aneurysmal subarachnoid hemorrhage (Tseng *et al.*, 2005), and chronic subepidural hemorrhage (Wang *et al.*, 2014). Statin treatment is associated with attenuated oxidative stress biomarkers and subsequent unfavorable complications (Kinlay *et al.*, 2008; Tseng *et al.*, 2005; Wemmelund *et al.*, 2014). The growth rate of aortic aneurysm is reduced after statin therapy (Karrowni *et al.*, 2011).

Statin regulates cerebrovascular stability by interfering with post-translational modification of cell signaling molecules (Eisa-Beygi *et al.*, 2013). Isoprenoids are an essential lipid attachment for the Rho family of GTPases. Rho GTPase-mediated signaling pathways participate in the development of the vascular network (Eisa-Beygi *et al.*, 2013). Malfunction of the Rho-GTPase pathway impairs the stabilization and maturation of nascent vessels by preventing the association of mural cells with endothelial cells, giving rise to pronounced vascular malformation and multiple hemorrhages in the brain (Buchner *et al.*, 2007; Liu *et al.*, 2007). *In vivo* and *in vitro* evidence suggest that cerebral cavernous malformations are associated with RhoA hyperactivation, and therefore, overactivation of downstream signaling via Rho kinase and increased stress fiber assembly, which leads to disrupted endothelial cell–cell junctions and loss of vascular stability (Borikova *et al.*, 2010; Stockton *et al.*, 2010). Statin inhibits isoprenoids, and thus may improve the vascular stability and prevent cerebral hemorrhage in cavernous malformations. In addition, a retrospective review of the medication in patients with dural arteriovenous fistula, a cerebrovascular malformation with arteriovenous shunt similar to bAVM, indicate that patients with statin treatment have a lower rate of hemorrhagic presentation (Gross *et al.*, 2013). However, Rho–GTPase signaling has not been found to be associated with bAVM pathology. There is no evidence associating statin use with hemorrhage risk in bAVM patients. Thus, the putative molecular mechanisms or the benefit of statins in a bAVM context requires further studies.

Increased tissue perfusion promotes vascular dysplasia in a bAVM animal model (Hao *et al.*, 2008b). Flow shear forces due to high blood flow in arteries and veins affect vascular sprouting and proliferation (Asano *et al.*, 2005; Song & Munn, 2011). Intraluminal shear stress regulates expression of angiogenic genes in endothelial cells (Abumiya *et al.*, 2002). Fluid shear stress rapidly increase levels of VEGF mRNA expression in the brain microvasculature (Malek *et al.*, 2000). In human bAVM with endovascular embolization, blood flow increases in non-occluded portions of an AVM after partial embolization, resulting in increased angiogenic activity in treated bAVM tissue (Buell *et al.*, 2014) that may enhance abnormal angiogenesis.

Vasodilators, including beta-blockers and angiotensin-converting enzyme-inhibitors, have shown promise in relieving vascular malformation (Gross *et al.*, 2013). In 2008, treatment using propranolol (2mg/kg/day, for 4–10 months), a non-selective beta-blocker, resulted in a regression of severe cutaneous infantile hemangiomas in 8 out of 11 infants (Leaute-Labreze *et al.*, 2008). This observation has caused a worldwide revolution in treating complicated infantile hemangioma. Prospective control trials have also confirmed that oral propranolol at 2 mg/kg/day has a consistent and rapid therapeutic effect on superficial infantile hemangioma without significant adverse events (Leaute-Labreze *et al.*, 2008). Potential explanations for the therapeutic effect of propranolol on infantile capillary hemangiomas include vasoconstriction (Leaute-Labreze *et al.*, 2008), decreased VEGF and bFGF expression (D'Angelo *et al.*, 1997), and capillary endothelial cell apoptosis (Sommers Smith & Smith, 2002). Although serum VEGF levels did not decrease in children after 3-month propranolol treatment, VEGF and VEGFR1 mRNA levels were reduced in hemangioma tissue after propranolol treatment (Przewratil *et al.*, 2016). A recent case report showed a successful rescue of intracranial and spinal hemangiomas with propranolol (Miquel *et al.*, 2014). Hence, the effect of propranolol should be evaluated in bAVM patients.

5.3 Summary

Developing medical therapies for bAVM relies on understanding the pathogenesis and pathophysiology of bAVMs. Human and animal studies suggest that a combination of angiogenic stimulation (response-to-injury), inflammation, BM-derived cell participation, and hemodynamic changes may be required for bAVM formation and development. Animal studies further indicate that specific therapies targeting the aforementioned are effective in attenuating bAVM. Phase I clinical trials have completed safety assessment of some drugs and have validated the feasibility of further efficacy studies. Advances in animal studies are introducing more potential therapeutic approaches, while novel imaging tools and clinical trials are helping apply our increased knowledge about bAVM pathogenesis to patient care.

Acknowledgments

This research is supported by grants from the U.S. National Institutes of Health (R01 NS027713, R01 HL122774, and R21 NS083788), the Michael Ryan Zodda Foundation, and the UCSF Research Evaluation and Allocation Committee (REAC) to H.S.; and by a Young Investigator Award from Cure HHT Foundation to W.Z.

References

Abdalla, S.A., Pece-Barbara, N., *et al.* (2000). Analysis of ALK-1 and endoglin in newborns from families with hereditary hemorrhagic telangiectasia type 2. *Hum Mol Genet* **9**(8): 1227–1237.

Abumiya, T., Sasaguri, T., *et al.* (2002). Shear stress induces expression of vascular endothelial growth factor receptor Flk-1/KDR through the CT-rich Sp1 binding site. *Arterioscler Thromb Vasc Biol* **22**(6): 907–913.

Achrol, A.S., Pawlikowska, L., *et al.* (2006a). Tumor necrosis factor-alpha-238G>A promoter polymorphism is associated with increased risk of new hemorrhage in the natural course of patients with brain arteriovenous malformations. *Stroke* **37**(1): 231–234.

Achrol, A.S., Pawlikowska, L., *et al.* (2006b). TNFa-238G>A promoter polymorphism is associated with increased risk of new hemorrhage in the natural course of patients with brain arteriovenous malformations [Abstract]. *Stroke* **37**(2): 638–639.

Adam, P.J., Clesham, G.J., *et al.* (1998). Expression of endoglin mRNA and protein in human vascular smooth muscle cells. *Biochem Biophys Res Commun* **247**(1): 33–37.

Akers, A.L., Johnson, E., *et al.* (2009). Biallelic somatic and germline mutations in cerebral cavernous malformations (CCM): Evidence for a two-hit mechanism of CCM pathogenesis. *Hum Mol Genet* **18**(5): 919–930.

Alam, M.A., Sami, S., *et al.* (2011). Successful treatment of bleeding gastro-intestinal angiodysplasia in hereditary haemorrhagic telangiectasia with thalidomide. *BMJ Case Rep.*

Albert, M.A., Danielson, E., *et al.* (2001). Effect of statin therapy on C-reactive protein levels: The pravastatin inflammation/CRP evaluation (PRINCE): A randomized trial and cohort study. *JAMA* **286**(1): 64–70.

Allinson, K.R., Carvalho, R.L., *et al.* (2007). Generation of a floxed allele of the mouse Endoglin gene. *Genesis* **45**(6): 391–395.

Amarenco, P., Labreuche, J., *et al.* (2004). Statins in stroke prevention and carotid atherosclerosis: Systematic review and up-to-date meta-analysis. *Stroke* **35**(12): 2902–2909.

Amiridze, N.S., Darwish, R., *et al.* (2008). Treatment of arteriovenous malformations with hydrocoils in a swine model. *Interv Neuroradiol* **14**(2): 165–171.

Andreoli, C.M., Miller, J.W. (2007). Anti-vascular endothelial growth factor therapy for ocular neovascular disease. *Curr Opin Ophthalmol* **18**(6): 502–508.

Aragon-Ching, J.B., Li, H., *et al.* (2007). Thalidomide analogues as anticancer drugs. *Recent Pat Anticancer Drug Discov* **2**(2): 167–174.

Ardelean, D.S., Yin, M., *et al.* (2014). Anti-VEGF therapy reduces intestinal inflammation in Endoglin heterozygous mice subjected to experimental colitis. *Angiogenesis* **17**(3): 641–659.

Armulik, A., Genove, G., *et al.* (2010). Pericytes regulate the blood-brain barrier. *Nature* **468**(7323): 557–561.

Arthur, H.M., Ure, J., *et al.* (2000). Endoglin, an ancillary TGFbeta receptor, is required for extraembryonic angiogenesis and plays a key role in heart development. *Dev Biol* **217**(1): 42–53.

Asano, Y., Ichioka, S., *et al.* (2005). Sprouting from arteriovenous shunt vessels with increased blood flow. *Med Biol Eng Comput* **43**(1): 126–130.

Aziz, M.M., Takagi, Y., *et al.* (2010). Activation of nuclear factor kappaB in cerebral arteriovenous malformations. *Neurosurgery* **67**(6): 1669–1680.

Badea, T.C., Wang, Y., *et al.* (2003). A noninvasive genetic/pharmacologic strategy for visualizing cell morphology and clonal relationships in the mouse. *J Neurosci* **23**(6): 2314–2322.

Barbara, N.P., Wrana, J.L., *et al.* (1999). Endoglin is an accessory protein that interacts with the signaling receptor complex of multiple members of the transforming growth factor-beta superfamily. *J Biol Chem* **274**(2): 584–594.

Bederson, J.B., Wiestler, O.D., *et al.* (1991). Intracranial venous hypertension and the effects of venous outflow obstruction in a rat model of arteriovenous fistula. *Neurosurgery* **29**(3): 341–350.

Bell, R.D., Winkler, E.A., *et al.* (2010). Pericytes control key neurovascular functions and neuronal phenotype in the adult brain and during brain aging. *Neuron* **68**(3): 409–427.

Benzinou, M., Clermont, F.F., *et al.* (2012). Mouse and human strategies identify PTPN14 as a modifier of angiogenesis and hereditary haemorrhagic telangiectasia. *Nat Commun* 3: 616.

Bonelli, R.M., Heuberger, C., *et al.* (2003). Minocycline for Huntington's disease: An open label study. *Neurology* **60**(5): 883–884.

Bonyadi, M., Rusholme, S.A., *et al.* (1997). Mapping of a major genetic modifier of embryonic lethality in TGF beta 1 knockout mice. *Nat Genet* **15**(2): 207–211.

Borikova, A.L., Dibble, C.F., *et al.* (2010). Rho kinase inhibition rescues the endothelial cell cerebral cavernous malformation phenotype. *J Biol Chem* **285**(16): 11760–11764.

Bourdeau, A., Cymerman, U., *et al.* (2000a). Endoglin expression is reduced in normal vessels but still detectable in arteriovenous malformations of patients with hereditary hemorrhagic telangiectasia type 1. *Am J Pathol* **156**(3): 911–923.

Bourdeau, A., Faughnan, M.E., *et al.* (2000b). Endoglin-deficient mice, a unique model to study hereditary hemorrhagic telangiectasia. *Trends Cardiovasc Med* **10**(7): 279–285.

Bourdeau, A., Faughnan, M.E., *et al.* (2001). Potential role of modifier genes influencing transforming growth factor- beta1 levels in the development of vascular defects in endoglin heterozygous mice with hereditary hemorrhagic telangiectasia. *Am J Pathol* **158**(6): 2011–2020.

Braverman, I.M., Keh, A., *et al.* (1990). Ultrastructure and three-dimensional organization of the telangiectases of hereditary hemorrhagic telangiectasia. *J Invest Dermatol* **95**(4): 422–427.

Brundula, V., Rewcastle, N.B., *et al.* (2002). Targeting leukocyte MMPs and transmigration: Minocycline as a potential therapy for multiple sclerosis. *Brain* **125**(Pt 6): 1297–1308.

Buchner, D.A., Su, F., *et al.* (2007). pak2a mutations cause cerebral hemorrhage in redhead zebrafish. *Proc Natl Acad Sci USA* **104**(35): 13996–14001.

Buell, T.J., Ding, D., *et al.* (2014). Embolization-induced angiogenesis in cerebral arteriovenous malformations. *J Clin Neurosci* **21**(11): 1866–1871.

Carlson, T.R., Yan, Y., *et al.* (2005). Endothelial expression of constitutively active Notch4 elicits reversible arteriovenous malformations in adult mice. *Proc Natl Acad Sci U S A* **102**(28): 9884–9889.

Chalouhi, N., Starke, R.M., *et al.* (2016). Differential sex response to aspirin in decreasing aneurysm rupture in humans and mice. *Hypertension* **68**(2): 411–417.

Chaloupka, J.C., Vinuela, F., *et al.* (1994). An *in vivo* arteriovenous malformation model in swine: Preliminary feasibility and natural history study. *AJNR Am J Neuroradiol* **15**(5): 945–950.

Chan, M.F., Li, J., *et al.* (2013). Protective effects of matrix metalloproteinase-12 following corneal injury. *J Cell Sci* **126**(Pt 17): 3948–3960.

Chen, C.H., Hsu, H.H., *et al.* (2012). Long-term therapy with thalidomide in hereditary hemorrhagic Telangiectasia: Case report and literature review. *J Clin Pharmacol* **52**(9): 1436–1440.

Chen, W., Guo, Y., *et al.* (2013). Reduced mural cell coverage and impaired vessel integrity after angiogenic stimulation in the *Alk1*-deficient brain. *Arterioscler Thromb Vasc Biol* **33**(2): 305–310.

Chen, W., Sun, Z., *et al.* (2014). De novo cerebrovascular malformation in the adult mouse after endothelial Alk1 deletion and angiogenic stimulation. *Stroke* **45**(3): 900–902.

Chen, Y., Fan, Y., *et al.* (2006a). MMP-9 expression is associated with leukocytic but not endothelial markers in brain arteriovenous malformations. *Front Biosci* **11**: 3121–3128.

Chen, Y., Pawlikowska, L., *et al.* (2006b). Interleukin-6 involvement in brain arteriovenous malformations. *Ann Neurol* **59**(1): 72–80.

Chen, Y., Zhu, W., *et al.* (2008). Evidence of inflammatory cell involvement in brain arteriovenous malformations. *Neurosurgery* **62**(6): 1340–1349.

Choi, E.J., Chen, W., *et al.* (2014). Novel brain arteriovenous malformation mouse models for type 1 hereditary hemorrhagic telangiectasia. *PLoS One* **9**(2): e88511.

Choi, E.J., Walker, E.J., *et al.* (2013). Endoglin deficiency in bone marrow is sufficient to cause cerebrovascular dysplasia in the adult mouse after vascular endothelial growth factor stimulation. *Stroke* **44**(3): 795–798.

Choi, E.J., Walker, E.J., *et al.* (2012). Minimal homozygous endothelial deletion of Eng with VEGF stimulation is sufficient to cause cerebrovascular dysplasia in the adult mouse. *Cerebrovasc Dis* **33**(6): 540–547.

Chung, A.S., Ferrara, N. (2011). Developmental and pathological angiogenesis. *Annu Rev Cell Dev Biol* **27**: 563–584.

Cirulli, A., Liso, A., *et al.* (2003). Vascular endothelial growth factor serum levels are elevated in patients with hereditary hemorrhagic telangiectasia. *Acta Haematol* **110**(1): 29–32.

Claxton, S., Kostourou, V., *et al.* (2008). Efficient, inducible Cre-recombinase activation in vascular endothelium. *Genesis* **46**(2): 74–80.

Cole, S.G., Begbie, M.E., *et al.* (2005). A new locus for hereditary haemorrhagic telangiectasia (HHT3) maps to chromosome 5. *J Med Genet* **42**(7): 577–582.

Conley, B.A., Smith, J.D., *et al.* (2000). Endoglin, a TGF-beta receptor-associated protein, is expressed by smooth muscle cells in human atherosclerotic plaques. *Atherosclerosis* **153**(2): 323–335.

Corti, P., Young, S., *et al.* (2011). Interaction between alk1 and blood flow in the development of arteriovenous malformations. *Development* **138**(8): 1573–1582.

Curci, J.A., Mao, D., *et al.* (2000). Preoperative treatment with doxycycline reduces aortic wall expression and activation of matrix metalloproteinases in patients with abdominal aortic aneurysms. *J Vasc Surg* **31**(2): 325–342.

Davidson, B.E.N., Remst, D.F., *et al.* (2009). Increase in ALK1/ALK5 ratio as a cause for elevated MMP-13 expression in osteoarthritis in humans and mice. *J Immunol Methods* **182**(12): 7937–7945.

D'Amato, R.J., Loughnan, M.S., *et al.* (1994). Thalidomide is an inhibitor of angiogenesis. *Proc Natl Acad Sci USA* **91**(9): 4082–4085.

D'Angelo, G., Lee, H., *et al.* (1997). cAMP-dependent protein kinase inhibits the mitogenic action of vascular endothelial growth factor and fibroblast growth factor in capillary endothelial cells by blocking Raf activation. *J Cell Biochem* **67**(3): 353–366.

De Salles, A.A.F., Solberg, T.D., *et al.* (1996). Arteriovenous malformation animal model for radiosurgery: The rete mirabile. *AJNR Am J Neuroradiol* **17**(8): 1451–1458.

Dingenouts, C.K., Goumans, M.J., *et al.* (2015). Mononuclear cells and vascular repair in HHT. *Front Genet* 6: 114.

Dredge, K., Horsfall, R., *et al.* (2005). Orally administered lenalidomide (CC-5013) is anti-angiogenic *in vivo* and inhibits endothelial cell migration and Akt phosphorylation *in vitro*. *Microvasc Res* **69**(1–2): 56–63.

Dredge, K., Marriott, J.B., *et al.* (2002). Novel thalidomide analogues display anti-angiogenic activity independently of immunomodulatory effects. *Br J Cancer* **87**(10): 1166–1172.

Dupuis-Girod, S., Bailly, S., *et al.* (2010). Hereditary hemorrhagic telangiectasia (HHT): From molecular biology to patient care. *J Thromb Haemost* 8(7): 1447–1456.

Dupuis-Girod, S., Ginon, I., *et al.* (2012). Bevacizumab in patients with hereditary hemorrhagic telangiectasia and severe hepatic vascular malformations and high cardiac output. *JAMA* **307**(9): 948–955.

Eerola, I., Boon, L.M., *et al.* (2003). Capillary malformation–arteriovenous malformation, a new clinical and genetic disorder caused by RASA1 mutations. *Am J Hum Genet* **73**(6): 1240–1249.

Eisa-Beygi, S., Hatch, G., *et al.* (2013). The 3-hydroxy-3-methylglutaryl-CoA reductase (HMGCR) pathway regulates developmental cerebral-vascular stability via prenylation-dependent signalling pathway. *Dev Biol* **373**(2): 258–266.

Ferreira, R., Santos, T., *et al.* (2014). MicroRNA-18a improves human cerebral arteriovenous malformation endothelial cell function. *Stroke* **45**(1): 293–297.

Flieger, D., Hainke, S., *et al.* (2006). Dramatic improvement in hereditary hemorrhagic telangiectasia after treatment with the vascular endothelial growth factor (VEGF) antagonist bevacizumab. *Ann Hematol* **85**(9): 631–632.

Folkman, J., Klagsbrun, M. (1987). Angiogenic factors. *Science* **235**(4787): 442–447.

Foo, S.S., Turner, C.J., *et al.* (2006). Ephrin-B2 controls cell motility and adhesion during blood-vessel-wall assembly. *Cell* **124**(1): 161–173.

Franks, M.E., Macpherson, G.R., *et al.* (2004). Thalidomide. *Lancet* **363**(9423): 1802–1811.

Frenzel, T., Lee, C.Z., *et al.* (2008). Feasibility of minocycline and doxycycline use as potential vasculostatic therapy for brain vascular malformations: Pilot study of adverse events and tolerance. *Cerebrovasc Dis* **25**(1–2): 157–163.

Friedrich, G., Soriano, P. (1991). Promoter traps in embryonic stem cells: A genetic screen to identify and mutate developmental genes in mice. *Genes Dev* **5**(9): 1513–1523.

Gale, N.W., Dominguez, M.G., *et al.* (2004). Haploinsufficiency of delta-like 4 ligand results in embryonic lethality due to major defects in arterial and vascular development. *Proc Natl Acad Sci U S A* **101**(45): 15949–15954.

Gallione, C.J., Repetto, G.M., *et al.* (2004). A combined syndrome of juvenile polyposis and hereditary haemorrhagic telangiectasia associated with mutations in MADH4 (SMAD4). *Lancet* **363**(9412): 852–859.

Garrido-Martin, E.M., Nguyen, H.L., *et al.* (2014). Common and distinctive pathogenetic features of arteriovenous malformations in hereditary hemorrhagic telangiectasia 1 and hereditary hemorrhagic telangiectasia 2 animal models — brief report. *Arterioscler Thromb Vasc Biol* **34**(10): 2232–2236.

Gong, Z.P., Qiao, N.D., *et al.* (2011). Polymorphisms of VEGFA gene and susceptibility to hemorrhage risk of brain arteriovenous malformations in a Chinese population. *Acta Pharmacol Sin* **32**(8): 1071–1077.

Gore, A.V., Monzo, K., *et al.* (2012). Vascular development in the zebrafish. *Cold Spring Harb Perspect Med* **2**(5): a006684.

Gougos, A., Letarte, M. (1988). Identification of a human endothelial cell antigen with monoclonal antibody 44G4 produced against a pre-B leukemic cell line. *J Immunol* **141**(6): 1925–1933.

Gougos, A., Letarte, M. (1990). Primary structure of endoglin, an RGD-containing glycoprotein of human endothelial cells. *J Biol Chem* **265**(15): 8361–8364.

Goumans, M.J., Lebrin, F., *et al.* (2003). Controlling the angiogenic switch: A balance between two distinct TGF-b receptor signaling pathways. *Trends Cardiovasc Med* **13**(7): 301–307.

Govani, F.S., Shovlin, C.L. (2009). Hereditary haemorrhagic telangiectasia: A clinical and scientific review. *Eur J Hum Genet* **17**(7): 860–871.

Gross, B.A., Lai, P.M., *et al.* (2013). Angiotensin-converting enzyme-inhibitors, statins and the risk of hemorrhage from cerebral dural arteriovenous fistulae. *J Clin Neurosci* **20**(9): 1228–1231.

Guerrero-Esteo, M., Sanchez-Elsner, T., *et al.* (2002). Extracellular and cytoplasmic domains of endoglin interact with the transforming growth factor-beta receptors I and II. *J Biol Chem* **277**(32): 29197–29209.

Guo, Y., Qumu, S.W., *et al.* (2013). Human brain arteriovenous malformations are associated with interruptions in elastic fibers and changes in collagen content. *Turk Neurosurg* **23**(1): 10–15.

Guo, Y., Saunders, T., *et al.* (2012). Silent intralesional microhemorrhage as a risk factor for brain arteriovenous malformation rupture. *Stroke* **43**(5): 1240–1246.

Guo, Y., Tihan, T., *et al.* (2014). Distinctive distribution of lymphocytes in unruptured and previously untreated brain arterviovenous malformation. *Neuroimmunol Neuroinflamm* **1**(3): 147–152.

Hai, J., Ding, M., *et al.* (2002). A new rat model of chronic cerebral hypoperfusion associated with arteriovenous malformations. *J Neurosurg* **97**(5): 1198–1202.

Hai, J., Lin, Q., *et al.* (2004). Chronic cerebral hypoperfusion and reperfusion injury of restoration of normal perfusion pressure contributes to the neuropathological changes in rat brain. *Brain Res Mol Brain Res* **126**(2): 137–145.

Han, C., Choe, S.W., *et al.* (2014). VEGF neutralization can prevent and normalize arteriovenous malformations in an animal model for hereditary hemorrhagic telangiectasia 2. *Angiogenesis* **17**(4): 823–830.

Hao, Q., Liu, J., *et al.* (2008a). Contribution of bone marrow-derived cells associated with brain angiogenesis is primarily through leucocytes and macrophages. *Arterioscler Thromb Vasc Biol* **28**(12): 2151–2157.

Hao, Q., Su, H., *et al.* (2008b). Increased tissue perfusion promotes capillary dysplasia in the ALK1-deficient mouse brain following VEGF stimulation. *Am J Physiol Heart Circ Physiol* **295**(6): H2250–2256.

Hao, Q., Su, H., *et al.* (2011). Bone marrow-derived cells contribute to vascular endothelial growth factor-induced angiogenesis in the adult mouse brain by supplying matrix metalloproteinase-9. *Stroke* **42**(2): 453–458.

Hao, Q., Zhu, Y., *et al.* (2010). VEGF induces more severe cerebrovascular dysplasia in Endoglin$^{+/-}$ than in Alk1$^{+/-}$ mice. *Transl Stroke Res* **1**(3): 197–201.

Hasan, D.M., Amans, M., *et al.* (2012). Ferumoxytol-enhanced MRI to image inflammation within human brain arteriovenous malformations: A pilot investigation. *Transl Stroke Res* **3**(Suppl 1): 166–173.

Hasan, D.M., Chalouhi, N., *et al.* (2013). Evidence that acetylsalicylic acid attenuates inflammation in the wall of human cerebral aneurysms: Preliminary results. *J Am Heart Assoc* **2**(1): e000019.

Hasan, D.M., Mahaney, K.B., *et al.* (2011). Aspirin as a promising agent for decreasing incidence of cerebral aneurysm rupture. *Stroke* **42**(11): 3156–3162.

Hashimoto, T., Emala, C.W., *et al.* (2000). Abnormal pattern of Tie-2 and vascular endothelial growth factor receptor expression in human cerebral arteriovenous malformations. *Neurosurgery* **47**(4): 910–918; discussion 918–919.

Hashimoto, T., Lam, T., *et al.* (2001). Abnormal balance in the angiopoietin-tie2 system in human brain arteriovenous malformations. *Circ Res* **89**(2): 111–113.

Hashimoto, T., Lawton, M.T., *et al.* (2004). Gene microarray analysis of human brain arteriovenous malformations. *Neurosurgery* **54**(2): 410–423; discussion 423–425.

Hashimoto, T., Matsumoto, M., *et al.* (2005a). Suppression of MMP-9 by doxycycline in brain arteriovenous malformations. *BMC Neurol* **5**(1): 1.

Hashimoto, T., Wen, G., *et al.* (2003). Abnormal expression of matrix metalloproteinases and tissue inhibitors of metalloproteinases in brain arteriovenous malformations. *Stroke* **34**(4): 925–931.

Hashimoto, T., Wu, Y., *et al.* (2005b). Co-expression of angiogenic factors in brain arteriovenous malformations. *Neurosurgery* **56**(5): 1058–1065; discussion 1058–1065.

Hatva, E., Jaaskelainen, J., *et al.* (1996). Tie endothelial cell-specific receptor tyrosine kinase is upregulated in the vasculature of arteriovenous malformations. *J Neuropathol Exp Neurol* **55**(11): 1124–1133.

Hellstrom, M., Gerhardt, H., *et al.* (2001). Lack of pericytes leads to endothelial hyperplasia and abnormal vascular morphogenesis. *J Cell Biol* **153**(3): 543–553.

Herbert, S.P., Huisken, J., *et al.* (2009). Arterial-venous segregation by selective cell sprouting: An alternative mode of blood vessel formation. *Science* **326**(5950): 294–298.

Isoda, K., Fukuda, H., *et al.* (1981). Arteriovenous malformation of the brain — histological study and micrometric measurement of abnormal vessels. *Acta Pathol Jpn* **31**(5): 883–893.

Jahan, R., Solberg, T.D., *et al.* (2006). Stereotactic radiosurgery of the rete mirabile in swine: A longitudinal study of histopathological changes. *Neurosurgery* **58**(3): 551–558; discussion 551–558.

Jarriault, S., Brou, C., *et al.* (1995). Signalling downstream of activated mammalian Notch. *Nature* **377**(6547): 355–358.

Jerkic, M., Kabir, M.G., *et al.* (2011). Pulmonary hypertension in adult Alk1 heterozygous mice due to oxidative stress. *Cardiovasc Res* **92**(3): 375–384.

Jerkic, M., Sotov, V., *et al.* (2012). Oxidative stress contributes to endothelial dysfunction in mouse models of hereditary hemorrhagic telangiectasia. *Oxid Med Cell Longev* 2012: 686972.

Kaplitt, M.G., Leone, P., *et al.* (1994). Long-term gene expression and phenotypic correction using adeno-associated virus vectors in the mammalian brain. *Nat Genet* **8**(2): 148–154.

Karrowni, W., Dughman, S., *et al.* (2011). Statin therapy reduces growth of abdominal aortic aneurysms. *J Investig Med* **59**(8): 1239–1243.

Kendall, R.L., Thomas, K.A. (1993). Inhibition of vascular endothelial cell growth factor activity by an endogenously encoded soluble receptor. *Proc Natl Acad Sci U S A* **90**(22): 10705–10709.

Kilic, T., Pamir, M.N., *et al.* (2000). Expression of structural proteins and angiogenic factors in cerebrovascular anomalies. *Neurosurgery* **46**(5): 1179–1191; discussion 1191–1172.

Kim, H., Hysi, P.G., *et al.* (2009). Common variants in interleukin-1-beta gene are associated with intracranial hemorrhage and susceptibility to brain arteriovenous malformation. *Cerebrovasc Dis* **27**(2): 176–182.

Kim, H., Pawlikowska, L., *et al.* (2011a). Genetics and vascular biology of brain vascular malformations (Chapter 12). In: Mohr, J.P., Wolf, P.A., *et al.*, (eds.) *Stroke: Pathophysiology, Diagnosis, and Management*, 5th edn. Philadelphia, PA: Churchill Livingstone Elsevier.

Kim, H., Su, H., *et al.* (2011b). Brain arteriovenous malformation pathogenesis: A response-to-injury paradigm. *Acta Neurochir Suppl* **111**: 83–92.

Kinlay, S., Schwartz, G.G., *et al.* (2008). Inflammation, statin therapy, and risk of stroke after an acute coronary syndrome in the MIRACL study. *Arterioscler Thromb Vasc Biol* **28**(1): 142–147.

Koizumi, T., Shiraishi, T., *et al.* (2002). Expression of vascular endothelial growth factors and their receptors in and around intracranial arteriovenous malformations. *Neurosurgery* **50**(1): 117–124; discussion 124–116.

Kojima, T., Miyachi, S., *et al.* (2007). The relationship between venous hypertension and expression of vascular endothelial growth factor: Hemodynamic and immunohistochemical examinations in a rat venous hypertension model. *Surg Neurol* **68**(3): 277–284; discussion 284.

Koo, T., Okada, T., *et al.* (2011). Long-term functional adeno-associated virus-microdystrophin expression in the dystrophic CXMDj dog. *J Gene Med* **13**(9): 497–506.

Krebs, L.T., Shutter, J.R., *et al.* (2004). Haploinsufficient lethality and formation of arteriovenous malformations in Notch pathway mutants. *Genes Dev* **18**(20): 2469–2473.

Krebs, L.T., Starling, C., *et al.* (2010). Notch 1 activation in mice causes arteriovenous malformations phenocopied by EphrinB2 and EphB4 mutants. *Genesis* **48**(3): 146–150.

Kume, T., Jiang, H., *et al.* (2001). The murine winged helix transcription factors, Foxc1 and Foxc2, are both required for cardiovascular development and somitogenesis. *Genes Dev* **15**(18): 2470–2482.

Lan, Y., Liu, B., *et al.* (2007). Essential role of endothelial Smad4 in vascular remodeling and integrity. *Mol Cell Biol* **27**(21): 7683–7692.

Lastres, P., Bellon, T., *et al.* (1992). Regulated expression on human macrophages of endoglin, an Arg-Gly-Asp-containing surface antigen. *Eur J Immunol* **22**(2): 393–397.

Lawson, N.D., Scheer, N., *et al.* (2001). Notch signaling is required for arterial-venous differentiation during embryonic vascular development. *Development* **128**(19): 3675–3683.

Lawton, M.T., Stewart, C.L., *et al.* (2004). The transgenic arteriovenous fistula in the rat: An experimental model of gene therapy for brain arteriovenous malformations. *Neurosurgery* **54**(6): 1463–1471; discussion 1471.

Leaute-Labreze, C., Dumas de la Roque, E., *et al.* (2008). Propranolol for severe hemangiomas of infancy. *N Engl J Med* **358**(24): 2649–2651.

Leblanc, G.G., Golanov, E., *et al.* (2009). Biology of vascular malformations of the brain. *Stroke* **40**(12): e694–702.

Lebrin, F., Srun, S., *et al.* (2010). Thalidomide stimulates vessel maturation and reduces epistaxis in individuals with hereditary hemorrhagic telangiectasia. *Nat Med* **16**(4): 420–428.

Lee, C.Z., Xu, B., *et al.* (2004). Doxycycline suppresses cerebral matrix metalloproteinase-9 and angiogenesis induced by focal hyperstimulation of vascular endothelial growth factor in a mouse model. *Stroke* **35**(7): 1715–1719.

Lee, C.Z., Xue, Z., *et al.* (2007). Matrix metalloproteinase-9 inhibition attenuates vascular endothelial growth factor-induced intracranial hemorrhage. *Stroke* **38**(9): 2563–2568.

Letteboer, T.G., Benzinou, M., *et al.* (2015). Genetic variation in the functional ENG allele inherited from the non-affected parent associates with presence of pulmonary arteriovenous malformation in hereditary hemorrhagic telangiectasia 1 (HHT1) and may influence expression of PTPN14. *Front Genet* 6: 67.

Li, C., Hampson, I.N., *et al.* (2000). CD105 antagonizes the inhibitory signaling of transforming growth factor beta1 on human vascular endothelial cells. *FASEB J* **14**(1): 55–64.

Li, H., Ezra, D.G., *et al.* (2013). Doxycycline prevents matrix remodeling and contraction by trichiasis-derived conjunctival fibroblasts. *Invest Ophthalmol Vis Sci* **54**(7): 4675–4682.

Liao, J.K., Laufs, U., *et al.* (2005). Pleiotropic effects of statins. *Annu Rev Pharmacol Toxicol* 45: 89–118.

Lim, M., Haddix, T., *et al.* (2005). Characterization of the integrin alphavbeta3 in arteriovenous malformations and cavernous malformations. *Cerebrovasc Dis* **20**(1): 23–27.

Limaye, N., Wouters, V., *et al.* (2009). Somatic mutations in angiopoietin receptor gene TEK cause solitary and multiple sporadic venous malformations. *Nat Genet* **41**(1): 118–124.

Lindahl, P., Johansson, B.R., *et al.* (1997). Pericyte loss and microaneurysm formation in PDGF-B-deficient mice. *Science* **277**(5323): 242–245.

Liu, J., Fraser, S.D., *et al.* (2007). A betaPix Pak2a signaling pathway regulates cerebral vascular stability in zebrafish. *Proc Natl Acad Sci U S A* **104**(35): 13990–13995.

Lukason, M., Dufresne, E., *et al.* (2011). Inhibition of choroidal neovascularization in a nonhuman primate model by intravitreal administration of an AAV2 vector expressing a novel anti-VEGF molecule. *Mol Ther* **19**(2): 260–265.

Lupu, A., Stefanescu, C., *et al.* (2013). Bevacizumab as rescue treatment for severe recurrent gastrointestinal bleeding in hereditary hemorrhagic telangiectasia. *J Clin Gastroenterol* **47**(3): 256–257.

Ma, L., Guo, Y., *et al.* (2015). The role of macrophage in the pathogenesis of brain arteriovenous malformation. *Int J Hematol Res* **1**(2): 52–56.

Ma, L., Shen, F., *et al.* (2016). Integrin β8 deletion enhances vascular dysplasia and hemorrhage in the brain of adult Alk1 heterozygous mice. *Transl Stroke Res* **7**(6): 488–496.

Maclachlan, T.K., Lukason, M., *et al.* (2011). Preclinical safety evaluation of AAV2-sFLT01- a gene therapy for age-related macular degeneration. *Mol Ther* **19**(2): 326–334.

Maestraggi, Q., Bouattour, M., *et al.* (2015). Bevacizumab to treat cholangiopathy in hereditary hemorrhagic telangiectasia: Be cautious: A case report. *Medicine* **94**(46): e1966.

Mahmoud, M., Allinson, K.R., *et al.* (2010). Pathogenesis of arteriovenous malformations in the absence of endoglin. *Circ Res* **106**(8): 1425–1433.

Malek, A.M., Lee, I., *et al.* (2000). Identification of vascular endothelial growth factor (VEGF) as a flow-regulated mediator of angiogenesis [Abstract]. *Neurosurgery* **47**(2): 499–500.

Massoud, T.F., Ji, C., *et al.* (1994). An experimental arteriovenous malformation model in swine: Anatomic basis and construction technique. *AJNR Am J Neuroradiol* **15**(8): 1537–1545.

Milton, I., Ouyang, D., *et al.* (2012). Age-dependent lethality in novel transgenic mouse models of central nervous system arteriovenous malformations. *Stroke* **43**(5): 1432–1435.

Milton, I., Seki, T. (2011). Transgenic mouse model of hemorrhagic arteriovenous malformation in brain and spinal cord [Abstract]. *Stroke* **42**(3): e91 (#163).

Miniati, D., Jelin, E.B., *et al.* (2010). Constitutively active endothelial Notch4 causes lung arteriovenous shunts in mice. *Am J Physiol Lung Cell Mol Physiol* **298**(2): L169–177.

Miquel, J., Bruneau, B., *et al.* (2014). Successful treatment of multifocal intracerebral and spinal hemangiomas with propranolol. *J Am Acad Dermatol* **70**(4): e83–84.

Mitchell, A., Adams, L.A., *et al.* (2008). Bevacizumab reverses need for liver transplantation in hereditary hemorrhagic telangiectasia. *Liver Transpl* **14**(2): 210–213.

Miyasaka, Y., Tokiwa, K., *et al.* (1994). The effects of a carotid-jugular fistula on cerebral blood flow in the cat: An experimental study in the acute period. *Surg Neurol* **41**(5): 396–398.

Mohr, J.P., Hartmann, A., *et al.* (2014). Viewpoints on the ARUBA trial. *AJNR Am J Neuroradiol* **36**(4): 615–617.

Morgan, M.K., Anderson, R.E., *et al.* (1989a). The effects of hyperventilation on cerebral blood flow in the rat with an open and closed carotid-jugular fistula. *Neurosurgery* **25**(4): 606–611; discussion 611–612.

Morgan, M.K., Anderson, R.E., *et al.* (1989b). A model of the pathophysiology of cerebral arteriovenous malformations by a carotid-jugular fistula in the rat. *Brain Res* **496**(1–2): 241–250.

Mouchtouris, N., Jabbour, P.M., *et al.* (2015). Biology of cerebral arteriovenous malformations with a focus on inflammation. *J Cereb Blood Flow Metab* **35**(2): 167–175.

Munroe, P.B., Olgunturk, R.O., *et al.* (1999). Mutations in the gene encoding the human matrix Gla protein cause Keutel syndrome. *Nat Genet* **21**(1): 142–144.

Murphy, P.A., Kim, T.N., *et al.* (2014). Constitutively active Notch4 receptor elicits brain arteriovenous malformations through enlargement of capillary-like vessels. *Proc Natl Acad Sci U S A* **111**(50): 18007–18012.

Murphy, P.A., Kim, T.N., *et al.* (2012). Notch4 normalization reduced blood vessel size in arteriovenous malformations. *Sci Transl Med* **4**(117): 117ra118.

Murphy, P.A., Lam, M.T., *et al.* (2008). Endothelial Notch4 signaling induces hallmarks of brain arteriovenous malformations in mice. *Proc Natl Acad Sci U S A* **105**(31): 10901–10906.

Murphy, P.A., Lu, G., *et al.* (2009). Endothelial Notch signaling is upregulated in human brain arteriovenous malformations and a mouse model of the disease. *Lab Invest* **89**(9): 971–982.

Nielsen, C.M., Cuervo, H., *et al.* (2014). Deletion of Rbpj from postnatal endothelium leads to abnormal arteriovenous shunting in mice. *Development* **141**(19): 3782–3792.

Niu, H., Cao, Y., *et al.* (2012). Relationships between hemorrhage, angioarchitectural factors and collagen of arteriovenous malformations. *Neurosci Bull* **28**(5): 595–605.

Numazawa, S., Sasaki, T., *et al.* (2005). Experimental model of intracranial arteriovenous shunting in the acute stage. *Neurol Med Chir (Tokyo)* **45**(6): 288–293.

Oh, S.P., Seki, T., *et al.* (2000). Activin receptor-like kinase 1 modulates transforming growth factor- beta 1 signaling in the regulation of angiogenesis. *Proc Natl Acad Sci USA* **97**(6): 2626–2631.

Park, S.O., Lee, Y.J., *et al.* (2008). ALK5- and TGFBR2-independent role of ALK1 in the pathogenesis of hereditary hemorrhagic telangiectasia type 2 (HHT2). *Blood* **111**(2): 633–642.

Park, S.O., Wankhede, M., *et al.* (2009). Real-time imaging of de novo arteriovenous malformation in a mouse model of hereditary hemorrhagic telangiectasia. *J Clin Invest* **119**(11): 3487–3496.

Pawlikowska, L., Tran, M.N., *et al.* (2004). Polymorphisms in genes involved in inflammatory and angiogenic pathways and the risk of hemorrhagic presentation of brain arteriovenous malformations. *Stroke* **35**(10): 2294–2300.

Pechan, P., Rubin, H., *et al.* (2009). Novel anti-VEGF chimeric molecules delivered by AAV vectors for inhibition of retinal neovascularization. *Gene Ther* **16**(1): 10–16.

Pepper, M.S. (1997). Manipulating angiogenesis. From basic science to the bedside. *Arterioscler Thromb Vasc Biol* **17**(4): 605–619.

Pietila, T.A., Zabramski, J.M., *et al.* (2000). Animal model for cerebral arteriovenous malformation. *Acta Neurochir* **142**(11): 1231–1240.

Post, S., Smits, A.M., *et al.* (2010). Impaired recruitment of HHT-1 mononuclear cells to the ischaemic heart is due to an altered CXCR4/CD26 balance. *Cardiovasc Res* **85**(3): 494–502.

Power, C., Henry, S., *et al.* (2003). Intracerebral hemorrhage induces macrophage activation and matrix metalloproteinases. *Ann Neurol* **53**(6): 731–742.

Przewratil, P., Kobos, J., *et al.* (2016). Serum and tissue profile of VEGF and its receptors VGFR1/R2 in children with infantile hemangiomas on systemic propranolol treatment. *Immunol Lett* **175**: 44–49.

Qian, Z., Climent, S., *et al.* (1999). A simplified arteriovenous malformation model in sheep: Feasibility study. *AJNR Am J Neuroradiol* **20**(5): 765–770.

Qu, R., Silver, M.M., *et al.* (1998). Distribution of endoglin in early human development reveals high levels on endocardial cushion tissue mesenchyme during valve formation. *Cell Tissue Res* **292**(2): 333–343.

Quach, H., Kalff, A., *et al.* (2012). Lenalidomide in multiple myeloma: Current status and future potential. *Am J Hematol* **87**(12): 1089–1095.

Rangel-Castilla, L., Cress, M.C., *et al.* (2015). Feasibility, safety, and periprocedural complications of pipeline embolization for intracranial neurysm treatment under conscious sedation: University at Buffalo neurosurgery experience. *Neurosurgery* **11** (Suppl 3): 426–430.

Riss, D., Burian, M., *et al.* (2015). Intranasal submucosal bevacizumab for epistaxis in hereditary hemorrhagic telangiectasia: A double-blind, randomized, placebo-controlled trial. *Head Neck* **37**(6): 783–787.

Rivera, V.M., Gao, G.P., *et al.* (2005). Long-term pharmacologically regulated expression of erythropoietin in primates following AAV-mediated gene transfer. *Blood* **105**(4): 1424–1430.

Rivera, V.M., Ye, X., *et al.* (1999). Long-term regulated expression of growth hormone in mice after intramuscular gene transfer. *Proc Natl Acad Sci U S A* **96**(15): 8657–8662.

Riviere, C., Danos, O., *et al.* (2006). Long-term expression and repeated administration of AAV type 1, 2 and 5 vectors in skeletal muscle of immunocompetent adult mice. *Gene Ther* **13**(17): 1300–1308.

Rochon, E.R., Menon, P.G., *et al.* (2016). Alk1 controls arterial endothelial cell migration in lumenized vessels. *Development* **143**(14): 2593–2602.

Roelen, B.A., van Rooijen, M.A., *et al.* (1997). Expression of ALK-1, a type 1 serine/threonine kinase receptor, coincides with sites of vasculogenesis and angiogenesis in early mouse development. *Dev Dyn* **209**(4): 418–430.

Roman, B.L., Pham, V.N., *et al.* (2002). Disruption of acvrl1 increases endothelial cell number in zebrafish cranial vessels. *Development* **129**(12): 3009–3019.

Rothbart, D., Awad, I.A., *et al.* (1996). Expression of angiogenic factors and stuctural proteins in central nervous system vascular malformations. *Neurosurgery* **38**(5): 915–924; discussion 924–925.

Sadick, H., Naim, R., *et al.* (2005a). Plasma level and tissue expression of angiogenic factors in patients with hereditary hemorrhagic telangiectasia. *Int J Mol Med* **15**(4): 591–596.

Sadick, H., Riedel, F., *et al.* (2005b). Patients with hereditary hemorrhagic telangiectasia have increased plasma levels of vascular endothelial growth factor and transforming growth factor-beta1 as well as high ALK1 tissue expression. *Haematologica* **90**(6): 818–828.

Sanchez-Mejia, R.O., Ona, V.O., *et al.* (2001). Minocycline reduces traumatic brain injury-mediated caspase-1 activation, tissue damage, and neurological dysfunction. *Neurosurgery* **48**(6): 1393–1399; discussion 1399–1401.

Sanchez de Miguel, L., de Frutos, T., *et al.* (1999). Aspirin inhibits inducible nitric oxide synthase expression and tumour necrosis factor-alpha release by cultured smooth muscle cells. *Eur J Clin Invest* **29**(2): 93–99.

Saunders, T.L. (2011). Inducible transgenic mouse models. *Methods Mol Biol* **693**: 103–115.

Schwartz, G.G., Olsson, A.G., *et al.* (2001). Effects of atorvastatin on early recurrent ischemic events in acute coronary syndromes: The MIRACL study: A randomized controlled trial. *JAMA* **285**(13): 1711–1718.

Seki, T., Yun, J., *et al.* (2003). Arterial endothelium-specific activin receptor-like kinase 1 expression suggests its role in arterialization and vascular remodeling. *Circ Res* **93**(7): 682–689.

Seo, S., Fujita, H., *et al.* (2006). The forkhead transcription factors, Foxc1 and Foxc2, are required for arterial specification and lymphatic sprouting during vascular development. *Dev Biol* **294**(2): 458–470.

Shen, F., Degos, V., *et al.* (2014). Endoglin deficiency impairs stroke recovery. *Stroke* **45**(7): 2101–2106.

Shen, F., Mao, L., *et al.* (2015). Inhibition of pathological brain angiogenesis through systemic delivery of AAV vector expressing soluble FLT1. *Gene Ther* **22**(11): 893–900.

Shen, Z., Li, C., *et al.* (2012). Smooth muscle protein 22 alpha-Cre is expressed in myeloid cells in mice. *Biochem Biophys Res Commun* **422**(4): 639–642.

Shibuya, M. (2006). Vascular endothelial growth factor receptor-1 (VEGFR-1/Flt-1): A dual regulator for angiogenesis. *Angiogenesis* **9**(4): 225–230; discussion 231.

Shovlin, C.L. (2010). Hereditary haemorrhagic telangiectasia: Pathophysiology, diagnosis and treatment. *Blood Rev* **24**(6): 203–219.

Simonds, J., Miller, F., *et al.* (2009). The effect of bevacizumab (Avastin) treatment on epistaxis in hereditary hemorrhagic telangiectasia. *Laryngoscope* **119**(5): 988–992.

Sommers Smith, S.K., Smith, D.M. (2002). Beta blockade induces apoptosis in cultured capillary endothelial cells. *In Vitro Cell Dev Biol Anim* **38**(5): 298–304.

Song, J.W., Munn, L.L. (2011). Fluid forces control endothelial sprouting. *Proc Natl Acad Sci U S A* **108**(37): 15342–15347.

Sonstein, W.J., Kader, A., *et al.* (1996). Expression of vascular endothelial growth factor in pediatric and adult cerebral arteriovenous malformations: An immunocytochemical study. *J Neurosurg* **85**(5): 838–845.

Sorensen, L.K., Brooke, B.S., *et al.* (2003). Loss of distinct arterial and venous boundaries in mice lacking endoglin, a vascular-specific TGFbeta coreceptor. *Dev Biol* **261**(1): 235–250.

Spetzler, R.F., Martin, N.A., *et al.* (1987). Surgical management of large AVM's by staged embolization and operative excision. *J Neurosurg* **67**(1): 17–28.

Spetzler, R.F., Wilson, C.B., *et al.* (1978). Normal perfusion pressure breakthrough theory. *Clin Neurosurg* 25: 651–672.

Srinivasan, S., Hanes, M.A., *et al.* (2003). A mouse model for hereditary hemorrhagic telangiectasia (HHT) type 2. *Hum Mol Genet* **12**(5): 473–482.

St-Jacques, S., Cymerman, U., *et al.* (1994). Molecular characterization and in situ localization of murine endoglin reveal that it is a transforming growth factor-beta binding protein of endothelial and stromal cells. *Endocrinology* **134**(6): 2645–2657.

Starke, R.M., Chalouhi, N., *et al.* (2015). Potential role of aspirin in the prevention of aneurysmal subarachnoid hemorrhage. *Cerebrovasc Dis* **39**(5–6): 332–342.

Stockton, R.A., Shenkar, R., *et al.* (2010). Cerebral cavernous malformations proteins inhibit Rho kinase to stabilize vascular integrity. *J Exp Med* **207**(4): 881–896.

Storer, K.P., Tu, J., *et al.* (2008). Inflammatory molecule expression in cerebral arteriovenous malformations. *J Clin Neurosci* **15**(2): 179–184.

Su, H., Kim, H., *et al.* (2010). Reduced expression of integrin alphavbeta8 is associated with brain arteriovenous malformation pathogenesis. *Am J Pathol* **176**(2): 1018–1027.

Sumii, T., Lo, E.H. (2002). Involvement of matrix metalloproteinase in thrombolysis-associated hemorrhagic transformation after embolic focal ischemia in rats. *Stroke* **33**(3): 831–836.

Sure, U., Battenberg, E., *et al.* (2004). Hypoxia-inducible factor and vascular endothelial growth factor are expressed more frequently in embolized than in nonembolized cerebral arteriovenous malformations. *Neurosurgery* **55**(3): 663–669; discussion 669–670.

Sure, U., Butz, N., *et al.* (2001a). Endothelial proliferation, neoangiogenesis, and potential *de novo* generation of cerebrovascular malformations. *J Neurosurg* **94**(6): 972–977.

Sure, U., Butz, N., *et al.* (2001b). Treatment-induced neoangiogenesis in cerebral arteriovenous malformations. *Clin Neurol Neurosurg* **103**(1): 29–32.

Takagi, H., Matsui, M., *et al.* (2010). A meta-analysis of clinical studies of statins for prevention of abdominal aortic aneurysm expansion. *J Vasc Surg* **52**(6): 1675–1681.

Tang, Y., Lee, K.S., *et al.* (2005). Epistatic interactions between modifier genes confer strain-specific redundancy for Tgfb1 in developmental angiogenesis. *Genomics* **85**(1): 60–70.

Tang, Y., McKinnon, M.L., *et al.* (2003). Genetic modifiers interact with maternal determinants in vascular development of Tgfb1(-/-) mice. *Hum Mol Genet* **12**(13): 1579–1589.

Tanvetyanon, T., Murtagh, R., *et al.* (2009). Rupture of a cerebral arteriovenous malformation in a patient treated with bevacizumab. *J Thorac Oncol* **4**(2): 268–269.

Tillet, E., Bailly, S. (2015). Emerging roles of BMP9 and BMP10 in hereditary hemorrhagic telangiectasia. *Front Genet* **5**: 456.

Tokiwa, K., Miyasaka, Y., *et al.* (1995). The effects of a carotid-jugular fistula on cerebral blood flow in the cat: An experimental study in the chronic period. *Neurol Res* **17**(4): 297–300.

Toporsian, M., Jerkic, M., *et al.* (2010). Spontaneous adult-onset pulmonary arterial hypertension attributable to increased endothelial oxidative stress in a murine model of hereditary hemorrhagic telangiectasia. *Arterioscler Thromb Vasc Biol* **30**(3): 509–517.

Torsney, E., Charlton, R., *et al.* (2003). Mouse model for hereditary hemorrhagic telangiectasia has a generalized vascular abnormality. *Circulation* **107**(12): 1653–1657.

Tseng, M.Y., Czosnyka, M., *et al.* (2005). Effects of acute treatment with pravastatin on cerebral vasospasm, autoregulation, and delayed ischemic deficits after aneurysmal subarachnoid hemorrhage: A phase II randomized placebo-controlled trial. *Stroke* **36**(8): 1627–1632.

Urness, L.D., Sorensen, L.K., *et al.* (2000). Arteriovenous malformations in mice lacking activin receptor-like kinase-1. *Nat Genet* **26**(3): 328–331.

van Laake, L.W., van den Driesche, S., *et al.* (2006). Endoglin has a crucial role in blood cell-mediated vascular repair. *Circulation* **114**(21): 2288–2297.

Walker, E.J., Su, H., *et al.* (2011). Arteriovenous malformation in the adult mouse brain resembling the human disease. *Ann Neurol* **69**(6): 954–962.

Walker, E.J., Su, H., *et al.* (2012). Bevacizumab attenuates VEGF-induced angiogenesis and vascular malformations in the adult mouse brain. *Stroke* **43**(7): 1925–1930.

Wang, D., Li, T., *et al.* (2014). Effects of atorvastatin on chronic subdural hematoma: A preliminary report from three medical centers. *J Neurol Sci* **336**(1–2): 237–242.

Weber, C., Erl, W., *et al.* (1995). Aspirin inhibits nuclear factor-kappa B mobilization and monocyte adhesion in stimulated human endothelial cells. *Circulation* **91**(7): 1914–1917.

Wemmelund, H., Hogh, A., *et al.* (2014). Statin use and rupture of abdominal aortic aneurysm. *Br J Surg* **101**(8): 966–975.

Williams, B.J., Park, D.M., *et al.* (2012). Bevacizumab used for the treatment of severe, refractory perilesional edema due to an arteriovenous malformation treated with stereotactic radiosurgery. *J Neurosurg* **116**(5): 972–977.

Wooderchak-Donahue, W.L., McDonald, J., *et al.* (2013). BMP9 mutations cause a vascular-anomaly syndrome with phenotypic overlap with hereditary hemorrhagic telangiectasia. *Am J Hum Genet* **93**(3): 530–537.

Xu, B., Wu, Y.Q., *et al.* (2004). Vascular endothelial growth factor induces abnormal microvasculature in the endoglin heterozygous mouse brain. *J Cereb Blood Flow Metab* **24**(2): 237–244.

Yang, X., Li, C., *et al.* (2002). Generation of Smad4/Dpc4 conditional knockout mice. *Genesis* **32**(2): 80–81.

Yang, X., Li, C., *et al.* (1998). The tumor suppressor SMAD4/DPC4 is essential for epiblast proliferation and mesoderm induction in mice. *Proc Natl Acad Sci U S A* **95**(7): 3667–3672.

Yao, Y., Jumabay, M., *et al.* (2011). Matrix Gla protein deficiency causes arteriovenous malformations in mice. *J Clin Invest* **121**(8): 2993–3004.

Yao, Y., Yao, J., *et al.* (2013). Reducing Jagged 1 and 2 levels prevents cerebral arteriovenous malformations in matrix Gla protein deficiency. *Proc Natl Acad Sci U S A* **110**(47): 19071–19076.

Yao, Y., Zebboudj, A.F., *et al.* (2006). Regulation of bone morphogenetic protein-4 by matrix GLA protein in vascular endothelial cells involves activin-like kinase receptor 1. *J Biol Chem* **281**(45): 33921–33930.

Yassari, R., Sayama, T., *et al.* (2004). Angiographic, hemodynamic and histological characterization of an arteriovenous fistula in rats. *Acta Neurochir* **146**(5): 495–504.

Yiqin, Y., Meilin, X., *et al.* (2009). Aspirin inhibits MMP-2 and MMP-9 expression and activity through PPARalpha/gamma and TIMP-1-mediated mechanisms in cultured mouse celiac macrophages. *Inflammation* **32**(4): 233–241.

Yotsui, T., Yasuda, O., *et al.* (2007). Aspirin prevents adhesion of T lymphoblasts to vascular smooth muscle cells. *FEBS Lett* **581**(3): 427–432.

Young, W.L., Yang, G.Y. (2004). Are there genetic influences on sporadic brain arteriovenous malformations? *Stroke* **35**(11 Suppl 1): 2740–2745.

Yrjanheikki, J., Tikka, T., *et al.* (1999). A tetracycline derivative, minocycline, reduces inflammation and protects against focal cerebral ischemia with a wide therapeutic window. *Proc Natl Acad Sci U S A* **96**(23): 13496–13500.

Zhang, R., Han, Z., *et al.* (2016a). Persistent infiltration and pro-inflammatory differentiation of monocytes cause unresolved inflammation in brain arteriovenous malformation. *Angiogenesis* **19**(4): 451–461.

Zhang, R., Zhu, W., *et al.* (2016b). Vascular integrity in the pathogenesis of brain arteriovenous malformation. *Acta Neurochir Suppl* 121: 29–35.

ZhuGe, Q., Zhong, M., *et al.* (2009). Notch1 signaling is activated in brain arteriovenous malformation in humans. *Brain* **132**(Pt 12): 3231–3241.

Zucco, L., Zhang, Q., *et al.* (2014). Circulating angiogenic cell dysfunction in patients with hereditary hemorrhagic telangiectasia. *PLoS One* **9**(2): e89927.

Biology of Brain Aneurysms 6

W. Caleb Rutledge and Tomoki Hashimoto*[†]

**Department of Neurological Surgery,*
University of California, San Francisco, San Francisco, California, USA
[†]Department of Anesthesia and Perioperative Care,
University of California, San Francisco, San Francisco, California, USA

Brain aneurysms are common, with a prevalence of about 3–5% in the general population (Vlak *et al.*, 2011). The overall rupture rate is about 1% per year, but depends on patient and aneurysm characteristics, including patient age and sex, and aneurysm location and size (Greving *et al.*, 2014; Juvela *et al.*, 2008; UCAS Japan Investigators *et al.*, 2012; Wiebers *et al.*, 2003). Despite advances in the detection and management of unruptured and ruptured aneurysms, subarachnoid hemorrhage (SAH) from aneurysm rupture still has significant morbidity and mortality, with a 30-day mortality up to 45% and moderate-to-severe disability in as many as 30% of survivors (Johnston *et al.*, 1998; Nieuwkamp *et al.*, 2009; Wiebers *et al.*, 2003). Surgical or endovascular treatment of aneurysms can eliminate the risk of rupture, but both are invasive treatments with non-negligible procedural-related morbidity and mortality. As such, in aneurysms with an exceedingly low risk of rupture, the risks of treatment may exceed the risk of rupture in the aneurysm's natural history (Wiebers *et al.*, 2003). Thus, there is considerable interest in developing new pharmacologic treatments to stabilize aneurysms and prevent rupture.

Little is known about the cause of brain aneurysms or why some rupture. Although there are rare associations with familial conditions, such as Autosomal Dominant Polycystic Kidney Disease and Ehlers-Danlos syndrome, most brain aneurysms are sporadic. There are well-established clinical risk factors for aneurysm rupture, including cigarette smoking, hypertension, and family history of aneurysm rupture, but the mechanisms of aneurysm pathogenesis are not well defined (Brinjikji *et al.*, 2016; Juvela 2011; Juvela *et al.*, 2008). There is, however, growing evidence that inflammation in response to endothelial cell injury and dysfunction is a critical contributor in aneurysm pathogenesis (Chalouhi *et al.*, 2012).

Moreover, observational studies in humans suggest that anti-inflammatory therapies may stabilize aneurysms and prevent rupture. In a nested case–control study from the International Study of Unruptured Intracranial Aneurysms (ISUIA) cohort, patients who took aspirin frequently had significantly lower odds of hemorrhage compared with those who did not, suggesting that aspirin may have a protective effect via its anti-inflammatory properties (Hasan *et al.*, 2011).

6.1 Histopathology of Normal Intracranial Arteries and Aneurysms

Intracranial arteries are composed of three layers: intima, media, and adventitia. The intima is the innermost layer, a monolayer of endothelial cells and a subendothelial layer of connective tissue. The internal elastic lamina separates the intima from the media. The media is composed of smooth muscle cells and collagen fibers (predominately type III collagen fibers) that provide tensile strength to blood vessels. In systemic arteries, an external elastic lamina separates the media from the outermost layer, the adventitia, but is absent in intracranial arteries. The adventitia is made up of collagen fibers that are predominantly type I collagen fibers, as well as elastin, fibroblasts, nerves, and the vaso vasorum. Connective tissue in the artery wall is primarily composed of extracellular matrix that provides structural support and participates in important biologic functions.

In contrast to the ordered structure of the normal arterial wall, aneurysms have a disorganized structure. A key feature is loss of the internal elastic lamina and disruption of the media (Abruzzo *et al.*, 1998; Scanarini *et al.*, 1978). There is loss of smooth muscle cells and thinning of the media. Other important histopathological features are intimal hyperplasia, loss of cellular components, disorganization of muscle fiber structure, and changes in the composition of the vessel wall, including collagen, fibronectin, and laminin expression (Chalouhi *et al.*, 2012).

6.2 Aneurysm Pathogenesis

It has been long postulated that aneurysms arise as a result of a disrupted balance between local hemodynamic stress and arterial wall strength. Hemodynamic stress causes physiologic flow-induced vascular remodeling, an adaptive process of the vascular wall to reduce wall shear stress. With sustained hemodynamic stress, such as in hypertension, pathologic vascular remodeling may occur. Inflammatory cells, including macrophages and proteinases like matrix metalloproteinases (MMPs), seem to play critical roles in flow-induced vascular remodeling in both physiologic and pathologic states. Nuki *et al.* showed that macrophage depletion reduced flow-induced remodeling in a rat model (Nuki *et al.*, 2009a). Subsequently, Ota *et al.* showed that pharmacologic inhibition of MMPs produced by macrophages, especially MMP-9, also reduced flow-induced remodeling in the same rat model (Ota *et al.*, 2009). MMP-9 is the main gelatinase produced by macrophages and digests the extracellular matrix. Pathologic flow-induced remodeling then may lead to formation and ultimately rupture of intracranial aneurysms.

Most brain aneurysms occur in the circle of Willis, arising at arterial bifurcations or branch points where there is disruption in laminar flow, intravascular turbulence, and abnormal wall shear stress (Nixon *et al.*, 2010). Perturbations in flow and abnormal wall shear stress at these branch points may cause endothelial cell injury and dysfunction, and initiate an inflammatory cascade resulting in pathological vascular wall remodeling (Cebral *et al.*, 2017; Chalouhi *et al.*, 2012; Fukuda *et al.*, 2000; Kataoka *et al.*, 1999; Sho *et al.*, 2002; Tada *et al.*, 2010; Tamura *et al.*, 2009).

Inflammatory cells play key roles in cardiovascular pathologies, such as atherosclerosis and abdominal aortic aneurysms, and inflammation has also recently emerged as a key contributor in the pathophysiology of intracranial aneurysms. Observational studie have shown the presence of inflammatory cells and markers in intracranial aneurysm tissue and serum samples, including macrophages, macrophage-derived cytokines, matrix metalloproteinases, mast cells, and mast cell-derived cytokines (Aoki *et al.*, 2007; Baker *et al.*, 1995; Chyatte *et al.*, 1999; Chyatte & Lewis, 1997; Frosen *et al.*, 2004; Kataoka *et al.*, 1999; Kim *et al.*, 1997; Nuki *et al.*, 2009b; Todor *et al.*, 1998).

6.3 Macrophage Infiltration in Aneurysms

Macrophages and macrophage-derived cytokines are present in the wall of both unruptured and ruptured aneurysms (Aoki *et al.*, 2007; Bruno *et al.*, 1998; Chyatte *et al.*, 1999; Frosen *et al.*, 2004; Kataoka *et al.*, 1999; Kim *et al.*, 1997). Macrophages destabilize the vascular wall by secreting pro-inflammatory proteinases and MMPs that digest the extracellular matrix and remodel the vascular wall. Compared with unruptured aneurysms, macrophage infiltration appears to increase in ruptured aneurysms (Frosen *et al.*, 2004). Macrophages and MMPs are also known to be involved in physiologic flow-induced remodeling (Nuki *et al.*, 2009a; Ota *et al.*, 2009).

Aoki *et al.* showed that expression of macrophages was associated with aneurysm growth in mice (Aoki *et al.*, 2007). Also in a mouse model of intracranial aneurysm, Kanematsu *et al.* found macrophage infiltration in the aneurysm wall, and that pharmacologic depletion of macrophages decreased the incidence of aneurysm formation (Kanematsu *et al.*, 2011). Aoki *et al.* also noted that critical aneurysmal changes, including loss of the elastic lamina, were less frequent in monocyte chemotactic protein-1 (MCP-1) knockout mice. Kanematsu *et al.* showed that MCP-1 knockout mice had a lower incidence of aneurysm formation (Kanematsu *et al.*, 2011). MCP-1 is critical for macrophage function (Aoki *et al.*, 2009b).

Interestingly, in humans, macrophages exist as different subtypes with opposite effects on inflammation. Hasan *et al.* showed a critical imbalance of these macrophage subsets in ruptured aneurysms in humans. M1 and M2 macrophages are the major subsets of macrophages in humans, and these have opposite effects on inflammation: M1 macrophages produce inflammatory cytokines and a pro-inflammatory state, while M2 macrophages reduce inflammation. Hasan found that M1 expression significantly increased compared with M2 expression in ruptured aneurysms (Hasan *et al.*, 2012).

Nowicki *et al.* also found an increased pro-inflammatory M1/M2 ratio in cerebral aneurysm formation over time. Neutrophil blockade produced a decreased or an anti-inflammatory M1/M2 ratio, suggesting that neutrophil-dependent inflammation produces a pro-inflammatory phenotype leading to aneurysm formation (Nowicki *et al.*, 2017).

6.4 Matrix Metalloproteinases in Aneurysms

In addition to macrophage infiltration, Aoki *et al.* found that macrophage-derived MMPs are associated with aneurysm progression (Aoki *et al.*, 2007). MMPs are proteinases involved in remodeling of vascular walls. In their rat model, MMP expression increased with progression of aneurysms, and selective inhibition of MMPs blocked aneurysm progression (Aoki *et al.*, 2007). Makino *et al.* found that doxycycline and minocycline, non-specific inhibitors for MMPs, prevent aneurysmal rupture in mice (Makino *et al.*, 2012).

6.5 Mast Cells in Aneurysms

The role mast cells play in allergy and anaphylaxis has been well described; however, they are also involved in the pathogenesis of aneurysms. In an experimental rat model, Ishibashi *et al.* found that mast cells also significantly increased in aneurysm walls. Degranulation of mast cells induced expression and activation of pro-inflammatory cytokines, including MMP-2 and MMP-9. Inhibition of mast cell degranulation had anti-inflammatory effects with decreased nuclear factor-k B activation, macrophage infiltration, and expression of MCP-1, MMPs, and cytokines (Ishibashi *et al.*, 2010). Hasan *et al.* also demonstrated mast cell infiltration in both ruptured and unruptured aneurysms, but more so in ruptured cases (Hasan *et al.*, 2012).

Liu *et al.* showed that administration of human mesenchymal stem cell-derived microvesicles reduced aneurysm rupture rate in a mouse model of intracranial aneurysm, likely by suppressing mast cell activation via a prostaglandin E2-dependent mechanism (Liu *et al.*, 2016).

6.6 Pharmacologic Treatments Targeting Inflammation

Anti-inflammatory therapy is emerging as a potential therapy for preventing aneurysmal rupture. In addition to aspirin that may exert its anti-inflammatory effects through inactivation of cyclooxygenases, statins, and angiotensin-receptor blockers have also been investigated as possible therapies for preventing aneurysm formation and subsequent rupture.

6.6.1 *Statins*

In addition to their cholesterol-lowering effect, statins improve endothelial function and suppress inflammation and oxidative stress. Aoki *et al.* found that simvastatin prevented development of cerebral aneurysms in rats by inhibiting inflammatory reactions in aneurysmal walls (Aoki *et al.*, 2008). Similarly, pitavastatin prevented aneurysm progression in rats by inhibiting NF-kB activation and chronic inflammation in aneurysmal walls (Aoki *et al.*, 2009a). Moreover, Kimura *et al.* showed that pravastatin prevented aneurysm formation in rats (Kimura *et al.*, 2010). Interestingly, however, Tada *et al.* found that statins may exert bidirectional effects on cerebral aneurysms. At a low dose, pravastatin reduced endothelial damage and inhibited aneurysm formation, but at high doses, statins promoted growth and even rupture of cerebral aneurysms in rats (Tada *et al.*, 2011). In a large case–control study that included 1,200 patients, there was no association between statin use and aneurysm formation (Marbacher *et al.*, 2012). Despite conflicting data, statins appear to modulate the inflammatory cascade in aneurysm pathogenesis. Further investigation is therefore needed.

6.6.2 *Peroxisome proliferator-activated receptor gamma agonists*

Shimada *et al.* showed that peroxisome proliferator-activated receptor-gamma (PPARγ) has a protective role in the development of intracranial aneurysms (Shimada *et al.*, 2015). Activation of the nuclear hormone cell receptor, PPARγ, present in many cell types (including endothelial cells, smooth muscle cells, and inflammatory cells), modulates the inflammatory response and production of inflammatory cytokines (Jiang *et al.*, 1998; Welch *et al.*, 2003). In a mouse model of intracranial aneurysms, the PPARγ agonist pioglitazone reduced the aneurysm rupture rate as well as the production of macrophage-related cytokines. The protective effect of pioglitazone required macrophage PPARγ.

6.6.3 *Renin–angiotensin system*

Activation of the renin–angiotensin system (RAS) is known to cause vascular inflammation and remodeling. In an intracranial aneurysm mouse model, Tada *et al.* showed that inhibition of RAS can prevent aneurysm rupture independent of blood pressure reduction (Tada *et al.*, 2014). Similarly, Kimura *et al.* showed that an angiotensin II receptor blocker inhibited aneurysm formation (Kimura *et al.*, 2010). However, the downstream mechanisms by which RAS mediates aneurysm formation and/or rupture are unknown. Epidermal growth factor receptor (EGFR) transactivation by angiotensin may be a critical transduction step. EGFR transactivation by angiotensin is mediated by activation of the metalloprotease ADAM17 (Ohtsu *et al.*, 2006). While the mechanism by which ADAM17 expression and activity are increased in aneurysm rupture is unknown, endoplasmic reticulum (ER) stress may contribute to ADAM17 upregulation.

6.7 Conclusion

Inflammation plays a critical role in the pathogenesis of cerebral aneurysms. Brain aneurysms result from endothelial cell injury and dysfunction, which triggers an inflammatory response leading to degradation of the vascular wall. Endothelial cell injury and dysfunction are related to hemodynamic and wall shear stress. The inflammatory process is largely mediated by macrophages and macrophage-derived MMPs that digest the extracellular matrix, weakening the vessel/aneurysm wall. Anti-inflammatory therapies are emerging as a potential therapy for preventing aneurysmal rupture.

References

Abruzzo, T., Shengelaia, G.G., *et al.* (1998). Histologic and morphologic comparison of experimental aneurysms with human intracranial aneurysms. *AJNR Am J Neuroradiol* **19**(7): 1309–1314.

Aoki, T., Kataoka, H., *et al.* (2007). Macrophage-derived matrix metalloproteinase-2 and -9 promote the progression of cerebral aneurysms in rats. *Stroke* **38**(1):162–169.

Aoki, T., Kataoka, H., *et al.* (2008). Simvastatin suppresses the progression of experimentally induced cerebral aneurysms in rats. *Stroke* **39**(4): 1276–1285.

Aoki, T., Kataoka, H., *et al.* (2009a). Pitavastatin suppresses formation and progression of cerebral aneurysms through inhibition of the nuclear factor kappaB pathway. *Neurosurgery* **64**(2): 357–365.

Aoki, T., Kataoka, H., *et al.* (2009b). Impact of monocyte chemoattractant protein-1 deficiency on cerebral aneurysm formation. *Stroke* **40**(3): 942–951.

Baker, C.J., Fiore, A., *et al.* (1995). Serum elastase and alpha-1-antitrypsin levels in patients with ruptured and unruptured cerebral aneurysms. *Neurosurgery* **37**(1): 56–61.

Brinjikji, W., Zhu, Y.Q., *et al.* (2016). Risk factors for growth of intracranial aneurysms: A systematic review and meta-analysis. *AJNR Am J Neuroradiol* **37**(4): 615–620.

Bruno, G., Todor, R., *et al.* (1998). Vascular extracellular matrix remodeling in cerebral aneurysms. *J Neurosurg* **89**(3): 431–440.

Cebral, J., Ollikainen, E., *et al.* (2017). Flow conditions in the intracranial aneurysm lumen are associated with inflammation and degenerative changes of the aneurysm wall. *AJNR Am J Neuroradiol* **38**(1): 119–126.

Chalouhi, N., Ali, M.S., *et al.* (2012). Biology of intracranial aneurysms: Role of inflammation. *J Cereb Blood Flow Metab* **32**(9): 1659–1676.

Chyatte, D., Bruno, G., *et al.* (1999). Inflammation and intracranial aneurysms. *Neurosurgery* **45**(5): 1137–1146.

Chyatte, D., Lewis, I. (1997). Gelatinase activity and the occurrence of cerebral aneurysms. *Stroke* **28**(4): 799–804.

Frosen, J., Piippo, A., *et al.* (2004). Remodeling of saccular cerebral artery aneurysm wall is associated with rupture: Histological analysis of 24 unruptured and 42 ruptured cases. *Stroke* **35**(10): 2287–2293.

Fukuda, S., Hashimoto, N., *et al.* (2000). Prevention of rat cerebral aneurysm formation by inhibition of nitric oxide synthase. *Circulation* **101**(21): 2532–2538.

Greving, J.P., Wermer, M.J., *et al.* (2014). Development of the PHASES score for prediction of risk of rupture of intracranial aneurysms: A pooled analysis of six prospective cohort studies. *Lancet Neurol* **13**(1): 59–66.

Hasan, D., Chalouhi, N., *et al.* (2012). Macrophage imbalance (M1 vs. M2) and upregulation of mast cells in wall of ruptured human cerebral aneurysms: Preliminary results. *J Neuroinflammation* **9**: 222.

Hasan, D.M., Mahaney, K.B., *et al.* (2011). Aspirin as a promising agent for decreasing incidence of cerebral aneurysm rupture. *Stroke* **42**(11): 3156–3162.

Ishibashi, R., Aoki, T., *et al.* (2010). Contribution of mast cells to cerebral aneurysm formation. *Curr Neurovasc Res* **7**(2): 113–124.

Jiang, C., Ting, A.T., *et al.* (1998). PPAR-gamma agonists inhibit production of monocyte inflammatory cytokines. *Nature* **391**(6662): 82–86.

Johnston, S.C., Selvin, S., *et al.* (1998). The burden, trends, and demographics of mortality from subarachnoid hemorrhage. *Neurology* **50**(5): 1413–1418.

Juvela, S. (2011). Prevalence of and risk factors for intracranial aneurysms. *Lancet Neurol* **10**(7): 595–597.

Juvela, S., Porras, M., *et al.* (2008). Natural history of unruptured intracranial aneurysms: Probability of and risk factors for aneurysm rupture. *J Neurosurg* **108**(5): 1052–1060.

Kanematsu, Y., Kanematsu, M., *et al.* (2011). Critical roles of macrophages in the formation of intracranial aneurysm. *Stroke* **42**(1): 173–178.

Kataoka, K., Taneda, M., *et al.* (1999). Structural fragility and inflammatory response of ruptured cerebral aneurysms. A comparative study between ruptured and unruptured cerebral aneurysms. *Stroke* **30**(7): 1396–1401.

Kim, S.C., Singh, M., *et al.* (1997). Matrix metalloproteinase-9 in cerebral aneurysms. *Neurosurgery* **41**(3): 642–666.

Kimura, N., Shimizu, H., *et al.* (2010). Effect of olmesartan and pravastatin on experimental cerebral aneurysms in rats. *Brain Res* **1322**: 144–152.

Liu, J., Kuwabara, A., *et al.* (2016). Human mesenchymal stem cell-derived microvesicles prevent the rupture of intracranial aneurysm in part by suppression of mast cell activation via a PGE2-dependent mechanism. *Stem Cells* **34**(12): 2943–2955.

Makino, H., Tada, Y., *et al.* (2012). Pharmacological stabilization of intracranial aneurysms in mice: A feasibility study. *Stroke* **43**(9): 2450–2456.

Marbacher, S., Schlappi, J.A., *et al.* (2012). Do statins reduce the risk of aneurysm development? A case–control study. *J Neurosurg* **116**(3): 638–642.

Nieuwkamp, D.J., Setz, L.E., *et al.* (2009). Changes in case fatality of aneurysmal subarachnoid haemorrhage over time, according to age, sex, and region: A meta-analysis. *Lancet Neurol* **8**(7): 635–642.

Nixon, A.M., Gunel, M., *et al.* (2010). The critical role of hemodynamics in the development of cerebral vascular disease. *J Neurosurg* **112**(6): 1240–1253.

Nowicki, K.W., Hosaka, K., *et al.* (2017). M1 macrophages are required for murine cerebral aneurysm formation. *J Neurointerv Surg*. doi: 10.1136/neurintsurg-2016-012911.

Nuki, Y., Matsumoto, M.M., *et al.* (2009a). Roles of macrophages in flow-induced outward vascular remodeling. *J Cereb Blood Flow Metab* **29**(3): 495–503.

Nuki, Y., Tsou, T.L., *et al.* (2009b). Elastase-induced intracranial aneurysms in hypertensive mice. *Hypertension* **54**(6): 1337–1344.

Ohtsu, H., Dempsey, P.J., *et al.* (2006). ADAM17 mediates epidermal growth factor receptor transactivation and vascular smooth muscle cell hypertrophy induced by angiotensin II. *Arterioscler Thromb Vasc Biol* **26**(9): e133–137.

Ota, R., Kurihara, C., *et al.* (2009). Roles of matrix metalloproteinases in flow-induced outward vascular remodeling. *J Cereb Blood Flow Metab* **29**(9): 1547–1558.

Scanarini, M., Mingrino, S., *et al.* (1978). Histological and ultrastructural study of intracranial saccular aneurysmal wall. *Acta Neurochir* **43**(3-4): 171–182.

Shimada, K., Furukawa, H., *et al.* (2015). Protective role of peroxisome proliferator-activated receptor-gamma in the development of intracranial aneurysm rupture. *Stroke* **46**(6): 1664–1672.

Sho, E., Sho, M., *et al.* (2002). Arterial enlargement in response to high flow requires early expression of matrix metalloproteinases to degrade extracellular matrix. *Exp Mol Pathol* **73**(2): 142–153.

Tada, Y., Kitazato, K.T., *et al.* (2011). Statins promote the growth of experimentally induced cerebral aneurysms in estrogen-deficient rats. *Stroke* **42**(8): 2286–2293.

Tada, Y., Wada, K., *et al.* (2014). Roles of hypertension in the rupture of intracranial aneurysms. *Stroke* **45**(2): 579–586.

Tada, Y., Yagi, K., *et al.* (2010). Reduction of endothelial tight junction proteins is related to cerebral aneurysm formation in rats. *J Hypertens* **28**(9): 1883–1891.

Tamura, T., Jamous, M.A., *et al.* (2009). Endothelial damage due to impaired nitric oxide bioavailability triggers cerebral aneurysm formation in female rats. *J Hypertens* **27**(6): 1284–1292.

Todor, D.R., Lewis, I., *et al.* (1998). Identification of a serum gelatinase associated with the occurrence of cerebral aneurysms as pro-matrix metalloproteinase-2. *Stroke* **29**(8): 1580–1583.

UCAS Japan Investigators, Morita, A., *et al.* (2012). The natural course of unruptured cerebral aneurysms in a Japanese cohort. *N Engl J Med* **366**(26): 2474–2482.

Vlak, M.H., Algra, A., *et al.* (2011). Prevalence of unruptured intracranial aneurysms, with emphasis on sex, age, comorbidity, country, and time period: A systematic review and meta-analysis. *Lancet Neurol* **10**(7): 626–636.

Welch, J.S., Ricote, M., *et al.* (2003). PPARgamma and PPARdelta negatively regulate specific subsets of lipopolysaccharide and IFN-gamma target genes in macrophages. *Proc Natl Acad Sci USA* **100**(11): 6712–6717.

Wiebers, D.O., Whisnant, J.P., *et al.* (2003). Unruptured intracranial aneurysms: Natural history, clinical outcome, and risks of surgical and endovascular treatment. *Lancet* **362**(9378): 103–110.

Intracranial Aneurysms: Imaging, Hemodynamics, and Remodeling

7

David Saloner

Department of Radiology and Biomedical Imaging,
University of California, San Francisco, San Francisco, California, USA

Aneurysms are localized dilatations resulting from weakening of the structure of the vessel wall that can, if left untreated, grow and rupture. Data on the prevalence of intracranial aneurysms are highly variable, with reports providing estimates ranging between 2 to 6% of the adult population (Chan *et al.*, 2016; Etminan & Rinkel, 2016; Juvela, 2011; Loewenstein *et al.*, 2012; Nasr & Brown, 2016; Rinkel *et al.*, 1998; Seibert *et al.*, 2011; Thien *et al.*, 2016; Wardlaw & White, 2000; Wills *et al.*, 2003). This variability reflects the challenges associated with the evaluation techniques used. The most sensitive *in vivo* imaging methods (catheter angiography) has a selection bias, as it is invasive and only used when there is a pre-existing clinical question (Beck *et al.*, 2006). Reports of prevalence data from other *in vivo* imaging methods, e.g., MR angiography, with a reported prevalence of 1.8% (Vernooij *et al.*, 2007), is likely an underestimate as it did not include the use of contrast agents, the most sensitive and reliable approach that is available for aneurysm detection by MRA (Thien *et al.*, 2016; Wang *et al.*, 2002). Similarly, retrospective autopsy studies reporting an incidence of approximately 4% (Juvela, 2011) face difficulties in providing an accurate assessment of vascular anatomy that comprehensively covers the intracranial circulation in non-pressurized blood vessels.

7.1 Origin and History of Intracranial Aneurysms

The anatomic origin and natural history of intracranial aneurysms vary. Aneurysms can be spherically shaped, such as saccular aneurysms (Figure 1) that commonly develop at the point of bifurcation of large cerebral arteries, or they can be fusiform aneurysms (Figure 2) that are extended concentric dilatations of the vessel wall. Intracranial aneurysms are predominantly saccular, and non-saccular aneurysms — fusiform, dolichoectatic, or dissecting — are rare with an incidence of less than 0.1% (Anson *et al.*, 1996). Fusiform aneurysms often originate from atherosclerotic lesions (Hademenos & Massoud, 1998; Nakayama *et al.*, 1999), and saccular aneurysms are the results of congenital defects of the media or focal degeneration of the internal elastic lamella (Byrne & Guglielmi, 1998; Sekhar & Heros, 1981). While the underlying pathophysiology is poorly understood (Krex *et al.*, 2001), genetic factors have been associated with systemic pathology of the vascular system and the formation of intracranial aneurysms. Arteries in the intracranial circulation lack the external elastic lamina, are generally surrounded by cerebrovascular fluid, and have less supportive perivascular tissue than in the extracranial circulation. Intracranial aneurysms often occur at

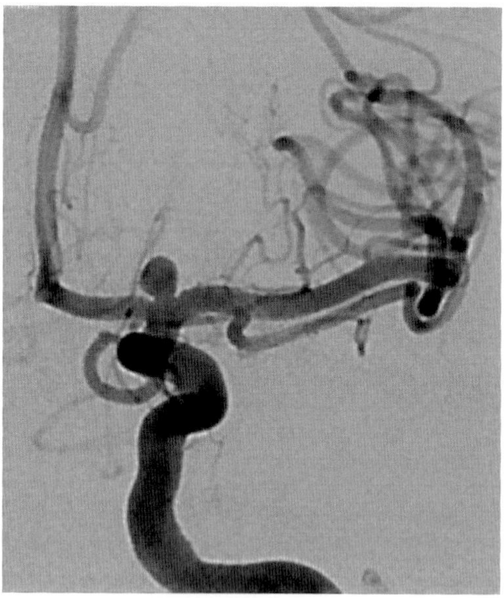

Figure 1. Catheter angiogram of a saccular aneurysm at the terminus of the internal carotid artery.

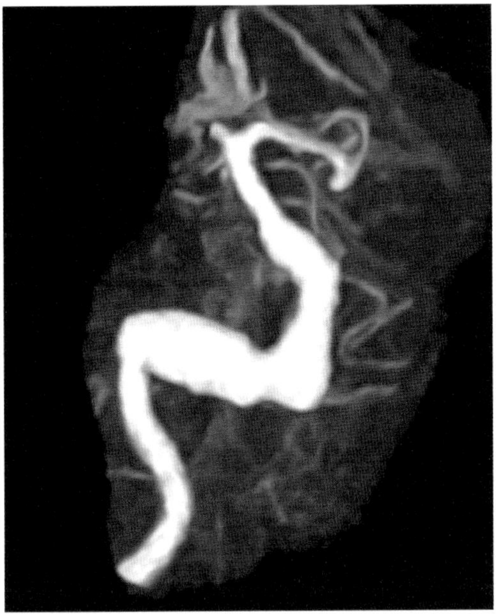

Figure 2. MR angiogram of a fusiform aneurysm of the vertebral artery extending into the basilar artery.

arterial branch points around the circle of Willis where there are structural variations in the collagen matrix (Finlay *et al.*, 1998; Rowe *et al.*, 2003). In addition, hemodynamics at locations of geometric complexity, such as at bifurcations, deviates from simple uniform laminar flow, and it is for this reason that hemodynamic factors are considered to be candidate descriptors for the development of abnormal morphology (Kondo *et al.*, 1997; Rowe *et al.*, 2003; Shojima *et al.*, 2004; Stehbens, 1990).

The most common location for saccular aneurysms is in the anterior circulation (90%) of the circle of Willis (Bonneville *et al.*, 2006). Of the aneurysms of the anterior circulation, those in the anterior communicating territory are most common (30–35%), followed by the internal carotid artery (30%). In the posterior circulation, aneurysms are most common at the basilar apex and represent approximately 10% of all intracranial aneurysms (Bonneville *et al.*, 2006).

Given the relationship of aneurysm size to rupture risk, as demonstrated in large clinical trials, categorization by aneurysm size is also common. Aneurysms are generally classified as small (< 10 mm), large (10–25 mm), or giant (>25 mm), with measurements typically made on the basis of maximal diameter assessment.

Aneurysms smaller than 12 mm in dome size account for more than 75% of unruptured aneurysms (Wiebers *et al.*, 2003).

7.2 Treatment Options

The primary rationale to treat brain vascular malformations is prophylaxis against new or recurrent intracranial hemorrhage. Cerebral aneurysms account for roughly 10% of all strokes (Bederson *et al.*, 2000; Hademenos & Massoud, 1998; Rinkel *et al.*, 1998; Wardlaw & White, 2000). The 30-day mortality rate after hemorrhage is about 45%, with about half the survivors sustaining permanent brain damage. The detection rate is estimated at 10 per 100,000. Therefore, some 15,000 Americans per year will suffer aneurysmal subarachnoid hemorrhage (SAH). Of all the risk factors, aneurysm size is regarded as the best predictor for rupture, putatively because aneurysms experience larger wall strains as they grow (Humphrey & Na, 2002; Juvela *et al.*, 2000; Steiger *et al.*, 1989). However, until recently, reliable high-resolution non-invasive imaging of intracranial aneurysms was not available, and given that patients with diagnosed aneurysms typically undergo immediate treatment, there is little literature that exists on longitudinal monitoring of intracranial aneurysms, and reports on the prospective determination of the growth and progression to rupture of intracranial aneurysms are limited. Optimal management of patients with unruptured intracranial aneurysms, therefore, remains controversial (ISUIA Investigators, 1998).

The large multicenter International Study of Unruptured Intracranial Aneurysms (ISUIA) supplied data on the rupture rate of aneurysms (Wiebers *et al.*, 2003). A pivotal finding of the ISUIA study is that patients with unruptured aneurysms smaller than 7 mm in diameter are at lower risk if their aneurysm remains untreated than if they undergo surgical or endovascular repair. This dramatically alters treatment options for patients with aneurysms that are below the threshold for intervention, and such individuals can now be followed with serial imaging to determine whether their aneurysms remain stable in size over time.

On the other hand, the natural history of untreated giant intracranial aneurysms is dismal. There is a high risk of progressive neurological deficits from cerebral compression, hemorrhage, and distal thromboembolism, causing stroke or death (Kodama & Suzuki, 1982; McCormick & Acosta-Rua, 1970; Michel, 1974; Peerless *et al.*, 1990; Pia & Zierski, 1982). The mortality rate of untreated giant aneurysms reported by Peerless *et al.* was 68% after 2 years and 85% after 5 years (Peerless *et al.*, 1990). All survivors had marked neurological dysfunction. In one series, a third of the patients exhibited symptoms and signs of mass effect and neural compression reflecting the aneurysm location (Lawton & Spetzler, 1999).

Giant aneurysms in the anterior circulation produced pain and dysfunction of extraocular movement and vision, and in the posterior circulation produced cranial-nerve dysfunction, bulbar palsy, and limb weakness. It is therefore considered that the poor prognosis of patients with giant aneurysms warrants aggressive treatment.

In general, aneurysms in the anterior circulation are accessible enough to be managed with clipping. Aneurysms that have a well-defined neck are good candidates for transcatheter treatments such as stenting and/or coiling. Despite advances in the therapy of cerebral aneurysms (Bederson *et al.*, 2000), there is a small but significant number that cannot be adequately controlled by conventional surgical or endovascular techniques. This includes, for example, fusiform aneurysms of the basilar artery involving a number of critical perforators that must not be occluded, and neither clipping nor coiling is an attractive alternative. Clinicians are therefore extremely reluctant to treat giant fusiform aneurysms and will delay intervention until there is an imperative to act, such as loss of function secondary to mass effect.

It had previously been the standard of care to treat smaller aneurysms by endovascular coiling or surgical clipping, depending on anatomy and patient-specific factors. As mentioned above, the results of ISUIA demonstrate that aneurysms smaller than 7 mm in size do not show a benefit from treatment, and there is now a trend to follow these patients with imaging, provided the aneurysms do not bleed.

7.3 Geometric Morphology of Intracranial Aneurysms

While many intracranial aneurysms have shapes that are approximately spherical, many do not, and even if they do, there are many factors suggesting that characterizing all aneurysms based on a diameter measure alone is a major simplification of the risk of future growth and/or rupture. While the neck of the aneurysm is an important descriptor in determining whether a given aneurysm might be able to retain coils that might be endovascularly deployed to promote thrombo-occlusion of the aneurysmal sac, the neck is also important in determining the inflow conditions into the aneurysm, and the pressure field within the sac. Aneurysms with identical shapes might be subject to very different biomechanical forces if they are sited at different locations relative to the parent vessel, e.g., a sidewall aneurysm versus an aneurysm at the tip of a bifurcation. While much less common in intracranial aneurysms than in, e.g., aneurysms of the abdominal aorta, the presence of intraluminal thrombus alters the relationship of endothelial cells to the media of the wall, and therefore modifies the relationship between hemodynamics and aneurysm progression in ways that have not yet been determined. Furthermore, the biomechanical stress and strain forces in the wall are highly sensitive to local features of the wall such as wall thickness. The paucity of information in this regard is related to the inability of conventional catheter-based angiographic methods to depict extra-lumenal features. Until recently, all *in vivo* imaging methods assess intracranial aneurysmal disease in terms of the lumenal morphology, and this has been a major restriction in determining those features of the true disease process, namely, the morphology and composition of the vessel wall, that drive aneurysm progression and rupture. Over the last few years, novel MRI methods have been developed that now provide the spatial resolution to identify features of the aneurysm wall, and the response of the wall to contrast agents, which could provide an insight into processes such as inflammation that are relevant to disease progression. Despite these important advances, it is improbable that, in the foreseeable future, any *in vivo* imaging modality will provide the kind of resolution needed to provide direct visualization of key structural elements of aneurysmal disease such as density, orientation, and status of the constituent collagen fibers.

Given the wide variety of morphologic descriptors, investigators have attempted to utilize shape descriptors available from 3D approaches and to include these in the analysis of likely aneurysmal risk factors. However, the findings from several investigations are inconsistent. A study of 109 unruptured and 105 ruptured aneurysms found that the aspect ratio — the ratio of the height of the aneurysm to the width of the neck — and the volume-to-neck ratio were most correlated with a risk of rupture (Ryu *et al.*, 2011), while another study that included 79 middle cerebral artery aneurysms (40 ruptured and 39 unruptured) also found that aspect ratio was correlated with rupture risk, and they determined that flow angle, and parent–daughter angle are more strongly associated with ruptured MCA aneurysms than size. Raghavan *et al.* (2005) found that the undulation index, a measure of the unevenness of the aneurysm surface, and the non-sphericity index, are strong indicators of aneurysm rupture risk. It is feasible that these two indices are surrogate markers of the presence of intraluminal thrombus.

There has been extensive investigation of the biomechanical forces that are present in the aneurysm wall. In particular, the curvature of the aneurysm and the thickness of the wall are key parameters in these considerations. Humphrey has advocated a Fluid-Solid-Growth model that accounts for both the hemodynamic forces from blood flow and the biomechanical forces in the constituents of the wall (Di Achille & Humphrey, 2012). As noted, an approach that uses generic information on aneurysms is inadequate for describing the patient-specific risk. However, individualized patient-specific estimation of rupture risk is limited by a lack of knowledge on key features of the vessel wall. Not only is it difficult to determine these features *in vivo*, histopathologic measures are also unreliable as the tissue is

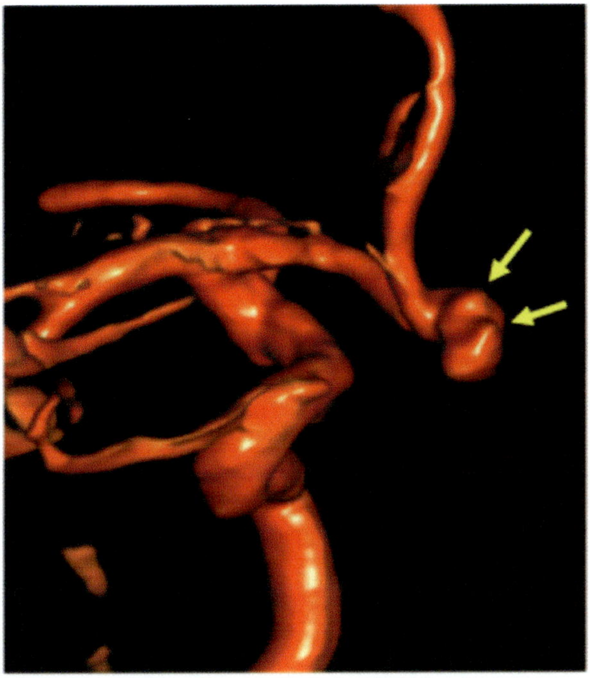

Figure 3. DSA of a giant basilar artery aneurysm showing blebs on the aneurysm.

no longer pressurized. Studies are inconsistent in the measured thickness of the wall between aneurysms and within single aneurysms. While some of this variation might be methodological, much of it is likely inherent variability within and between aneurysms. Despite this variability, most studies report that the wall is thinnest in the dome, and that the wall can be extremely thin at sites of a bleb (Frosen *et al.*, 2004; MacDonald *et al.*, 2000; Scott *et al.*, 1972; Suzuki & Ohara, 1978) (Figure 3). Aneurysm tissue resected at surgery and autopsy studies of unruptured aneurysms (MacDonald *et al.*, 2000; Scott *et al.*, 1972) report thicknesses that varied between 16 and 400 μm.

7.4 *In vivo* Imaging of the Lumen of Intracranial Aneurysms

The ability to monitor the degree and extent of aneurysmal disease is critically important in determining appropriate medical and/or surgical interventions. Several invasive and non-invasive imaging modalities have been developed to study the presentation of vascular disease. These include X-ray angiography (XRA), MRI, and multidetector CT (Angelini, 2001; Back *et al.*, 2000; Johnson *et al.*, 2000; Magarelli *et al.*, 1998), each of which has unique strengths and limitations. While ultrasound is an inexpensive and non-invasive modality, it has limited application in evaluating intracranial vessels given its inability to penetrate the skull. Consideration of *in vivo* imaging capabilities must be assessed relative to the physiologic and structural features of aneurysms that are of clinical relevance. These include: (1) the 3D contours of the vascular lumen; (2) the structure and composition of the vessel wall and intraluminal thrombus; (3) the physiological status of the wall such as inflammatory activity; (4) the relationship of the aneurysm to the adjacent brain parenchyma; (5) dynamics of flow through the aneurysm; and (6) the velocity field of intraluminal blood in 3D space and through the cardiac cycle. In addition, the expense and risk associated with each modality is also of importance.

2D projectional catheter-based X-ray digital subtraction angiography (DSA) has long been the gold standard for evaluation of the vascular lumen and provides the highest resolution imaging of aneurysmal disease with resolution approaching 0.2 mm. Rotational DSA is now widely available and offers multiple views of the anatomy, *albeit* with somewhat reduced resolution given the algorithms used to reconstruct the different views. While conventional display provides multiple projection views (either as maximum intensity projections or as surface-rendered volume displays), the data can also be reformatted in cross-sectional views resembling Computerized Tomography (CT) views. In DSA, selective injection of each arterial bed provides clear delineation not only of the aneurysm but also the feeding and draining arteries, and the relation of the aneurysm to the parent vessel. Furthermore, visualization of the dynamics of contrast wash-in and wash-out provide insights into flow dynamics, such as the identification of regions of relative stasis. In cases where endovascular treatment is possible, DSA benefits from the possibility of providing treatment in the same session as when diagnostic imaging is performed. Despite these advantages, DSA is limited in visualization of features of the vessel wall, only providing some ability to determine the presence of calcifications. An example of this is in fusiform aneurysms that are often accompanied by extensive intralumenal thrombus not apparent on DSA. The burden of intralumenal thrombus can have significant impact in terms of mass effect and in the potential for thromboembolic events. Not only is DSA unable to assess the vessel wall, it also has no ability to determine flow velocities and requires invasive catheterization and the injection of iodinated contrast, and it is therefore undesirable as a monitoring modality for repeated studies.

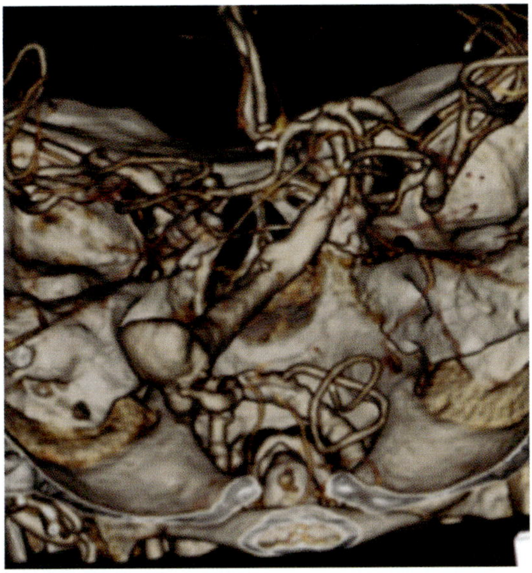

Figure 4. Surface rendered CTA of a fusiform aneurysm of the basilar artery showing the relationship of vessels to bony structures.

Modalities that offer a true 3D depiction of the vessel are appealing for delineating features of the lumen and the wall. This is currently routinely available with multidetector CT and with MRI. Multidetector CT provides a 3D overview of the vasculature that is relatively robust and is insensitive to artifacts from blood flow. Acquisition is rapid with extensive coverage from the origin of the great vessels at the arch and including the entire intracranial volume of interest possible in less than 10 seconds. In-plane resolution can be as low as 0.4 mm with 0.6 mm-thick acquired slices. Compared with MRI, CTA is particularly helpful in the setting of acute stroke where there is a need to simultaneously identify the presence of an aneurysm and rapidly determine whether there is a cerebral bleed in the patient with unknown contraindications to MRI, such as a pacemaker. Although CTA provides high-image quality, images are difficult to evaluate when the lesion is proximal to bone or when there is calcification in the wall that abuts the lumen. While bony structures can confound vascular visualization, including them in display is helpful in placing the lesion in its anatomic milieu (Figure 4). CTA has little ability to visualize the wall or differentiate the composition of intraluminal thrombus, and is also subject to artifact from pre-existing metal clips. Given the associated radiation dose and iodinated contrast agent effects, its appeal as a monitoring modality is reduced. Finally, CTA is also unable to provide information on flow velocities.

MRI has attractive features for the comprehensive evaluation of intracranial vascular disease and has the ability to assess multiple features of intracranial aneurysms. This includes angiographic methods that delineate the vascular lumen. In addition, MR can be adjusted to provide excellent contrast between different tissue types and to define the relationship of the aneurysm to the adjacent soft tissue. With appropriate settings, MR has recently been demonstrated to provide visualization of the aneurysm wall, and has a unique ability to provide a comprehensive map of the flow velocities of intraluminal blood. MR is also particularly well-suited for longitudinal studies, given that it does not require catheterization, use of iodinated contrast agents, or application of ionizing radiation. The major limitation of MRI is that it is a relatively slow modality. Not only does each imaging series require several minutes, and in some cases 10 minutes or longer, a comprehensive study requires time of the order of 45 minutes. There are contraindications to MRI such as pacemakers, and these studies are poorly tolerated by the claustrophobic patient. Furthermore, MR image quality is locally degraded by the presence of surgically placed metal clips. In many cases, MR images are acquired with modest resolution as increased resolution requires prohibitively lengthy acquisition times.

There are different approaches to achieving MR angiographic studies of intracranial vessels. MR acquisitions build up data for image formation by repeated sampling of the magnetization in a volume of interest. One class of techniques that can be used relies on the inflow of blood into the region of interrogation; it therefore enters with full magnetization strength that is in high contrast to signal from stationary tissue effectively suppressed by the serial magnetization sampling, which is inherent to the method. This method is termed time-of-flight (TOF) and can be performed as sequential 2D slices to build up a volumetric coverage, or, more effectively, as an explicit 3D volume acquisition. Although MR methods are comparatively slow, it is relatively easy to immobilize the head over the lengthy examination time (3 to 6 minutes). 3D imaging of the contours of the vascular lumen using MRA is now routinely obtained with good resolution (0.4 mm × 0.4 mm × 0.6 mm) at 3T. However, flow patterns in giant (and even relatively small aneurysms or ulcerations) often show strong recirculation patterns resulting in long residence time of blood in these structures with an associated saturation of magnetization and poor depiction on TOF imaging.

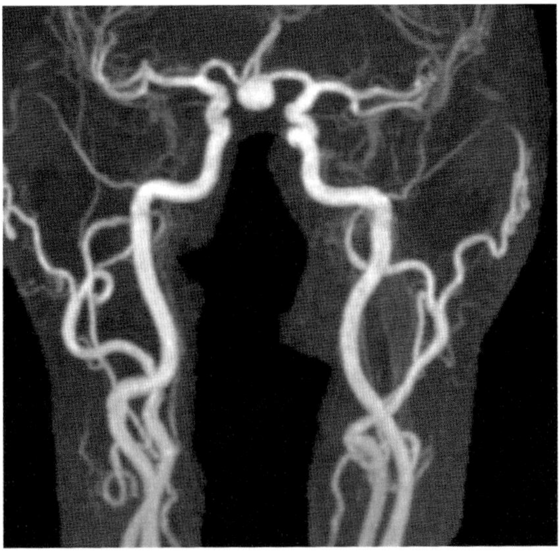

Figure 5. CE MR of an anterior communicating artery aneurysm.

In recent years there has been widespread use of gadolinium-based contrast agents for MR vascular studies(Aoki *et al.*, 2000; Huston *et al.*, 1999; Huston *et al.*, 2001; Kim *et al.*, 1998; Korosec *et al.*, 1999; Maki *et al.*, 1998; Prince *et al.*, 1993; Wilman *et al.*, 1998). The principal advantages of contrast-enhanced MRA (CE-MRA) are: (1) reduction of patient motion in the short acquisition time (<25 seconds) compared with TOF studies that take of the order of 3–6 minutes; (2) reduction of disordered flow-related effects with a short echo time of <2 ms; and (3) absence of saturation effects, which is of particular importance for aneurysms. The spatial resolution of CE-MRA is limited by the time window available during the passage of contrast agent and is of the order of 0.7 mm × 0.7 mm × 0.7 mm. Image quality is subject to bolus timing effects. In monitoring a known localized aneurysm, high spatial resolution acquisition, coincident with maximum contrast-enhancement of the blood is paramount (Figure 5). Acquisition timing is of particular importance in territories such as the aneurysms of the distal internal carotid artery that abut the venous drainage in the cavernous sinus. If acquisition of the arterial phase is delayed, contrast enhancement is noted in the cavernous sinus and complicates delineation of the boundary of aneurysms of the internal carotid artery.

7.5 *In vivo* Imaging of the Vessel Wall and the Velocity Field

Most investigators studying disease of the vessel wall have used methods that suppress blood signal, termed black blood methods, because of their good soft tissue signal properties and the fact that flow signal is eliminated and special techniques for retaining blood signal (as is needed for MRA) need not be invoked. A class of techniques, referred to as balanced steady-state methods, has shown promise for evaluation of the vessel wall but do not yield a black lumen impairing the contrast between wall and lumen. However, they can be applied in 3D mode and have good contrast to neighboring structures, particularly CSF. An approach that has recently been implemented and is very promising is a method that uses a train of variable flip angles to build up image data in tolerable time frames. These methods provide images with good soft tissue signal (generally utilized with T1-weighting) and have excellent black blood properties. They yield images with isotropic spatial resolution of 0.5mm in each direction and permit visualization of the vessel wall (Figure 6). The voxel size is comparable to the thickness of the aneurysm wall, and the extent to which these techniques are able to differentiate between aneurysms with different risk profiles has not yet been established. The role of these methods in guiding clinical decision-making in the management of intra-cranial aneurysms remains to be determined. Preliminary studies have been performed to assess the extent to which different aneurysms display enhancement following the administration of Gd-based contrast agents in the hope that this will be an indicator of the inflammatory burden and might correlate with the risk of aneurysm progression or rupture (Edjlali *et al.*, 2014).

Another class of contrast agents, the ultrasmall paramagnetic iron oxides (USPIOs), has also been studied. These agents are scavenged by activated macrophages and when concentrated in cells appear on many MR imaging sequences as regions of hypointensity. This signal change is best detected several days after USPIO administration as the agent has an extended intravascular half-life and the high vascular signal obscures all peri-vascular signal for 72 hours or longer (Hasan *et al.*, 2012).

MR imaging has demonstrated exciting capabilities in quantitating flow velocities *in vivo* (Drangova *et al.*, 1998; Markl *et al.*, 2003a, 2003b; Pelc, 1995; Sommer *et al.*, 1998). The method is not limited by anatomic features (such as bowel gas, depth of vessel or bony structures) as is Doppler ultrasound (DUS). MR velocimetry (MRV) methods do not have the temporal or spatial resolution of US, and these studies tend to require lengthy acquisition times (10 minutes or longer for 3D studies.) MRI is the only imaging modality that can provide cardiac cycle resolved flow velocities on a pixel-by-pixel basis and is the only variable

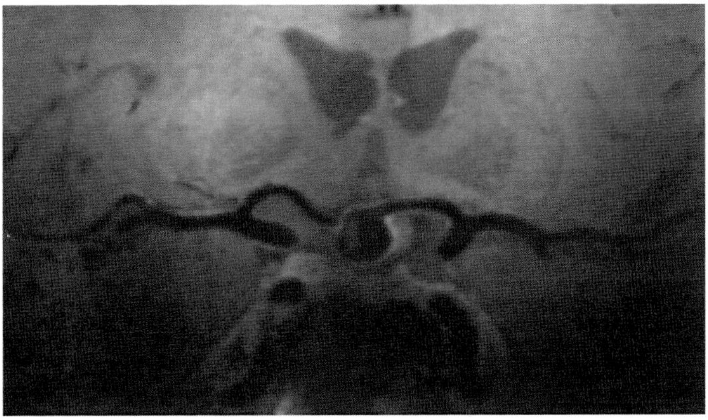

Figure 6. Black blood image of the wall of an anterior communicating artery aneurysm.

modality in applications such as measurement of flow in the intracranial circulation. Although enjoying relatively widespread use in a variety of clinical- and research-oriented studies, MRV studies are typically restricted to measuring a single component of the velocity vector. In quantitative applications, those measurements are made in a plane transverse to the target vessel, while qualitative visualization is often obtained in the longitudinal plane. There has been a rapid increase in the popularity and availability of so-called 4D methods, i.e., methods that measure the full velocity vector in 3D through the pulsatile cycle (Kvitting *et al.*, 2004; Markl *et al.*, 2003). While these methods provide good visualization of the prevailing flow patterns, such as the presence and location of jets, rotational vortices, and regions of slowly recirculating flow (Figure 7), they have limited spatial and temporal resolution; their ability to provide valuable information on quantitative metrics such as wall shear stress (WSS) remains to be validated.

7.6 Computational Fluid Dynamics

The flow of blood through vessels of known geometry can be numerically determined. These methods use the principles of physics that govern this flow, in much the same way as Newton's laws can be applied to determining the motion of solids. The governing equations for fluid flow are referred to as the Navier–Stokes equations, and methods for solving these are termed Computational Fluid Dynamics (CFD). In general, CFD methods require a prescription of the geometric and physiologic boundary conditions for the vasculature of interest. Specifically, what

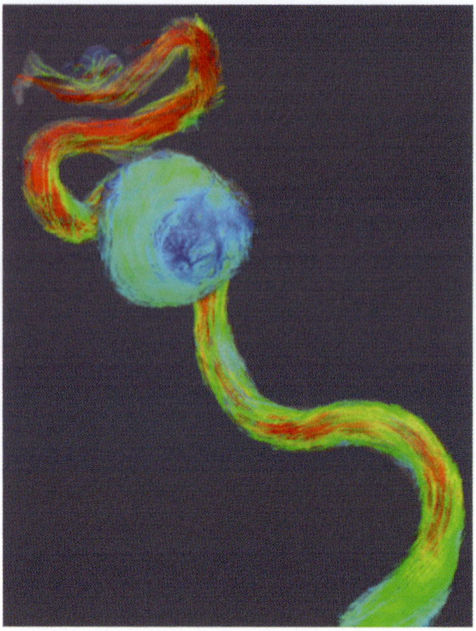

Figure 7. Depiction of velocity streamlines in a saccular aneurysm of the internal carotid artery.

is needed are the contours of the surface of the vascular lumen, the flow waveform at the inlet to the vasculature, and the pressure conditions in the outlet vessels (although in practice, in the intracranial vessels where there are no reflected waves, it is sufficient to prescribe a so-called zero pressure outlet condition.) With those values, the velocity field at every point in the entire lumenal space can be calculated at each point in the cardiac cycle. With that velocity field, secondary descriptors, such as pressure differences through the lumen, particle residence times, WSS, and gradients of WSS in time and space, can be derived.

CFD methods have been used to investigate the initiation and progression of vascular disease given the identification of hemodynamics as an important contributor in the disease process. Hemodynamics has long been implicated in the pathogenesis and pathophysiology of vascular disease (Giddens *et al.*, 1993; Glagov *et al.*, 1988; Zarins *et al.*, 1983). Caro *et al*, 1971 proposed that atherosclerosis originated in low shear regions. Other pathogenic pathways have been proposed (Davies, 1995; Eliasziw *et al.*, 1994). Ku *et al.* (1985) introduced OSI, the oscillating shear index, as an important parameter in vascular disease progression. The influence of spatial variation of wall shear (WSSG, wall shear stress gradient) on vascular dysfunction has also been investigated (Lei *et al.*, 1995). Fluid shear stress can exert profound effects on endothelial phenotype, and many early events in atherogenesis are thought to be regulated by hemodynamic factors (Berger & Jou, 2000; Chen *et al.*, 1999; Davies *et al.*, 1995; Fisher *et al.*, 2001; Gimbrone *et al.*, 1997; Glagov *et al.*, 1988). Pathological values of WSS can significantly affect the biology and phenotype of endothelial cells (DePaola *et al.*, 1992; Fry, 1968; Gimbrone *et al.*, 1997; Langille *et al.*, 1986) as well as activate platelets (Kroll *et al.*, 1996; Strony *et al.*, 1993). Activated proteolytic enzymes may act to degrade the extracellular matrix (ECM) and consequently lead to vessel wall destabilization; the matix metalloproteinases are particularly important for aneurysms (Bruno *et al.*, 1998; Chyatte & Lewis, 1997; Connolly *et al.*, 1997; Gaetani *et al.*, 1999; Kim *et al.*, 1997; Todor *et al.*, 1998). The abnormalities appear to be in protease activity rather than in the substrates, e.g., collagens or elastin.

CFD methods have benefited from rapid improvements in computer hardware and software. While early studies were restricted to simulations of flow in idealized geometries, it is now routine to perform calculations of patient-specific tortuous vessels with bifurcations, with realistic pulsatile flow patterns (Jou, 2002; Jou, 1998; Jou & Saloner, 1998; Jou *et al.*, 1996; Perktold *et al.*, 1991; Berger & Jou, 2000; Stroud *et al.*, 2000, 2002;Antiga *et al.*, 2003; Cebral *et al.*, 2001, 2002; Jou *et al.*, 2003; Leuprecht *et al.*, 2002; Long *et al.*, 2000; Steinman, 2004; Steinman *et al.*, 2003, 2002) (Figure 8). Although approaches differ in detail, there are

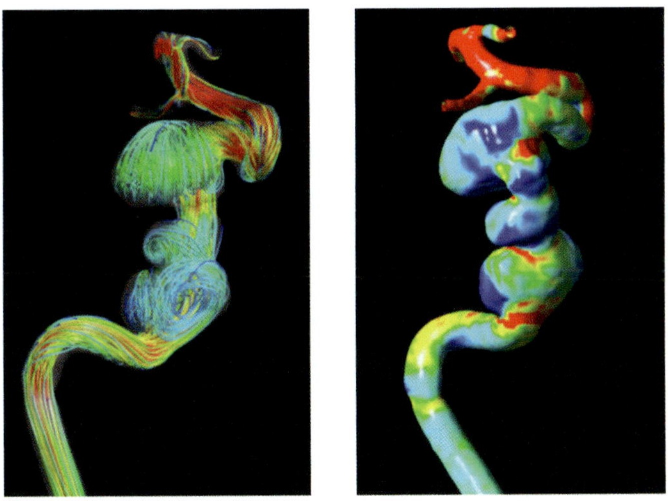

(a) (b)

Figure 8. CFD results for a tri-lobed aneurysm of the internal carotid artery. (a) Depiction of velocity streamlines, and (b) Calculated WSS.

commonalities. The principal steps taken in those approaches are to first segment out the vessels from a 3D image data set (MRA, DSA, or CTA) using thresholding and edge detection. A binary data set is created, and a mesh is generated to fit a "water-tight" surface onto the image data for the vessel surface. Depending on the specifics of the computational algorithm that is used, volume meshing is then performed to generate a grid throughout the vascular lumen. CFD calculations are then performed with suitable assumptions on inlet velocities (based on *in vivo* measurements of flow rates in the proximal arteries of the specific patient) and outlet pressures. Most models assume Newtonian flow. Some attempts have been made to implement *ad hoc* models for true blood viscosity, which is expected to be non-Newtonian in the slow recirculation in giant aneurysms. Typically, performing CFD studies in vessels with compliant walls that move through the cardiac cycle is substantially more challenging. Modeling the appropriate amount of distension is also challenging, as that requires detailed information on the elasto-mechanical behavior of the aneurysm. For CFD purposes, vessels are therefore generally considered to be rigid, and that decision is often justified by the observation that aneurysms have typically lost their elastin components and thus have little compliance.

Although commercial platforms are available for performing CFD simulations, substantial expertise is needed to ensure that meaningful results are obtained. Computation times can be extensive, although advanced hardware such as high performance computing clusters can be used to accelerate processing times. Validation of CFD models are a key component in determining the predictive capabilities of this approach. This is of particular importance when the simulations attempt to describe complex situations not previously well characterized. Experimental Fluid Dynamics measurements in well-defined vascular models that contain realistic characterization of physiological conditions can be directly compared with CFD studies made in identical conditions (Cebral *et al.*, 2002; Kohler *et al.*, 2001; Leuprecht *et al.*, 2002; Long *et al.*, 2002; Siegel *et al.*, 1997; Tang *et al.*, 2001; Taylor *et al.*, 1999; Zhao *et al.*, 2003). Those experiments provide a rigorous standard to which the numerical simulations can be compared. (Figure 9).

Laser Doppler velocimetry (LDV), which provides high spatial and temporal resolution (~10m, ~100 kHz) has been used to measure velocities with high precision in models of vascular disease. It is very tedious when used to map a velocity field, since the velocity can only be measured at a single point at one time. Particle image velocimetry (PIV) techniques are far more efficient because the velocity field over an entire plane of the model can be measured instantaneously (Adrian, 1991). However, PIV has lower spatial and temporal resolutions (>0.25 mm, and typically 15 Hz). Both LDV and PIV can only be used in transparent models. DUS and

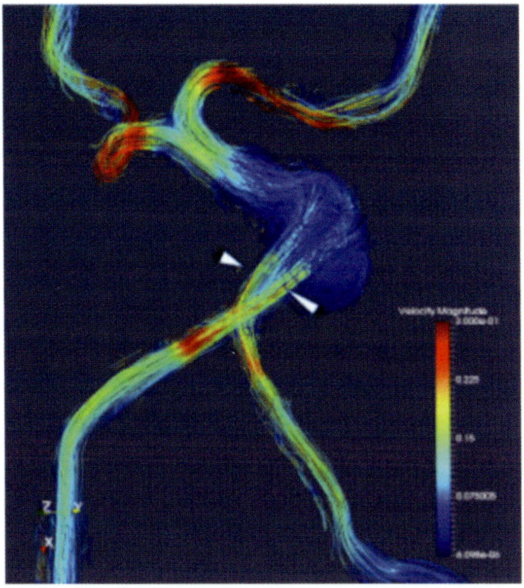

Figure 9. Streamline display of MR velocity field measured in an exact replica model of a basilar artery aneurysm.

magnetic resonance velocimetry (MRV) both have rather low spatial resolution compared with LDV and PIV (cf. 0.4 mm for DUS and 1 mm for MRV). However, DUS and MRV both have the advantage that they can be used *in vivo*, enabling direct comparison between *in vitro* and *in vivo* studies. Modeling or measuring wall and fluid shear stresses is challenging even in healthy blood vessels. Even more daunting is the characterization of shear stresses in the highly irregular lumen of a diseased blood vessel. Currently, no available technique can measure the real time gradient of velocities at the wall *in vivo*, and therefore shear stresses have been, at best, estimated.

7.7 Image-based Monitoring of Cerebral Aneurysms

Monitoring aneurysm features over time can be of great importance as it has been postulated that an increased rate of aneurysm growth could be indicative of an increased likelihood of rupture risk. One challenge for monitoring aneurysm size is the specific geometric metric that is used. Traditionally, as discussed above, the maximal luminal diameter has guided treatment decision-making. Establishing changes in that metric is particularly difficult for CA where the 2D projection views cannot be exactly reproduced in studies performed at different times, resulting in potential variations from changes in viewing angle. Little has been reported on the use of non-invasive 3D imaging methods for monitoring the progression of vascular disease. Different image analysis approaches can be pursued for this purpose depending on the question at hand. For stenotic disease, it is likely sufficient to be able to quantitate the severity of stenosis and monitor how that quantity changes over time. For aneurysmal disease, a similar approach would be to calculate and monitor the aneurysm volume.

Yasui *et al.* (1996) found that in patients with untreated intracranial aneurysms, the aneurysm diameters in a subset that went on to bleed had demonstrated growth from baseline. Furthermore, more than half of those aneurysms were smaller than 5 mm at initial assessment. This indicates that even given the low rate of rupture for small aneurysms found in ISUIA, that growth rates could be important. In a later study, the same group studied 166 unruptured aneurysms in 140 patients with CTA and identified growth in 6.4% of all aneurysms, with aneurysms of 2–4 mm, 5–9 mm, and 10-20 mm, having growth frequencies of 2.4, 9.1, and 50, respectively(Matsubara *et al.*, 2004). Villablanca *et al.* (2013) followed a cohort of subjects with untreated intracranial aneurysms with 3D CT angiography measurements. In this cohort, they reported that 18% of aneurysms of all sizes showed growth, where growth was defined on the basis of change in geometric factors that were over and above the measurement error (e.g., a change in aneurysm volume of greater than 5%.) This included a growth rate of 14% for aneurysms smaller than 7 mm. While they found that there is a twelve-fold higher risk of rupture for growing aneurysms (2.4% per patient-year) compared with stable aneurysms (0.2% per patient year), those results must be tempered by the small absolute number of ruptures in these two groups (3 in the group of growers, and 1 in the no-growth group.)

MR methods are well-suited to serial monitoring as they are minimally invasive and do not require catheterization, ionizing radiation, or the use of iodinated contrast agents. A retrospective study (Burns *et al.*, 2009) was performed on 165 patients with 191 UIAs who were followed with serial MR angiography, and it was discovered that after a median follow-up period of 47 months that 20 aneurysms (~10%) grew. Frequency of enlargement was 6.9, 25, and 83 for aneurysms <8 mm, 8–12 mm, and ≥13 mm, respectively. They found that the only independent predictor of enlargement was the original diameter of the aneurysm, and that other factors such as patient age or aneurysm location were not predictive of growth (Loewenstein *et al.*, 2012). Limitations of this study are that: (1) it included a mix of evolving methodologies (such as both 1.5T and 3T scanners); (2) it used

TOF methods that provide somewhat inconsistent results because of sensitivity to flow conditions; and (3) the metric used was a diameter measure (rather than a volumetric measure.)

In order to determine the relationship between hemodynamic descriptors and geometric change, it is important to have a clearer identification of where change occurs. This is conveniently done by co-registering the serial imaging studies and visually inspecting the images to identify geometric changes. (Figure 10) One co-registration algorithm approach is based on maximization of global normalized mutual information and compensates for volumes of different sizes (Studholme *et al.*, 1996). Affine transformation (rotation and translation) is used to co-register image sets. Another approach uses surface registration. Commercial software packages that are available, such as Rapidform™ (INUS Technology, Inc.), achieve co-registration by surface registration, again with an affine transformation. Rapidform™ uses an "Intelligent Registration" described as a "user-guided registration, fully automated global registration based on curvature or volume and rotary table registration." Since surface co-registration occurs after segmentation, this method can be prone to segmentation errors.

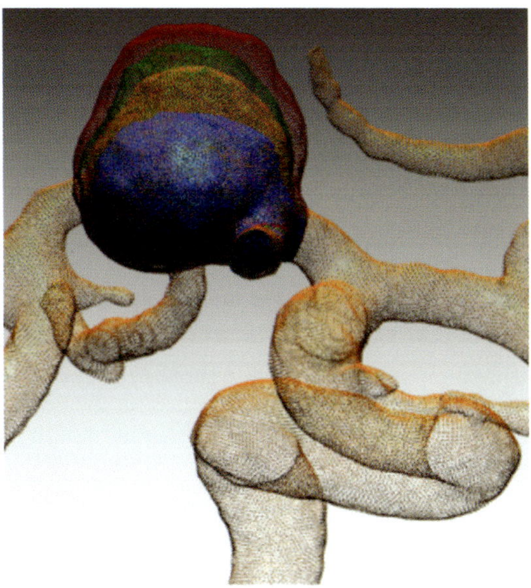

Figure 10. Co-registered serial studies of CE-MRA studies of an anterior communicating artery aneurysm showing incremental growth.

7.8 Relationship of Aneurysmal Changes to Hemodynamics

Many studies have indicated the possibility of a close relationship between hemodynamics and the initiation, growth, and rupture of intracranial aneurysms (Akkas, 1990; Burleson *et al.*, 1995; Burleson & Turitto, 1996; Fahrig *et al.*, 1999; Foutrakis *et al.*, 1999; Gonzalez *et al.*, 1992; Strother *et al.*, 1992; Turner *et al.*, 2001). With advances in vascular biology, the important role of shear stress — the frictional force generated by blood flowing over the endothelium — in vascular remodeling and angiogenesis is more clearly understood (Davies & Tripathi, 1993; Malek *et al.*, 1999). Aneurysm growth, therefore, is now seen as an active process (Burleson & Turitto, 1996), sharing the same mechanisms as normal vascular adaptation (Malek *et al.*, 1999). The relative importance of active remodeling and passive degradation, however, has not been determined with any degree of certainty. A major reason for this is the absence of methods for directly quantifying the biophysical forces on the aneurysm wall.

Another challenge in determining the specific links between hemodynamics and alterations in aneurysm progression is the broad range of hemodynamic descriptors that could be relevant. Biomechanical forces secondary to the morphology of the vascular territory of interest include absolute pressures, pressure drops, impingement forces that transfer momentum from fluid flow to the vessel wall, and frictional forces of blood flowing over the lumenal surface. All these forces have been referenced as modifiers of aneurysm progression. Furthermore these descriptors vary substantially in time (through the cardiac cycle) and space (on the aneurysm surface.) A candidate descriptor that has been investigated extensively is WSS, the near-wall gradient of velocities that reflects the force exerted by elements of the wall in bringing blood flowing through the vessel to zero velocity at the wall, to satisfy the no-slip condition. The endothelial cells that line the vessel wall are sensitive transducers of force. If WSS forces are outside a moderate range, endothelial cell signaling induces an upregulation of vessel wall degradation factors such as proteolytic enzymes that result in active vessel wall remodeling.

The pattern of intra-aneurysmal flow depends on the depth of the aneurysm and the width of the ostium. The location of inflow is affected by aneurysm geometry and relationship to the parent vessel. For lateral saccular aneurysms, inflow occurs at the distal ostium, forming a large vortex (Burleson *et al.*, 1995; Steiger *et al.*, 1987) that dominates the flow pattern in a region limited to less than 1.6 times the ostium width (Ujiie *et al.*, 1999), with little flow around the dome in larger aneurysms. Postulated mechanisms of intracranial aneurysm rupture are: excessive stretching of the aneurysmal wall (Awad & Barrow, 1995), resonance of wall motion and blood flow (Hung & Botwin, 1975; Mast & Pierce, 1995; Steiger & Reulen, 1986), or high wall stress (Hademenos & Massoud, 1998; Humphrey & Na, 2002; Sekhar & Heros, 1981). The majority of saccular aneurysms are speculated to rupture at the dome, not at the distal neck where the pressure and wall shear are at their maxima (Awad & Barrow, 1995). It is postulated that intra-aneurysmal flow impacts at the dome, and that this causes the rupture (Sekhar & Heros, 1981) (Gobin *et al.*, 1994; Strother, 1995); however, the exact cause of aneurysm rupture is still not clear. Since the size of an aneurysm is the only quantitative feature that can be measured on angiograms, it is still an accepted predictor for rupture. The mechanical behavior of an aneurysm is often simplified and described by Laplace's equation (Humphrey & Kyriacou, 1996; Steiger *et al.*, 1989), in which the aneurysm is modeled as a thin spherical shell undergoing uniform pressure. Finite element analysis permits the study of material asymmetry (Ryan & Humphrey, 1999), influence of aneurysm size and shape (Kyriacou & Humphrey, 1996), and stress distribution (Mower *et al.*, 1993).

It has also been demonstrated that not only arterial blood flow disturbance but hypertension in the brain vessels leads to increased mechanical stress on arterial walls (Swietaszczyk *et al.*, 2004). Prolonged elevation of blood pressure leads to excessive mechanical loading and causes remodeling of the arterial wall. The relationship between WSS and initiation of cerebral aneurysm formation has also been investigated in experimental models (Kondo *et al.*, 1997; Shojima *et al.*, 2004). The biomolecular responses to adverse flow conditions have also demonstrated that prolonged high WSS induces matrix metalloproteinase production and fragmentation of the internal elastic lamina at, or immediately adjacent to, the apex of bifurcations (Masuda *et al.*, 1999; Meng *et al.*, 2007; Shojima *et al.*, 2004). While the precise series of events is not completely understood, it is likely that not only the current biomechanical forces are of relevance but that extended exposure to adverse forces leads to a decrease in the structural integrity of the tissue and that this may be of importance in aneurysm formation and growth.

A CFD study on patients who underwent rotational 3D DSA studies and who had either ruptured or unruptured intracranial aneurysms has been reported (Cebral *et al.*, 2011, 2015; Sforza *et al.*, 2016). That study indicated that there was a tendency (not statistically significant) for ruptured aneurysms to have a greater number of intra-aneurysmal vortices than for unruptured aneurysms. It also found that there was a higher incidence of rupture in aneurysms where the CFD predicted a narrow impact zone. However, studies that include aneurysms that have already ruptured are of questionable validity because the post-rupture aneurysm likely has a different geometry (and hence predicted flow pattern) to that which existed prior to rupture. Nevertheless, establishment of this possible relationship would be extremely valuable — particularly in a prospective study and certainly warrants further investigation.

The UCSF group has provided data indicating that in patients with untreated aneurysms, aneurysms that grew had a large fraction of the surface area exposed to low WSS (Boussel *et al.*, 2008). Another result presented by the Tokyo group provides strong support to the hypothesis that low WSS might be a key parameter (Shojima *et al.*, 2004). They show, in a computational modeling study of 20 middle cerebral artery aneurysms, that WSS at the location of aneurysm rupture was lower than in unruptured aneurysms. Although speculative, it is not unreasonable that regions of low WSS are co-located with increased rates of growth; low WSS correlates with endothelial dysfunction, and the mechanotransduction of signals that lead to an overexpression of proteolytic enzmes, such as the matrix metallo-proteinases, leading to a potential for further weakening of the wall.

CFD methods have been used extensively for modeling flow conditions in idealized representations of vascular structures. Recently, there has been a greater effort to base these predictions on patient-specific conditions. This requires extraction of lumenal surfaces from imaging data and determination of flow velocities *in vivo*. Progression of disease is best monitored using co-registered longitudinal image data.

7.9 Summary

Intracranial aneurysmal disease is not uncommon, and it is likely that with the increasing utilization of non-invasive imaging methods, that the reported prevalence will increase. Advanced imaging methods provide new capabilities to provide a comprehensive overview of structural and functional features of these conditions. This will help to better understand the natural history of progression of disease, as current data indicate that many of these aneurysms might remain quiescent for extended periods. The challenge then is to find those features of aneurysms that will identify those that might grow more rapidly than others and those that are at increased risk of rupture, to permit improved staging of aneurysm treatment. CFD provides additional information on flow dynamics and the related hemodynamic biomechanical forces that might drive aneurysm progression. Coupling those capabilities with non-invasive imaging offers the potential for improvements in the management of intracranial aneurysms.

Acknowledgments

This research is supported by grants from the U.S. National Institutes of Health (R01 NS059944 and R01 HL114118) to D.S.

References

Akkas, N. (1990). Aneurysms as a biomechanical instability problem. In: Mosora, F. (ed.). *Biomechanical Transport Processes*. New York: Plenum Press.

Angelini, P. (2001). Is angiography the gold standard to establish the severity of a carotid lesion? Does duplex doppler ultrasound compete with it? *Catheter Cardiovasc Interv* **52**(1): 16–17.

Anson, J.A., Lawton, M.T., *et al.* (1996). Characteristics and surgical treatment of dolichoectatic and fusiform aneurysms. *J Neurosurg* **84**(2): 185–193.

Antiga. L., Ene-Iordache, B., *et al.* (2003). Computational geometry for patient-specific reconstruction and meshing of blood vessels from MR and CT angiography. *IEEE Trans Med Imaging* **22**(5): 674–684.

Aoki, S., Nakajima, H., *et al.* (2000). Dynamic contrast-enhanced mr angiography and mr imaging of the carotid artery: High-resolution sequences in different acquisition planes. *AJNR Am J Neuroradiol* **21**(2): 381–385.

Awad, I.A., Barrow, D.L. (1995). *Giant Intracranial Aneurysms*. Park Ridge, IL: AANS Publishers.

Back, M.R., Wilson, J.S., *et al.* (2000). Magnetic resonance angiography is an accurate imaging adjunct to duplex ultrasound scan in patient selection for carotid endarterectomy. *J Vasc Surg* **32**(3): 429–438.

Beck, J., Rohde, S., *et al.* (2006). Size and location of ruptured and unruptured intracranial aneurysms measured by 3-dimensional rotational angiography. *Surg Neurol* **65**(1): 18–25.

Bederson, J.B., Awad, I.A., *et al.* (2000). Recommendations for the management of patients with unruptured intracranial aneurysms: A statement for healthcare professionals from the Stroke Council of the American Heart Association. *Stroke* **31**(11): 2742–2750.

Berger, S.A., Jou, L.D. (2000). Flows in stenotic vessels. *Annu Rev Fluid Mech* **32**: 347–382.

Bonneville, F., Sourour, N., *et al.* (2006). Intracranial aneurysms: An overview. *Neuroimaging Clin N Am* **16**(3): 371–382.

Boussel, L., Rayz, V., *et al.* (2008). Aneurysm growth occurs at region of low wall shear stress: Patient-specific correlation of hemodynamics and growth in a longitudinal study. *Stroke* **39**(11): 2997–3002

Bruno, G., Todor, R., *et al.* (1998). Vascular extracellular matrix remodeling in cerebral aneurysms. *J Neurosurg* **89**(3): 431–440

Burleson, A., Strother, C., *et al.* (1995). Computer modeling of intracranial saccular and lateral aneurysms for the study of their hemodynamics. *Neurosurgery* **37**(4): 774–784.

Burleson, A.C., Turitto, V.T. (1996).Identification of quantifiable hemodynamic factors in the assessment of cerebral aneurysm behavior. On behalf of the subcommittee on biorheology of the scientific and standardization committee of the isth. *Thromb Haemost* **76**(1): 118–123.

Burns. J.D., Huston, J., 3rd *et al.* (2009).Intracranial aneurysm enlargement on serial magnetic resonance angiography: Frequency and risk factors. *Stroke* **40**(2): 406–411.

Byrne, J., Guglielmi, G. (1998). *Endovascular Treatment of Intracranial Aneurysms*. Berlin: Springer-Verlag.

Caro, C.G., Fitz-Gerald, J.M., *et al.* (1971). Atheroma and arterial wall shear. Observation, correlation and proposal of a shear dependent mass transfer mechanism for altherogenesis. *Proc R Soc Lond B Biol Sci* **177**(1046): 109–159.

Cebral, J.R., Duan, X., *et al.* (2015). Wall mechanical properties and hemodynamics of unruptured intracranial aneurysms. *AJNR Am J Neuroradiol* **36**(9): 1695–1703.

Cebral, J.R., Lohner, R., *et al.* (2001). Merging of intersecting triangulations for finite element modeling. *J Biomech* **34**(6): 815–819.

Cebral, J.R., Mut, F., *et al.* (2011). Quantitative characterization of the hemodynamic environment in ruptured and unruptured brain aneurysms. *AJNR Am J Neuroradiol* **32**(1): 145–151.

Cebral, J.R., Yim, P.J., *et al.* (2002). Blood flow modeling in carotid arteries with computational fluid dynamics and mr imaging. *Acad Radiol* **9**(11): 1286–1299.

Chan, D.Y., Abrigo, J.M., *et al.* (2016). Screening for intracranial aneurysms? Prevalence of unruptured intracranial aneurysms in hong kong chinese. *J Neurosurg* **124**(5): 1245–1249.

Chen, K.D., Li, Y.S., *et al.* (1999). Mechanotransduction in response to shear stress. Roles of receptor tyrosine kinases, integrins, and shc. *J Biol Chem* **274**(26): 18393–18400.

Chyatte, D., Lewis, I. (1997). Gelatinase activity and the occurrence of cerebral aneurysms. *Stroke* **28**(4): 799–804.

Connolly, E.S., Jr., Fiore, A.J., *et al.* (1997). Elastin degradation in the superficial temporal arteries of patients with intracranial aneurysms reflects changes in plasma elastase. *Neurosurgery* **40**(5): 903–908.

Davies, P.F. (1995). Flow-mediated endothelial mechanotransduction. *Physiol Rev* **75**(3): 519–560.

Davies, P.F., Barbee, K.A., *et al.* (1995). Hemodynamics and atherogenesis. Endothelial surface dynamics in flow signal transduction. *Ann N Y Acad Sci* **748**: 86–102.

Davies, P.F., Tripathi, S.C. (1993). Mechanical stress mechanisms and the cell. An endothelial paradigm. *Circ Res* **72**(2): 239–245.

DePaola, N., Gimbrone, M.A., Jr., *et al.* (1992). Vascular endothelium responds to fluid shear stress gradients. *Arterioscler Thromb* **12**(11): 1254–1257.

Di Achille, P., Humphrey, J.D. (2012). Toward large-scale computational fluid-solid-growth models of intracranial aneurysms. *Yale J Biol Med* **85**(2): 217–228.

Drangova, M., Zhu, Y., *et al.* (1998). *In vitro* verification of myocardial motion tracking from phase-contrast velocity data. *Magn Reson Imaging* **16**(8): 863–870.

Edjlali, M., Gentric, J.C., *et al.* (2014). Does aneurysmal wall enhancement on vessel wall MRI help to distinguish stable from unstable intracranial aneurysms? *Stroke* **45**(12): 3704–3706.

Eliasziw, M., Streifler, J.Y., *et al.* (1994). Significance of plaque ulceration in symptomatic patients with high-grade carotid stenosis. North american symptomatic caroti endarterectomy trial. *Stroke* **25**(2): 304–308.

Etminan, N., Rinkel, G.J. (2016). Unruptured intracranial aneurysms: Development, rupture and preventive management. *Nat Reviews Neurol.* **12**: 699–713

Fahrig, R., Nikolov, H., *et al.* (1999). A three-dimensional cerebrovascular flow phantom. *Med. Phys.* **26**(8): 1589–1599.

Finlay, H.M., Whittaker, P., *et al.* (1998). Collagen organization in the branching region of human brain arteries. *Stroke* **29**(8): 1595–1601.

Fisher, A.B., Chien, S., *et al.*, (2001). Endothelial cellular response to altered shear stress. *Am J Physiol Lung Cell Mol Physiol* **281**(3): L529–533.

Foutrakis, G., Yonas, H., *et al.* (1999). Saccular aneurysm formation in curved and bifurcating arteries. *AJNR* **20**: 1309–1317.

Frosen, J., Piippo, A., *et al.* (2004). Remodeling of saccular cerebral artery aneurysm wall is associated with rupture: Histological analysis of 24 unruptured and 42 ruptured cases. *Stroke* **35**(10): 2287–2293.

Fry, D.L. (1968). Acute vascular endothelial changes associated with increased blood velocity gradients. *Circ Res* **22**(2): 165–197.

Gaetani, P., Rodriguez y Baena, R., *et al.* (1999). Metalloproteases and intracranial vascular lesions. *Neurol Res* **21**(4): 385–390.

Giddens, D.P., Zarins, C.K., *et al.* (1993). The role of fluid mechanics in the localization and detection of atherosclerosis. *J Biomech Eng* **115**(4B): 588–594.

Gimbrone, M.A., Jr., Resnick, N., *et al.* (1997). Hemodynamics, endothelial gene expression, and atherogenesis. *Ann N Y Acad Sci* **811**: 1–10.

Glagov, S., Zarins, C., *et al.* (1988). Hemodynamics and atherosclerosis. Insights and perspectives gained from studies of human arteries. *Arch Pathol Lab Med* **112**(10): 1018–1031.

Gobin, Y., Counord, J., *et al.* (1994). *In vitro* study of haemodynamics in a giant aneurysm model: Influence of flow dynamics in the parent vessel and effects of coil embolization. *Neuroradiology* **36**: 530–536.

Gonzalez, C., Cho, Y., *et al.* (1992). Intracranial aneurysms: Flow analysis of their origin and progression. *AJNR Am J Neuroradiol* **13**: 181–188.

Hademenos, G., Massoud, T. (1998). *The Physics of Cerebrovascular Diseases*. New York: Springer-Verlag.

Hasan, D., Chalouhi, N., *et al.* (2012). Early change in ferumoxytol-enhanced magnetic resonance imaging signal suggests unstable human cerebral aneurysm: A pilot study. *Stroke* **43**(12): 3258–3265.

Humphrey, J., Kyriacou, S. (1996). The use of laplace's equation in aneurysm mechanics. *Neurol Res* **18**: 204–208.

Humphrey, J.D., Na, S. (2002). Elastodynamics and arterial wall stress. *Ann Biomed Eng* **30**: 509–523.

Hung, E.J., Botwin, M.R. (1975). Mechanics of rupture of cerebral saccular aneurysms. *J Biomechanics* 8: 385–392.

Huston, J., 3rd, Fain, S.B., *et al.* (1999). Carotid arteries: Maximizing arterial to venous contrast in fluoroscopically triggered contrast-enhanced mr angiography with elliptic centric view ordering. *Radiology* **211**(1): 265–273.

Huston, J., 3rd, Fain, S.B., *et al.* (2001). Carotid artery: Elliptic centric contrast-enhanced mr angiography compared with conventional angiography. *Radiology* **218**(1): 138–143.

ISUIA Investigators (1998). Unruptured intracranial aneurysms — risks of rupture and risks of surgical intervention. *N Engl J Med* **339**: 1725–1733.

Johnson, M.B., Wilkinson, I.D., *et al.* (2000). Comparison of doppler ultrasound, magnetic resonance angiographic techniques and catheter angiography in evaluation of carotid stenosis. *Clin Radiol* **55**(12): 912–920.

Jou. L. (2002). Flow reduction in stented saccular aneurysm. In: Batra, R.C., Henneke, E.G., (eds.). *Contemporary Research in Theoretical and Applied Mechanics, 14th United States National Congress of Theoretical and Applied Mechanics Proceedings*. Blacksburg, VA: Department of Engineering Science and Mechanics.

Jou, L.D., Berger, S.A. (1998). Numerical simulation of the flow in the carotid bifurcation. *Theoret Comput Fluid Dynamics* **10**: 239.

Jou L.D., Quick C.M., *et al.* (2003). Computational approach to quantifying hemodynamic forces in giant cerebral aneurysms. *AJNR Am J Neuroradiol* **24**(9): 1804–1810.

Jou L.D., Saloner, D. (1998). A numerical study of magnetic resonance images of pulsatile flow in a two dimensional carotid bifurcation: A numerical study of MR images. *Med Eng Phys* **20**(9): 643–652.

Jou, L.D., van Tyen, R., *et al.* (1996). Calculation of the magnetization distribution for fluid flow in curved vessels. *Magn Reson Med* **35**(4): 577–584.

Juvela, S. (2011). Prevalence of and risk factors for intracranial aneurysms. *The Lancet. Neurology* **10**(7): 595–597.

Juvela, S., Porras, M., *et al.* (2000). Natural history of unruptured intracranial aneurysms: Probability of and risk factors for aneurysm rupture. *J Neurosurg.* **93**(3): 379–387.

Kim, J.K., Farb, R.I., *et al.* (1998). Test bolus examination in the carotid artery at dynamic gadolinium-enhanced mr angiography. *Radiology* **206**(1): 283–289.

Kim, S.C., Singh, M., *et al.* (1997). Matrix metalloproteinase-9 in cerebral aneurysms. *Neurosurgery* **41**(3): 642–666.

Kodama, N., Suzuki, J. (1982). Surgical treatment of giant aneurysms. *Neurosurg Rev* **5**(4): 155–160.

Kohler, U., Marshall, I., *et al.* (2001). MRI measurement of wall shear stress vectors in bifurcation models and comparison with CFD predictions. *J Magn Reson Imaging* **14**(5): 563–573.

Kondo, S., Hashimoto, N., *et al.* (1997). Cerebral aneurysms arising at nonbranching sites. An experimental study. *Stroke* **28**(2): 398–403.

Korosec, F.R., Turski, P.A., *et al.* (1999). Contrast-enhanced MR angiography of the carotid bifurcation. *J Magn Reson Imaging* **10**(3): 317–325.

Krex, D., Schackert, H.K., *et al.* (2001). Genesis of cerebral aneurysms — an update. *Acta Neurochir* **143**(5): 429–448.

Kroll, M.H., Hellums, J.D., *et al.* (1996). Platelets and shear stress. *Blood* **88**(5): 1525–1541.

Ku, D.N., Giddens, D.P., *et al.* (1985). Pulsatile flow and atherosclerosis in the human carotid bifurcation. Positive correlation between plaque location and low oscillating shear stress. *Arteriosclerosis* **5**(3): 293–302.

Kvitting, J.P., Ebbers, T., *et al.* (2004). Flow patterns in the aortic root and the aorta studied with time-resolved, 3-dimensional, phase-contrast magnetic resonance imaging: Implications for aortic valve-sparing surgery. *J Thorac Cardiovasc Surg* **127**(6): 1602–1607.

Kyriacou, S.K., Humphrey, J.D. (1996). Influence of size, shape and properties on the mechanics of axisymmetric saccular aneurysms. *J Biomech* **29**(8): 1015–1022.

Langille, B.L., Reidy, M.A., *et al.* (1986). Injury and repair of endothelium at sites of flow disturbances near abdominal aortic coarctations in rabbits. *Arteriosclerosis* **6**(2): 146–154.

Lawton, M.T., Spetzler, R.F. (1999). Surgical strategies for giant intracranial aneurysms. *Acta Neurochir Suppl* **72**: 141–156.

Lei, M., Kleinstreuer, C., *et al.* (1995). Numerical investigation and prediction of atherogenic sites in branching arteries. *J Biomech Eng* **117**(3): 350–357.

Leuprecht, A., Perktold, K., *et al.* (2002). Combined cfd and mri study of blood flow in a human ascending aorta model. *Biorheology* **39**(3–4): 425–429.

Loewenstein, J.E., Gayle, S.C., *et al.* (2012). The natural history and treatment options for unruptured intracranial aneurysms. *Intl J Vasc Med* **2012**(2012): 898052.

Long, Q., Xu, X.Y., *et al.* (2000). Numerical study of blood flow in an anatomically realistic aorto-iliac bifurcation generated from mri data. *Magn Reson Med* **43**(4): 565–576.

Long, Q., Xu, X.Y., *et al.* (2002). Quantitative comparison of cfd predicted and mri measured velocity fields in a carotid bifurcation phantom. *Biorheology* **39**(3–4): 467–474.

MacDonald, D., Finlay, H., *et al.* (2000). Directional wall strength in saccular brain aneurysms from polarized light microscopy. *Ann Biomed Eng* **28**: 533–542.

Magarelli, N., Scarabino, T., *et al.* (1998). Carotid stenosis: A comparison between mr and spiral ct angiography. *Neuroradiology* **40**(6): 367–373.

Maki, J.H., Prince, M.R., *et al.* (1998). Optimizing three-dimensional gadolinium-enhanced magnetic resonance angiography. Original investigation. *Invest Radiol* **33**(9): 528–537.

Malek, A.M., Alper, S.L., *et al.* (1999). Hemodynamic shear stress and its role in atherosclerosis. *JAMA* **282**(21): 2035–2042.

Markl, M., Alley, M.T., *et al.* (2003). Balanced phase-contrast steady-state free precession (pc-ssfp): A novel technique for velocity encoding by gradient inversion. *Magn Reson Med* **49**(5): 945–952.

Markl, M., Chan, F.P., *et al.* (2003). Time-resolved three-dimensional phase-contrast mri. *J Magn Reson Imaging* **17**(4): 499–506.

Mast, T.D., Pierce, A.D. (1995). A theory of aneurysm sounds. *J Biomechanics* **28**(9): 1045–1053.

Masuda, H., Zhuang, Y.J., *et al.* (1999). Adaptive remodeling of internal elastic lamina and endothelial lining during flow-induced arterial enlargement. *Arterioscler Thromb Vasc Biol* **19**(10): 2298–2307.

Matsubara, S., Hadeishi, H., *et al.* (2004). Incidence and risk factors for the growth of unruptured cerebral aneurysms: Observation using serial computerized tomography angiography. *J Neurosurg* **101**(6): 908–914.

McCormick, W.F., Acosta-Rua, G.J. (1970). The size of intracranial saccular aneurysms. An autopsy study. *J Neurosurg* **33**(4): 422–427.

Meng, H., Wang, Z., *et al.* (2007). Complex hemodynamics at the apex of an arterial bifurcation induces vascular remodeling resembling cerebral aneurysm initiation. *Stroke* **38**(6): 1924–1931.

Michel, W. (1974). Posterior fossa aneurysms simulating tumors. *J Neurol Neurosurg Psychiatry* **37**: 218–223.

Mower, W., Baraff, L., *et al.* (1993). Stress distributions in vascular aneurysms: Factors affecting risk of aneurysm rupture. *J Surg. Res.* **55**: 155–161.

Nakayama, Y., Tanaka, A., *et al.* (1999). Giant fusiform aneurysm of basilar artery: Consideration of its pathogenesis. *Surg Neurol* **51**: 140–145.

Nasr, D.M., Brown, R.D., Jr. (2016). Management of unruptured intracranial aneurysms. *Curr Cardiol Rep* **18**(9): 86.

Peerless, S., Wallace, M., *et al.* (1990). Giant intracranial aneurysms. In: Youmans J. (ed.). *Neurological Surgery. A Comprehensive Reference Guide to the Diagnosis and Management of Neurological Problems*. Philadelphia: W.B. Saunders.

Pelc, N.J. (1995). Flow quantification and analysis methods. *Magn Reson Imaging Clin N Am* **3**(3): 413–424.

Perktold, K., Resch, M., *et al.* (1991). Three-dimensional numerical analysis of pulsatile flow and wall shear stress in the carotid artery bifurcation. *J Biomech* **24**(6): 409–420.

Pia, H.W., Zierski, J. (1982). Giant cerebral aneurysms. *Neurosurg Rev* **5**(4): 117–148.

Prince, M.R., Yucel, E.K., *et al.* (1993). Dynamic gadolinium-enhanced three-dimensional abdominal mr arteriography. *J Magn Reson Imaging* **3**(6): 877–881.

Raghavan, M.L., Ma, B., *et al.* (2005). Quantified aneurysm shape and rupture risk. *J Neurosurg* **102**(2): 355–362.

Rinkel, G.J., Djibuti, M., *et al.* (1998). Prevalence and risk of rupture of intracranial aneurysms: A systematic review. *Stroke* **29**(1): 251–256.

Rowe, A.J., Finlay, H.M., *et al.* (2003). Collagen biomechanics in cerebral arteries and bifurcations assessed by polarizing microscopy. *J Vasc Res* **40**(4): 406–415.

Ryan, J.M., Humphrey, J.D. (1999). Finite element based predictions of preferred material symmetries in saccular aneurysms. *Ann Biomed Eng* **27**: 641–647.

Ryu, C.W., Kwon, O.K., *et al.* (2011). Analysis of aneurysm rupture in relation to the geometric indices: Aspect ratio, volume, and volume-to-neck ratio. *Neuroradiology* **53**(11): 883–889.

Scott, S., Ferguson, G.G., *et al.* (1972). Comparison of the elastic properties of human intracranial arteries and aneurysms. *Can J Physiol Pharmacol* **50**(4): 328–332.

Seibert, B., Tummal,a R.P., *et al.* (2011). Intracranial aneurysms: Review of current treatment options and outcomes. *Front Neurol* **2**: 45.

Sekhar, L., Heros, R. (1981). Origin, growth, and rupture of saccular aneurysms: A review. *Neurosurgery* **8**: 248–260.

Sforza, D.M., Kono, K., *et al.* (2016). Hemodynamics in growing and stable cerebral aneurysms. *J Neurointerv Surg* **8**(4): 407–412.

Shojima, M., Oshima, M., *et al.* (2004). Magnitude and role of wall shear stress on cerebral aneurysm: Computational fluid dynamic study of 20 middle cerebral artery aneurysms. *Stroke* **35**(11): 2500–2505.

Siegel, J.M., Jr., Oshinski, J.N., *et al.* (1997). Computational simulation of turbulent signal loss in 2d time-of-flight magnetic resonance angiograms. *Magn Reson Med* **37**(4): 609–614.

Sommer, G., Corrigan, G., *et al.* (1998). Renal blood flow: Measurement *in vivo* with rapid spiral mr imaging. *Radiology* **208**(3): 729–734.

Stehbens, W.E. (1990). Pathology and pathogenesis of intracranial berry aneurysms. *Neurol Res* **12**(1): 29–34.

Steiger, H.J., Aaslid, R., *et al.* (1989). Growth of aneurysms can be understood as passive yield to blood pressure: An experimental study. *Acta Neurochir* **100**: 74–78.

Steiger, H., Poll, A., *et al.* (1987a). Basic flow structure in saccular aneurysms: A flow visualization study. *Heart Vessels* **3**: 55–65.

Steiger, H.J., Poll, A., *et al.* (1987b). Hemodynamic stress in lateral saccular aneurysms. *Acta Neurochir* **86**: 98–105.

Steiger, H.J., Reulen, H.J. (1986). Low frequency flow fluctuations in saccular aneurysms. *Acta Neurochir* **83**: 131–137.

Steinman, D.A. (2004). Image-based computational fluid dynamics: A new paradigm for monitoring hemodynamics and atherosclerosis. *Curr Drug Targets Cardiovasc Haematol Disord* **4**(2): 183–197.

Steinman, D.A., Milner, J.S., *et al.* (2003). Image-based computational simulation of flow dynamics in a giant intracranial aneurysm. *AJNR Am J Neuroradiol* **24**(4): 559–566.

Steinman, D.A., Thomas, J.B., *et al.* (2002). Reconstruction of carotid bifurcation hemodynamics and wall thickness using conputational fluid dynamics and mri. *Magn. Reson. Med.* **47**: 149–159.

Strony, J., Beaudoin, A., *et al.* (1993). Analysis of shear stress and hemodynamic factors in a model of coronary artery stenosis and thrombosis. *Am J Physiol* **265**(5 Pt 2): H1787–1796.

Strother, C. (1995). *In vitro* study of haemodynamics in a giant saccular aneurysm model: Influence of flow dynamics in the parent vessel and effects of coil embolization. *Neuroradiology* **37**: 159–161.

Strother, C., Graves, V., *et al.* (1992). Aneurysm hemodynamics: An experimental study. *AJNR Am J Neuroradiol* **13**: 1089–1095.

Stroud ,J.S., Berger, S.A., *et al.* (2000). Influence of stenosis morphology on flow through severely stenotic vessels: Implications for plaque rupture. *J Biomech* **33**(4): 443–455.

Stroud, J.S., Berger, S.A., *et al.* (2002). Numerical analysis of flow through a severely stenotic carotid artery bifurcation. *J Biomech Eng* **124**(1): 9–20.

Studholme, C., Hill, D.L., *et al.* (1996). Automated 3-d registration of MR and CT images of the head. *Med Image Anal* **1**(2): 163–175.

Suzuki, J., Ohara, H. (1978). Clinicopathological study of cerebral aneurysms. Origin, rupture, repair, and growth. *J Neurosurg* **48**(4): 505–514.

Swietaszczyk, C., Maciaczyk, J., *et al.* (2004). What is the origin of cerebral aneurysms?. *Przeglad lekarski* **61**(2): 115–119.

Tang, D., Yang, C., *et al.* (2001). Steady flow and wall compression in stenotic arteries: A three-dimensional thick-wall model with fluid-wall interactions. *J Biomech Eng* **123**(6): 548–557.

Taylor, C.A., Draney, M.T., *et al.* (1999). Predictive medicine: Computational techniques in therapeutic decision-making. *Comput Aided Surg* **4**(5): 231–247.

Thien, A., See, A.A., *et al.* (2016). Prevalence of asymptomatic unruptured intracranial aneurysms in a Southeast Asian population. *World Neurosurg.*

Todor, D.R., Lewis, I., *et al.* (1998). Identification of a serum gelatinase associated with the occurrence of cerebral aneurysms as pro-matrix metalloproteinase-2. *Stroke* **29**(8): 1580–1583.

Turner, C.L., Tebbs, S., *et al.* (2001). The influence of hemodynamic stress factors on intracranial aneurysm formation. *J Neurosurg.* **95**: 764–770.

Ujiie, H., Tachibana, H., *et al.* (1999). Effects of size and shape (aspect ratio) on the hemodynamics of saccular aneurysms: A possible index for surgical treatment of intracranial aneurysms. *Neurosurgery* **45**(1): 119–130.

Vernooij, M.W., Ikram, M.A., *et al.* (2007). Incidental findings on brain mri in the general population. *N Engl J Med* **357**(18): 1821–1828.

Villablanca, J.P., Duckwiler, G.R., *et al.* (2013). Natural history of asymptomatic unruptured cerebral aneurysms evaluated at CT angiography: Growth and rupture incidence and correlation with epidemiologic risk factors. *Radiology* **269**(1): 258–265.

Wang, M.C., Rubinstein, D., *et al.* (2002). Prevalence of intracranial aneurysms in first-degree relatives of patients with aneurysms. *Neurosurg Focus* **13**(3): e2.

Wardlaw, J., White, P. (2000). The detection and management of unruptured intracranial aneurysms. *Brain* **123**: 205–221.

Wiebers, D.O., Whisnant, J.P., *et al.* (2003). Unruptured intracranial aneurysms: Natural history, clinical outcome, and risks of surgical and endovascular treatment. *Lancet* **362**(9378): 103–110.

Wills S, Ronkainen A., *et al.* (2003). Familial intracranial aneurysms: An analysis of 346 multiplex finnish families. *Stroke* **34**(6): 1370–1374.

Wilman A.H., Riederer S.J., *et al.* (1998). Arterial phase carotid and vertebral artery imaging in 3d contrast-enhanced MR angiography by combining fluoroscopic triggering with an elliptical centric acquisition order. *Magn Reson Med* **40**(1): 24–35.

Yasui, N., Magarisawa, S., *et al.* (1996). Subarachnoid hemorrhage caused by previously diagnosed, previously unruptured intracranial aneurysms: A retrospective analysis of 25 cases. *Neurosurgery* **39**(6): 1096–1100.

Zarins, C.K., Giddens, D.P., *et al.* (1983). Carotid bifurcation atherosclerosis. Quantitative correlation of plaque localization with flow velocity profiles and wall shear stress. *Circ Res* **53**(4): 502–514.

Zhao, S.Z., Papathanasopoulou, P., *et al.* (2003). Comparative study of magnetic resonance imaging and image-based computational fluid dynamics for quantification of pulsatile flow in a carotid bifurcation phantom. *Ann Biomed Eng* **31**(8): 962–971.

Recent Advances in CADASIL Research

8

Suning Ping and Li-Ru Zhao

*Department of Neurosurgery,
State University of New York,
Upstate Medical University,
Syracuse, New York, USA*

Cerebral autosomal dominant arteriopathy with subcortical infarcts and leukoencephalopathy (CADASIL) is the most common form of hereditary disease leading to recurrent ischemic stroke and vascular dementia. CADASIL mainly affects middle-aged adults and leads to severe disability and mortality (Chabriat *et al.*, 1995; Joutel, 2011; Peters *et al.*, 2004a). This disease is caused by mutations of the *NOTCH3* gene, which results in progressive degeneration and loss of vascular smooth muscle cells (VSMCs) in small arteries (Joutel, 2011). Although the *NOTCH3* gene coding Notch3 receptor is not specifically expressed in the VSMCs of small arteries in the brain, the clinical symptoms and pathological changes of CADASIL are almost exclusively restricted to the brain. The affected blood vessels in the brain are seen in the pial arteries, small penetrating arteries and arterioles. Over the past two decades, numerous studies have revealed that granular osmiophilic material (GOM) deposits surrounding the VSMCs play a key role in the pathogenesis of CADASIL (Monet-Lepretre *et al.*, 2013). In addition, it has been proposed that impaired blood flow supply and endothelial cell (EC) dysfunction in the brain may also be involved in the pathological progression of this disease. Since the diagnostic approach for genetic identification of *NOTCH3* gene mutation was established in 2000 (Joutel *et al.*, 2000), the number of reported and diagnosed cases has rapidly increased worldwide. So far, however, there is no treatment that has been proven to stop or delay the progression of CADASIL. During the past two decades, major efforts have been made to explore CADASIL's possible pathological processes, while a few studies have targeted the development of treatment for restricting its progression. In this chapter, we provide an overview of CADASIL, including the discovery process, current understanding of its pathology and pathogenesis, clinical syndromes and neuroimaging features, and the advances in both basic research and clinical studies for this inherited small vascular disease.

8.1 Discovery

CADASIL is an inherited small vascular disease leading to recurrent ischemic stroke and vascular dementia. Clinical observation of a special type of stroke led to the discovery of this unique disease.

Globally, stroke is the leading cause of disabilities and the second leading cause of death in people over the age of 60 (Feigin *et al.*, 2014; Mozaffarian *et al.*, 2016). The majority of strokes worldwide are ischemic strokes (Andersen *et al.*, 2009; Gunarathne *et al.*, 2009; Steiner *et al.*, 2014; Tsai *et al.*, 2013, 2016). Generally, high blood pressure and other cardiovascular diseases are the main risk factors for stroke. In addition to ordinary stroke, over the past two decades, a genetic cause has been identified. This type of stroke is characterized by recurrent subcortical ischemia and progressive dementia occurring in young or middle-aged adults with no history of hypertension and/or cardiovascular diseases (Chabriat *et al.*, 1995; Joutel, 2011; Peters *et al.*, 2004a) and is recognized today as CADASIL. It has taken more than 20 years to identify this genetically caused form of ischemic stroke and vascular dementia.

The first case of "CADASIL" was reported in 1955 by van Bogaert (van Bogaert, 1955) who described this special case as "hereditary Binswanger's disease" with rapidly progressive subcortical encephalopathy and dementia. This case was later verified as CADASIL through genetic analysis of the original tissue samples (Tikka *et al.*, 2014). Thereafter, several unrelated families in different countries were found to manifest inherited stroke syndromes independently. Various names were employed to designate these CADASIL-suspended cases. In 1977, Sourander and Walinder reported a Swedish family with "multi-infarct dementia of autosomal dominant inheritance" (Sourander & Walinder, 1977). Stevens and colleagues later described a case of another Swedish family with similar clinical symptoms as "chronic familial vascular encephalopathy" (Stevens *et al.*, 1977). In 2007, however, the case reported by Sourander and Walinder was demonstrated as an erroneous attribution to CADASIL, because the DNA screening of the original brain samples showed no mutation within the entire exon encoding region of *NOTCH3* (Low *et al.*, 2007). In 1987, the first "CADASIL" family in Finland was identified, and the disease was named "hereditary multi-infarct dementia" (Sonninen & Savontaus, 1987). After the discovery of a genetic detection approach for CADASIL, genetic analysis for identifying patients was conducted in Finland. Nearly 100 patients and presymptomatic carriers were found in 15 families, indicating the high incidence of CADASIL in that country (Shahien *et al.*, 2011; Sonninen & Savontaus, 1987).

The breakthrough in CADASIL disease research occurred between 1993 and 1996. In 1993, Tournier-Lasserve and colleagues performed linkage analysis on two large unrelated French families with CADASIL and discovered the localization of the defective gene to chromosome 19q12 (Tournier-Lasserve *et al.*, 1993); it was in this report that the acronym CADASIL was used for the first time. Around 3 years later, using tissue samples from CADASIL patients, Joutel and colleagues identified a mutation in the NOTCH3 gene that was mapped to chromosome 19q13.1–13.26. Based on this observation, Joutel and colleagues proposed that the unique mutation in the *NOTCH3* gene could be crucially involved in CADASIL's pathophysiological mechanisms (Joutel *et al.*, 1996). This discovery provided a solid genetic principle for the diagnosis and pathogenic understanding of the disease.

8.2 Epidemiology

Using the genetic identification approach, researchers throughout the world have performed large genetic studies on CADASIL families and found a diversity of mutation rate in different ethnic groups. So far, the majority of CADASIL patients have been found in Caucasian families in the United Kingdom (Markus *et al.*, 2002), Italy (Dotti *et al.*, 2005), and Finland (Mykkanen *et al.*, 2004). There have also been some cases reported in East Asia and South America (Liao *et al.*, 2015; Vishnevetsky *et al.*, 2016). The first CADASIL case reported in North America was traced to 1997 by Hedera and Friedland (Hedera & Friedland, 1997). All four cases in two independent American families showed characteristics of CADASIL clinical symptoms, and dementia was a prominent and constant feature in all subjects (Hedera & Friedland, 1997). Based on a 2011 report, the incidence of CADASIL patients is less than 1/100,000 in Canada and the United States (Shahien *et al.*, 2011). Considering the high rate of stroke incidence, the reported cases of CADASIL are relatively sparse.

Epidemiological studies to date are incomplete, and the exact frequency and mortality rate worldwide is still uncertain. Although 61 has been reported as the average age of death (Dichgans *et al.*, 1998), women tend to have a longer disease duration time than men, but the mortality appears to be the same for both genders (Opherk *et al.*, 2004). It is worth noting that the number of reported cases of CADASIIL has rapidly increased over the years, which may be due to widely recognized clinical features and well-established genetic diagnosis methods for identifying the disease.

8.3 Pathology and Pathogenesis

8.3.1 *Unique features of genetic defects*

As mentioned earlier, the driven mutation gene of CADASIL disease is *NOTCH3*. The *NOTCH3* gene encodes a single-pass transmembrane receptor consisting of 2,321 amino acids (Joutel *et al.*, 1996). Notch3 receptor belongs to the evolutionary conserved Notch/LIN-12 family, which is composed of four receptors: Notch 1, 2, 3, and 4. The Notch3 receptor comprises an extracellular domain containing 34 epidermal growth factor repeats (EGFrs), three cysteine-rich Notch/Lin repeats, a single transmembrane domain, and an intracellular domain containing seven ankyrin repeats (Chabriat *et al.*, 2009). Notch3 is initially synthesized as a 280 kDa precursor, and then it undergoes proteolysis to form a mature protein. The mature Notch3 receptor is a heterodimeric transmembrane receptor, consisting of a 210 kDa extracellular domain (N3ECD) and a 97 kDa transmembrane intracellular domain (N3TMIC). The N3TMIC connects to the N3ECD in a non-covalent way (Joutel *et al.*, 2000). The NOTCH3 gene has 33 exons, while the CADASIL-related pathological mutations are all located in exons 2–24 encoding the N3ECD, with strong clustering in exon 3 and 4 encoding EGFr2-5. In these mutations, more than 95% are missense mutations. In addition, splice site mutations and small in-frame deletions are also found in some cases (Joutel, 2011). Remarkably, all these mutations lead to the odd number of cysteine residues in the EGFrs, and this unique mutation may be the cause of CADASIL (Joutel *et al.*, 1997). Given the 100% specificity of NOTCH3 mutation that changes the number of cysteine residues, DNA sequencing of exons 2–24 has become the gold standard for diagnosing CADASIL. Genetic testing is highly recommended for patients who have a typical clinical syndrome with characterized neuroimaging features, particularly if the patient has no hypertension or has a positive familial history (Tikka *et al.*, 2014).

8.3.2 *Pathology*

GOM is a significant pathological feature of CADASIL (Lewandowska *et al.*, 2011). Through electron microscopy (EM) examination of autopsies of CADASIL patients, GOM deposits were found around cerebral VSMCs of small arteries and arterioles (Baudrimont *et al.*, 1993; Ishiko *et al.*, 2006; Joutel *et al.*, 2000).

In 1993, using the EM approach to examine the brain autopsy of a CADASIL patient, Baudrimont and colleagues were the first to report the granular electron-dense extracellular material in the media of small and leptomeningeal arteries in the brain (Baudrimont *et al.*, 1993). In 1995, Ruchoux and colleagues first reported GOM deposits in small arteries of muscle and skin taken as biopsy samples from a CADASIL patient (Ruchoux *et al.*, 1995); the term GOM was used for the first time in this report. However, due to technological limitations, the components of GOM remained unclarified. In 2000, Joutel and colleagues revealed that the ectodomain of the Notch3 receptor accumulated at the cytoplasmic membrane of VSMCs in the blood vessels throughout the brain of CADASIL patients, and additionally, through immune-EM analysis, GOM deposits were also observed in the cytoplasmic membrane of VSMCs. Moreover, Joutel and colleagues pointed out that the deposited N3ECD was located close to but not within the GOM (Joutel *et al.*, 2000). In 2006, using a more sensitive method, immunogold EM, Ishiko and coworkers observed that N3ECD was actually a component of GOM deposits (Ishiko *et al.*, 2006), an observation later confirmed by others (Yamamoto *et al.*, 2013). Because the arteriopathy of CADASIL also occurs in dermal arterioles, GOM, the specific diagnostic feature of CADASIL, has also been found in skin biopsies (Joutel *et al.*, 2001). As GOM appears to be present in all CADASIL patients, EM analysis is considered a reliable diagnostic method. Due to the unavailability of EM examinations, immunohistochemical analysis of N3ECD deposits in skin biopsy was proffered as an alternative approach for assisting in the diagnosis (Ebke *et al.*, 1997).

Thus far, the mechanisms of GOM deposition remain unclear. One of the hypotheses raised by Joutel and colleagues was that the deposits may be due to the impairments in the clearance of accumulated N3ECD in CADASIL patients (Joutel *et al.*, 2000). Recently, however, it has been shown that other proteins are involved in GOM deposition. In 2013, Monet-Lepretre and colleagues revealed that two functionally important extracellular matrix (ECM) proteins, tissue inhibitor of metalloproteinase 3 (TIMP3) and vitronectin (VTN), were located close to the N3ECD (Monet-Lepretre *et al.*, 2013). In the same study, they used functional studies demonstrating that these two proteins are recruited, sequestered, and accumulated by mutated N3ECD, and that the TIMP3, VTN, and mutated N3ECD are formed as disulfide cross-linked detergent-insoluble aggregates in both CADASIL patients and transgenic mice. The effects of TIMP3 and VTN on the pathological progression of CADASIL have also been studied recently. Capone and colleagues (Capone *et al.*, 2016a) found that these two proteins are responsible for aspects of CADASIL phenotypes. Genetical knockdown or knockout of TIMP3 and VTN-ameliorated disease manifests in CADASIL transgenic mice. In addition, Kast and team members reported in 2014 that latent TGF-b binding protein 1 (LTBP-1) is a novel component of GOM deposits (Kast *et al.*, 2014).

In addition to GOM deposition, CADASIL's arteriopathy, another key feature of its pathology, is characterized by morphological alterations of VSMCs and loss of VSMCs in the small arterial wall (Ruchoux *et al.*, 1995). These observations are in agreement with an earlier report by Baudrimont and colleagues. who found in their studies, through hematoxylin and eosin (H&E) staining, striking lesions with concentric thickening of the arterial wall, hyaline degeneration of media, and fragmentation of the internal elastic lamella in the small and leptomeningeal arteries of a CADASIL patient's brain (Baudrimont *et al.*, 1993). A later study claimed that microscopic examinations of CADASIL patients' brain showed typical features of chronic small artery disease (Chabriat *et al.*, 2009). Cerebral arteries are characterized by thickening of the vessel wall, especially in small arteries from cerebral white matter. Fibrous connective tissues around the arteries accumulating in the tunica adventitia may lead to lumen stenosis or complete obliteration (Kalimo *et al.*, 2002). There are diffused lesions in hemispheric white matter. In some cases, microbleeds were observed in the occipital lobe, showing an atypical image for CADASIL (Trikamji *et al.*, 2016).

Besides GOM deposition and VSMCs degeneration, over the past decade, many other pathological changes have also been found in CADASIL patients, including cerebral cortex neurons and cerebral vascular ECs.

Emerging evidence shows that CADASIL also affects the central nervous system. Diffused lesions appear in the hemispheric white matter, and multiple small lacunar infarcts are found in the white matter and deep gray matter, as well as in the brain stem (Chabriat *et al.*, 2009). The belief has been that cortical neurons are unaffected by CADASIL; however, accumulating evidence has shown that neuron damage is also involved in the pathological progression of the disease. In 2006, Viswanathan and colleagues first reported the cortical neuronal apoptosis in autopsy brain tissues from CADASIL patients (Viswanathan *et al.*, 2006); this finding was confirmed in 2015 by Liu and coworkers in transgenic mice carrying human NOTCH3 mutant genes (TgNotch3^{R90C} mice), revealing apoptosis in the cortex of these transgenic mice (Liu *et al.*, 2015).

It has long been thought that CADASIL is a VSMC degeneration-related genetic disease. Over the past two decades, evidence accumulated from both clinical and basic medical research has shown that ECs are also involved in the pathological progression of CADASIL. In 1997, Ruchoux and Maurage reported the pathological changes of ECs in muscle and skin biopsies of CADASIL patients (Ruchoux & Maurage, 1997), observing that ECs from both capillaries and arterioles were irregularly shaped and that the cytoplasm of ECs was shrunken. ECs were irregularly outlined and had numerous processes extending into the blood vessel lumen. The EC nucleus was retracted and contained obvious nucleoli, and in addition, large cavities were found in some of the ECs. In 2015, Liu and colleagues observed the replacement of capillary ECs in the brain of $TgNotch3^{R90C}$ mice by bone marrow-derived progenitors, suggesting that endothelial lesion in the cerebral capillaries are involved in the pathological process of CADASIL (Liu *et al.*, 2015). However, the mechanisms of the interaction between VSMC degeneration and endothelial lesion remain undetermined.

8.3.3 *Pathogenesis*

Substantial efforts have been made to explore the pathogenesis of CADASIL in the past decade. However, research is still needed to help us understand how NOTCH3 mutation leads to small artery pathology and brain dysfunction. Here, we summarize studies from both clinical and basic medical research concerning mechanistic determination of the origin and development of CADASIL.

8.3.3.1 *Cerebral vasculature dysfunction*

Several studies have determined that GOM accumulates outside the VSMCs in the small arteries of CADASIL patients. However, the mechanisms of how the deposition leads to VSMC degeneration and vasculature impairments in CADASIL patients remain largely unknown. It has been proposed that the GOM deposits may cause the thickening of the vessel wall and ultimately lead to complete obliteration of arterioles (Capone *et al.*, 2016b), but how NOTCH3 mutation causes small arterial wall degeneration, vascular dysfunction, and brain dysfunction in CADASIL is a subject that needs further exploration.

Numerous studies using positron emission tomography (PET) or magnetic resonance imaging (MRI) have revealed that dysfunction of the cerebral vasculature occurs in CADASIL patients. Abnormal changes in cerebral blood flow (CBF) and metabolism have been observed in the white matter of CADASIL patients during the early stage of the disease (Tuominen *et al.*, 2004). Tuominen and colleagues observed the abnormal changes in early adulthood of CADASIL patients, including significantly decreased CBF and low regional cerebral metabolic rates of glucose in both frontal and occipital white matter. They also noted that decreased CBF was significantly associated with the severity of the disease (Tuominen *et al.*, 2004). In addition, Joutel and colleagues also found decreased resting CBF in the white matter of mice carrying CADASIL-related NOTCH3 mutations (Joutel *et al.*, 2010). These studies suggest that persistent subcortical ischemia occurs in CADASIL (Joutel *et al.*, 2010; Tuominen *et al.*, 2004). It has been proposed that the persistent subcortical ischemia may lead to the recurrent lacunar infarcts and microstructural alterations that correlate with cognitive decline (Chabriat *et al.*, 2009). Additionally, the impaired cerebral hemodynamics may be the result of both structural and functional impairments in brain arteries (Chabriat *et al.*, 2009). In the white matter, the stenosis or even the complete obliteration of small arteries or arterioles are most likely the dominant cause of ischemic lesions (Tikka *et al.*, 2014). However, if the walls of the arterioles are thick and covered with intensive fibrosis, and even if the arterioles are not stenotic, there is still a risk of occurring infarction. Due to the vascular fibrosis and degeneration of the VSMCs, these arterioles gradually lose their compliance and auto-regulation (Miao *et al.*, 2006). Emerging functional studies have demonstrated that delayed increases of CBF response to vasodilation stimulus happen in CADASIL patients (Campolo *et al.*, 2012; De Maria *et al.*, 2014; Peters *et al.*, 2008; Pfefferkorn *et al.*, 2001).

In addition to the VSMC impairments, EC dysfunction appears to also be involved in the dysregulation of CBF. In 2008, Peters and colleagues conducted a clinical study in which L-arginine, a substrate for nitric oxide synthase in ECs, was given to CADASIL patients (Peters *et al.*, 2008), and they found that there were impairments occurring in the patients' cerebral hemodynamics. After L-arginine treatment, the vasoreactivity significantly increased. These findings suggest that the EC dysfunction occurs as a condition of CADASIL. This study has not only examined the hypothesis that EC injuries are involved in the disease's pathogenesis but has also provided a new therapeutic strategy for treating CADASIL.

Since a transgenic mouse model of CADASIL, TgNotch3^{R90C} mice, was established in 2003 (Ruchoux *et al.*, 2003), numerous studies using this mouse model have been performed. Vascular alterations seen in CADASIL patients have been partly modeled in TgNotch3^{R90C} mice (Ruchoux *et al.*, 2003). These CADASIL-like mice display impaired CBF before the structural changes in the small arteries, suggesting that the vascular dysfunction may be the initial event for CADASIL rather than the pathological changes in vascular structures. In 2015, Liu and colleagues revealed that the vascular density decreased in aged

TgNotch3^{R90C} mice compared with the age-matched wild-type controls; cerebral EC injury was also observed in the TgNotch3^{R90C} mice (Liu *et al.*, 2015). Further, Ghost and colleagues observed a breakdown in the blood–brain barrier (BBB) of TgNotch3^{R90C} mice (Ghosh *et al.*, 2015). These observations indicate that EC injuries occur in the brain of TgNotch3^{R90C} mice. A recent clinical study showed that the high levels of von Willebrand factor and low levels of endothelial progenitor cells and circulating progenitor cells were found in CADASIL patients' blood (Pescini *et al.*, 2016). Since von Willebrand is only expressed in the ECs, Pescini and coworkers proposed that the elevated levels of von Willebrand concentration in the blood of CADASIL patients might indicate EC injuries (Pescini *et al.*, 2016). In addition, EC dysfunction may lead to the multifocally breached BBB, resulting in multiple microbleeds seen as a common appearance in the MRIs of CADASIL patients (Salamon, 2014). Similar findings were observed by Ghosh and colleagues in TgNotch3^{R90C} mice (Ghosh *et al.*, 2015). By determining the extravasation of plasma proteins, including albumin and fibrinogen in brain parenchyma, they found that the protein deposition significantly increased in aged TgNotch3^{R90C} mice compared with the age-matched control group. These results suggest BBB impairments in aged TgNotch3^{R90C} mice. This finding is in line with a previous clinical study that free extracellular fluid (vasogenic edema) increased in the brain of CADASIL patients (Tikka *et al.*, 2014).

Recently in our laboratory, we revealed that capillary occlusion and endothelial injury coexist in the brain of TgNotch3^{R90C} mice. In this study, to track bone marrow-derived cells, the bone marrow cells isolated from UBC-GFP mice were transplanted to CADASIL transgenic mice. We noted that the transplanted bone marrow-derived cells (GFP-positive cells) co-express an endothelial marker (CD31) in the brain capillaries, suggesting that the ECs are bone marrow-derived. We also observed that some capillaries are infilled by the bone marrow-derived cells (GFP-positive cells). Through confocal imaging, these infilled bone marrow-derived cells were identified as the thrombosis occluding the capillaries. Importantly, the colocalization of the GFP^{+}/CD31^{+} ECs and thrombosis in the capillaries is frequently seen in TgNotch3^{R90C} mice. These findings suggest that the thrombosis may be a result of impairments in the endothelium.

8.3.3.2 *Impaired N3ECD processing in VSMCs and ECs*

Notch receptors are central components of a highly conserved signal transduction pathway that regulates cell fate in cells. They are initially synthesized as a single transmembrane protein, and then the protein undergoes proteolytic processing by furin-like convertase (Logeat *et al.*, 1998). This process is also called S1 cleavage, which results in the formation of a mature heterodimeric receptor that is transported to the cell surface. On the cell surface, ligand and receptor interaction is generally the initial step for signal transduction. In the Notch3 signaling pathway, there are five canonical ligands of the Delta–Serrate–Lag (DSL) type, including Jag1, Jag2, Dll1, Dll 3, and Dll 4 (Kandachar & Roegiers, 2012).

The Notch3 receptor is predominantly expressed in the VSMCs of small arteries (Joutel *et al.*, 2000, Villa *et al.*, 2001) and in the pericytes of brain capillaries (Joutel *et al.*, 2010). In physiological conditions, ECs are involved in the activation of Notch3 signaling in the VSMCs (Liu *et al.*, 2009). As a result, the Notch3 receptor is proteolytically processed. The released N3TMIC translocates into the nucleus, where N3TMIC further interacts with the transcription factor RBP-Jk and co-activators, and activates the transcription of target genes (Andersson & Lendahl, 2014).

In CADASIL research, the Notch3 signaling pathway has not been fully characterized. In 2004, Peters and colleagues reported that the NIH3T3 cells and VSMCs transfected with mutant human NOTCH3 gene showed reductions in S1 cleavage, indicating the impairments of N3ECD processing and trafficking to the membrane (Peters *et al.*, 2004b). This observation is in agreement with an earlier study by Karlstrom and colleagues (Karlstrom *et al.*, 2002). In their study, murine wild-type NOTCH3 or mutant NOTCH3 was transfected to HEK 293 cells. They found that there was a lower proportion of mutant Notch3 expressed in the S1-cleaved configuration, together with reduced amounts of mutant Notch3 on the cell surface, compared with the wild-type NOTCH3. Together, these findings suggest that processing, trafficking, and localization of the Notch3 receptor are impaired when the NOTCH3 gene mutates. However, the impaired processing, trafficking, and localization of mutant Notch3 receptor did not affect the intercellular pathway after Notch3 receptor activation (Joutel *et al.*, 2004). In addition, in 2006, using the method of transfecting human NOTCH3 gene harboring four different mutations to HEK293 and SH-SY5Y cells, Low and colleagues found similar results as reported by Joutel and coworkers (Low *et al.*, 2006). Thus, the defective Notch3 receptor in processing and trafficking appears to be not the main cause of CADASIL's pathogenesis.

Ligand–receptor interaction is the key step for activation of Notch signaling in VSMCs. There are five canonical ligands expressing on the ECs for binding Notch receptor: Jag1, Jag2, Dll1, Dll 3, and Dll 4. The binding of the Notch receptor to its ligand leads to the removal of Notch extracellular domain (NECD) and endocytosis of the NECD into the ECs. Endocytosis of NECD by ECs is an important process for keeping normal communication between VSMCs and ECs. Continuous regulation of NECD occurs after the internalization by the ECs, including mono-ubiquitylation and de-ubiquitylation processes (Andersson *et al.*, 2011). The balance between mono-ubiquitylation and de-ubiquitylation of NECD in the ECs leads to the regulation of NECD degeneration and post-endocytic sorting events for the Notch receptor (McGill *et al.*, 2009). After internalization by endocytosis, NECD undergoes degradation through the endocytic–lysosomal pathway (Vaccari *et al.*, 2008). These studies suggest that NECD internalization by ECs via endocytosis is responsible for the clearance of NECD. In addition, in 2012, Watanabe and colleagues reported that the disturbed process of degradation of mutant N3ECD on the cell surface was due to impairment of endocytosis. In this study, HEK293 cells stably expressing mutant NOTCH3 were co-cultured with NOTCH3 ligand Jag1-expressing HEK293 cells. Their findings showed that mutant N3ECD degraded slowly and that the vesicles of mutant N3ECD in Jag1-expressing cells also decreased when compared with the wild-type N3ECD (Watanabe-Hosomi *et al.*, 2012). In both CADASIL patients and transgenic mice carrying CADASIL-associated NOTCH3 mutations, accumulation of N3ECD on the VSMC surface is the most common pathological feature. Based on the findings from both clinical and basic biomedical research, we may hypothesize that there are impaired interactions between VSMCs and ECs, especially impairments in endocytosis and degradation process in ECs in CADASIL cases. While the mechanisms of the impairments in endocytosis and degradation process in ECs remain unknown to date, it is possible that these impairments may be caused by EC dysfunction. On one hand, it has been proposed that N3ECD may act as a platform for the abnormal accumulation of other proteins, such as TIMP and VTN, and, like a snowball effect, it forms a larger surface to aggregate more proteins (Monet-Lepretre *et al.*, 2013). On the other hand, mutant EGFrs caused by an odd number of cysteine residues may lead to the misfolded Notch3, resulting in aberrant multimerization with cross-linking by disulfide bonds (Tikka *et al.*, 2014). Due to the increased N3ECD volume or protein binding site coverage, the endocytosis of N3ECD may be affected in CADASIL cases. However, further studies need to be conducted to determine the pathogenic mechanisms of CADASIL.

8.3.3.3 *GOM deposition and impaired biological functions*

It has been proposed that abnormal accumulation and aggregation of TIMP3 and VTN by N3ECD may lead to GOM deposition outside the VSMCs in the brain of CADASIL patients (Monet-Lepretre *et al.*, 2013). The snowball effect of abnormal protein accumulation may result in the dysregulation of the normal protein biological function (Olzscha *et al.*, 2011). TIMP3 and VTN have multiple biological activities that are directly relevant to CADASIL pathogenesis. In the brain, TIMP3 acts as an inhibitor to matrix metalloproteinase; the dysfunction of TIMP3 may contribute to the small vascular pathology. Homeostasis of the ECM is maintained by a balance between matrix metalloproteinases and TIMP3 proteins. The matrix metalloproteinases degrade ECM components, while the TIMP3 proteins act as their endogenous inhibitors. Abnormal aggregations of TIMP3 in GOM suppress the function of metalloproteinases, resulting in vessel fibrosis. Fibrotic thickening of the arteriolar walls has been demonstrated in cerebral small arteries in CADASIL patients (Dong *et al.*, 2012). In addition, TIMP3 promotes anti-angiogenic activity (Qi *et al.*, 2003) and suppresses vascular endothelial growth factor-mediated angiogenesis and pericyte recruitment (Nagase *et al.*, 2006; Stratman *et al.*, 2010). These TIMP3-induced pathologic effects may cause the reduction in capillary density, which has been observed in a CADASIL mouse model (Liu *et al.*, 2015). Similarly, abnormal recruitment and sequestration of VTN may also impair its functions. VTN is known as an abundantly circulating plasma protein (Preissner *et al.*, 1996) and modulates the balance between fibrinolytic system and ECM homeostasis by binding and stabilizing plasminogen activator inhibitor-1. Through this modulation process, VTN regulates cell adhesion, signaling and cytoskeletal reorganization (Preissner & Reuning, 2011). However, more studies need to be performed to further test and confirm the function of TIMP3 and VTN in the development of CADASIL.

In total, accumulating evidence suggests that VSMC loss, fibrosis, and ECM protein deposition in small artery walls are the major pathology alterations of CADASIL. These pathological changes may cause persistent reduction in CBF and predominant damage in the white matter, resulting in brain lesions. However, there are still numerous questions concerning the pathogenesis of CADASIL that remain unanswered. Advancing knowledge about the mechanisms of CADASIL initiation and development will provide insightful directions when searching for specific strategies to stop or delay its pathological progression.

8.4 Clinical Syndromes and Neuroimaging Features

8.4.1 *Clinical syndromes*

The clinical symptoms of CADASIL have been described in several excellent reviews (Chabriat *et al.*, 2009, Tikka *et al.*, 2014). In this section, we summarize up-to-date knowledge about its clinical manifestations. Although they vary among individuals, these four symptoms are widely recognized as the most common.

8.4.1.1 *Migraine with aura*

More than one-half of CADASIL patients report experiencing a migraine with aura, and up to 90% suffer from migraine by the age of 40 (Guey *et al.*, 2015). Generally, migraine with aura is the first manifestation of CADASIL. The onset average age is around 30, and it may appear even before the age of 10 (Granild-Jensen *et al.*, 2009). The mean age of female CADASIL patients suffering from migraine with aura is 26, which is earlier than male patients who show symptoms at 36 (Chabriat *et al.*, 2009). According to the studies of Singhal and colleagues, the high concentration of serum homocysteine is closely related to the early age of this symptom onset (Singhal *et al.*, 2004). When the symptom is present, most attacks start with typical visual or sensory aura lasting 20–30 minutes then followed by headaches for a few hours. Other patients report atypical attacks with basilar, hemiplegic, or prolonged aura, and some with confusion, fever, meningitis, or coma (Feuerhake *et al.*, 2002; Schon *et al.*, 2003).

8.4.1.2 *Recurrent ischemic strokes*

Ischemic stroke and transient ischemic attacks are the most frequent manifestations of CADASIL, occurring in up to 85% of patients (Tikka *et al.*, 2014). The age of first-ever stroke varies among individuals. The youngest patient was only 11 years old (Granild-Jensen *et al.*, 2009), and the oldest CADASIL patient suffered the first stroke at the age of 70 (Watanabe *et al.*, 2012). The average age for occurring symptomatic ischemic stroke is around 40–50. The strokes are always subcortical and typically present as lacunar syndromes, such as sensory-motor deficits and dysarthria-clumsy hand syndrome (Chabriat *et al.*, 2009). Majority of CADASIL patients have recurrent strokes over several years (Chabriat *et al.*, 1995; Haddad *et al.*, 2015). It is worth mentioning that CADASIL's stroke feature is described as diffused lacunar infarcts; major strokes with large infarcts rarely happen in CADASIL patients.

8.4.1.3 *Mood disturbance and apathy*

One in five CADASIL patients have a mood disturbance (Joutel, 2011), the most common and severe manifestation of which is depression. The depression always alternates with a manic phase and is often mistaken for bipolar mood disorders until the typical CADASIL features are seen on MRI (Park *et al.*, 2014; Valenti *et al.*, 2011).

Apathy is another common symptom and is characterized by the absence of motivation together with decreased voluntary behavior. Apathy is present in almost 40% of CADASIL patients, independent of depression. In 2009, Reyes and colleagues proposed that apathy could be influenced by many factors, including age, severity of cognitive impairment, depression, and load of subcortical tissue lesions (Reyes *et al.*, 2012).

8.4.1.4 *Cognitive impairment and dementia*

Cognitive impairment and dementia are the second most frequent clinical manifestation of CADASIL after recurrent ischemic stroke. The affected age varies from early 30 to 70, and the patterns of cognitive impairments differ between young patients and patients over 60 years of age (Buffon *et al.*, 2006; Charlton *et al.*, 2006). The major heterogeneous symptoms that are seen in CADASIL patients younger than 50 include impairments in executive functions, attention, and memory (Dichgans, 2009). However, past the age of 60, the cognitive profile of CADASIL patients is homogeneous with significant deficits in all-cognitive aspects. Dementia occurs in almost one-third of patients, and always occurs at the end stage of the disease. During the years preceding death, dementia is often accompanied by motor disability, pseudobulbar palsy, and sphincter incontinence (Chabriat *et al.*, 2009).

8.4.2 *Neuroimaging features*

Neuroimaging features of CADASIL have been well described with MRI, which has been used as a reliable approach for diagnosing CADASIL, particularly in patients with misleading manifestations. In neuroimaging, the earliest and most frequent change is leukoaraiosis (also named white matter hyperintensities, WMHs) (Salamon, 2014), observed as white matter areas of increased signal intensity on T2-weighted or fluid-attenuated inversion recovery MRI. Later, the WMHs become more diffuse and are located in the external capsule and the anterior part of the temporal lobes in a symmetrical manner. In addition to WMHs, another diagnostic-valued image are the lacunar infarcts that appear on T1-weighted MRI as punctiform or large areas of decreased signal (Chabriat *et al.*, 1998; Herve *et al.*, 2009).

CADASIL is a severe genetic disease affecting both young and middle-aged populations. Generally, patients initially present with migraines with aura followed by three other common symptoms: recurrent ischemic stroke, mood disturbance, and dementia. Recurrent ischemic stroke and vascular dementia are the most frequently occurring symptoms. WMHs and lacunar infarcts represent unique MRI features for CADASIL. The characteristics of the special clinical manifestations and unique MRI neuroimaging provide valuable information in diagnosing CADASIL, but genetic screening is still the gold standard for diagnosis.

8.5 Treatment

To date, no treatment has been proven effective to delay or cure CADASIL, and only symptomatic relief medications are available for patients.

The first recommendation for CADASIL patients is to control the factors that may worsen the clinical manifestations. Maintaining blood pressure appears to be important. Vasoconstrictors such as ergot derivations and triptans need to be avoided as these may increase stroke risk factors (Chabriat & Bousser, 2008). Any treatment that risks reducing blood pressure must be used with care because it may decrease cerebral perfusion. "No smoking" is highly recommended, since there is evidence that smoking increases the risk of stroke and dementia in patients with CADASIL (Adib-Samii *et al.*, 2010).

The current strategy for the clinical management of CADASIL is symptomatic relief. If the patient has severe migraines with aura, prophylactic drugs such as antiepileptic drugs or beta-blockers are used to prevent seizures from occurring (Forteza *et al.*, 2001; Weller *et al.*, 1998). In the case of hypercholesterolaemia, statins are given because their prevention effect in vascular disease is well recognized. In addition, anti-platelet drugs, instead of anticoagulants, are often prescribed because of the risk of intracerebral hemorrhage posed by anticoagulants (Tikka *et al.*, 2014). For CADASIL patients with cognitive decline, cholinesterase inhibitors are used to alleviate this symptom. In 2008, Dichgans and colleagues reported that donepezil (cholinesterase inhibitor)-treated patients showed moderate improvement in executive functions compared with the placebo-control group (Dichgans *et al.*, 2008).

In basic science research, only a few studies have explored the therapeutic strategies for CADASIL. Using TgNotch3^{R90C} mice, our research group recently demonstrated the efficacy of two hematopoietic growth factors, stem cell factor (SCF) and granulocyte-colony stimulating factor (G-CSF), in restricting pathological progression of CADASIL (Liu *et al.*, 2015). Our findings revealed that SCF in combination with G-CSF (SCF+G-CSF) improves spatial learning and memory in TgNotch3^{R90C} mice. SCF+G-CSF treatment also increases angiogenesis and neurogenesis, restricts CADASIL-associated degeneration of VSMCs in small arteries and pathological changes in cerebral ECs, and inhibits neural progenitor cell loss and cerebral apoptosis in the CADASIL mouse model. These findings suggest that SCF+G-CSF treatment delays the pathological progression of the disease.

Because CADASIL is an autosomal dominant genetic disease, preventing the expression of the mutant NOTCH3 may lead to controlling its pathogenesis. This hypothesis has been examined using two possible genetic approaches. The first silences the mutant gene with interfering RNA molecules, and the other mediates the skipping of the NOTCH3 exons harboring the defective DNA sequence. In 2012, Tikka and colleagues revealed the efficient silencing of NOTCH3 in cultured VSMCs originally derived from human umbilical artery (Tikka *et al.*, 2012). In 2016, Rutten and colleagues proposed a model for the selection of exon skip (Rutten *et al.*, 2016). In their *in vitro* studies, mutant NOTCH3 exons were specifically skipped without influencing mRNA transcription and protein expression in CADASIL patient-derived cerebral VSMCs. Their data also showed no destruction to the structure and function of Notch3 receptor. These genetic studies provide theoretical support for using genetic intervention to treat CADASIL. Whether this genetic approach also works *in vivo* remains to be addressed in future studies. Before performing clinical trials for these new approaches, there is still much work to do at the preclinical level for clarifying the mechanisms of and addressing safety concerns for the treatment.

8.6 Conclusion

Over the past two decades, there have been many breakthroughs and significant advances in the field of CADASIL research, including recognition of the disease; discovery of a defective gene; establishment of a genetic approach for diagnosis; and identification of major clinical manifestations, signature images in MRI, and pathological features. Although many questions remain unanswered, recent studies have contributed to a better understanding of CADASIL's pathogenesis and provided new insights into its pathogenic mechanisms. Impairments in CBF supply, BBB damage, and EC dysfunction are all involved in the pathogenesis of CADASIL, along with VSMC degeneration and GOM deposition. Still unclear, however, are how NOTCH3 mutation leads to VSMC degeneration, what interaction between VSMCs and ECs plays crucial roles in the degeneration and dysfunction of small arteries, and how GOM deposition affects neuron loss and cognitive decline.

Notably, in contrast to the major efforts that have led to a better understanding of the pathological features and possible pathogenic mechanisms of CADASIL, very little work has been done so far in developing treatment for this devastating disease. Future studies should more fully investigate the therapeutic interventions that will impede its pathological progression.

Acknowledgments

This research is supported by a grant from the American Heart Association (15GRNT25700284).

References

Adib-Samii, P., Brice, G., *et al.* (2010). Clinical spectrum of CADASIL and the effect of cardiovascular risk factors on phenotype: Study in 200 consecutively recruited individuals. *Stroke* **41**(4): 630–634.

Andersson, E.R. Lendahl, U. (2014). Therapeutic modulation of Notch signalling — are we there yet? *Nat Rev Drug Discov* **13**(5): 357–378.

Andersen, K.K., Olsen, T.S., *et al.* (2009). Hemorrhagic and ischemic strokes compared: Stroke severity, mortality, and risk factors. *Stroke* **40**(6): 2068–2072.

Andersson, E.R., Sandberg, R., *et al.* (2011). Notch signaling: Simplicity in design, versatility in function. *Development* **138**(17): 3593–3612.

Baudrimont, M., Dubas, F., *et al.* (1993). Autosomal dominant leukoencephalopathy and subcortical ischemic stroke. A clinicopathological study. *Stroke* **24**(1): 122–125.

Buffon, F., Porcher, R., *et al.* (2006). Cognitive profile in CADASIL. *J Neurol Neurosurg Psychiatry* **77**(2): 175–180.

Campolo, J., De Maria, R., *et al.* (2012). Impaired vasoreactivity in mildly disabled CADASIL patients. *J Neurol Neurosurg Psychiatry* **83**(3): 268–274.

Capone, C., Cognat, E., *et al.* (2016a). Reducing Timp3 or vitronectin ameliorates disease manifestations in CADASIL mice. *Ann Neurol* **79**(3): 387–403.

Capone, C., Dabertrand, F., *et al.* (2016b). Mechanistic insights into a TIMP3-sensitive pathway constitutively engaged in the regulation of cerebral hemodynamics. *Elife* **5**: e17536.

Chabriat, H., Bousser, M. G. (2008). Cerebral autosomal dominant arteriopathy with subcortical infarcts and leukoencephalopathy. *Handb Clin Neurol* **89**: 671–686.

Chabriat, H., Joutel, A., *et al.* (2009). Cadasil. *Lancet Neurol* **8**(7): 643–653.

Chabriat, H., Levy, C., *et al.* (1998). Patterns of MRI lesions in CADASIL. *Neurology* **51**(2): 452–457.

Chabriat, H., Vahedi, K., *et al.* (1995). Clinical spectrum of CADASIL: A study of 7 families. *Lancet* **346**(8980): 934–939.

Charlton, R.A., Morris, R.G., *et al.* (2006). The cognitive profiles of CADASIL and sporadic small vessel disease. *Neurology* **66**(10): 1523–1526.

De Maria, R., Campolo, J., *et al.* (2014). Effects of sapropterin on endothelium-dependent vasodilation in patients with CADASIL: A randomized controlled trial. *Stroke* **45**(10): 2959–2966.

Dichgans, M. (2009). Cognition in CADASIL. *Stroke* **40**(3 Suppl): S45–47.

Dichgans, M., Markus, H.S., *et al.* (2008). Donepezil in patients with subcortical vascular cognitive impairment: A randomised double-blind trial in CADASIL. *Lancet Neurol* **7**(4): 310–318.

Dichgans, M., Mayer, M., *et al.* (1998). The phenotypic spectrum of CADASIL: Clinical findings in 102 cases. *Ann Neurol* **44**(5): 731–739.

Dong, H., Blaivas, M., *et al.* (2012). Bidirectional encroachment of collagen into the tunica media in cerebral autosomal dominant arteriopathy with subcortical infarcts and leukoencephalopathy. *Brain Res* **1456**: 64–71.

Dotti, M.T., Federico, A., *et al.* (2005). The spectrum of Notch3 mutations in 28 Italian CADASIL families. *J Neurol Neurosurg Psychiatry* **76**(5): 736–738.

Ebke, M., Dichgans, M., *et al.* (1997). CADASIL: Skin biopsy allows diagnosis in early stages. *Acta Neurol Scand* **95**(6): 351–357.

Feigin, V.L., Forouzanfar, M.H., *et al.* (2014). Global and regional burden of stroke during 1990–2010: Findings from the Global Burden of Disease Study 2010. *Lancet* **383**(9913): 245–254.

Feuerhake, F., Volk, B., *et al.* (2002). Reversible coma with raised intracranial pressure: An unusual clinical manifestation of CADASIL. *Acta Neuropathol* **103**(2): 188–192.

Forteza, A.M., Brozman, B., *et al.* (2001). Acetazolamide for the treatment of migraine with aura in CADASIL. *Neurology* **57**(11): 2144–2145.

Ghosh, M., Balbi, M., *et al.* (2015). Pericytes are involved in the pathogenesis of cerebral autosomal dominant arteriopathy with subcortical infarcts and leukoencephalopathy. *Ann Neurol* **78**(6): 887–900.

Granild-Jensen, J., Jensen, U.B., *et al.* (2009). Cerebral autosomal dominant arteriopathy with subcortical infarcts and leukoencephalopathy resulting in stroke in an 11-year-old male. *Dev Med Child Neurol* **51**(9): 754–757.

Guey, S., Mawet, J., *et al.* (2015). Prevalence and characteristics of migraine in CADASIL. *Cephalalgia*.

Gunarathne, A., Patel, J.V., *et al.* (2009). Ischemic stroke in South Asians: A review of the epidemiology, pathophysiology, and ethnicity-related clinical features. *Stroke* **40**(6): e415–423.

Haddad, N., Ikard, C., *et al.* (2015). Recurrent status epilepticus as the primary neurological manifestation of CADASIL: A case report. *Epilepsy Behav Case Rep* **3**: 26–29.

Hedera, P., Friedland, R.P. (1997). Cerebral autosomal dominant arteriopathy with subcortical infarcts and leukoencephalopathy: Study of two American families with predominant dementia. *J Neurol Sci* **146**(1): 27–33.

Herve, D., Godin, O., *et al.* (2009). Three-dimensional MRI analysis of individual volume of Lacunes in CADASIL. *Stroke* **40**(1): 124–128.

Ishiko, A., Shimizu, A., *et al.* (2006). Notch3 ectodomain is a major component of granular osmiophilic material (GOM) in CADASIL. *Acta Neuropathol* **112**(3): 333–339.

Joutel, A. (2011). Pathogenesis of CADASIL: Transgenic and knock-out mice to probe function and dysfunction of the mutated gene, Notch3, in the cerebrovasculature. *Bioessays* **33**(1): 73–80.

Joutel, A., Andreux, F., *et al.* (2000). The ectodomain of the Notch3 receptor accumulates within the cerebrovasculature of CADASIL patients. *J Clin Invest* **105**(5): 597–605.

Joutel, A., Corpechot, C., *et al.* (1996). Notch3 mutations in CADASIL, a hereditary adult-onset condition causing stroke and dementia. *Nature* **383**(6602): 707–710.

Joutel, A., Dodick, D.D., *et al.* (2000). De novo mutation in the Notch3 gene causing CADASIL. *Ann Neurol* **47**(3): 388–391.

Joutel, A., Favrole, P., *et al.* (2001). Skin biopsy immunostaining with a Notch3 monoclonal antibody for CADASIL diagnosis. *Lancet* **358**(9298): 2049–2051.

Joutel, A., Monet, M., *et al.* (2004). Pathogenic mutations associated with cerebral autosomal dominant arteriopathy with subcortical infarcts and leukoencephalopathy differently affect Jagged1 binding and Notch3 activity via the RBP/JK signaling Pathway. *Am J Hum Genet* **74**(2): 338–347.

Joutel, A., Monet-Lepretre, M., *et al.* (2010). Cerebrovascular dysfunction and microcirculation rarefaction precede white matter lesions in a mouse genetic model of cerebral ischemic small vessel disease. *J Clin Invest* **120**(2): 433–445.

Joutel, A., Vahedi, K., *et al.* (1997). Strong clustering and stereotyped nature of Notch3 mutations in CADASIL patients. *Lancet* **350**(9090): 1511–1515.

Kalimo, H., Ruchoux, M.M., *et al.* (2002). CADASIL: A common form of hereditary arteriopathy causing brain infarcts and dementia. *Brain Pathol* **12**(3): 371–384.

Kandachar, V., Roegiers, F. (2012). Endocytosis and control of Notch signaling. *Curr Opin Cell Biol* **24**(4): 534–540.

Karlstrom, H., Beatus, P., *et al.* (2002). A CADASIL-mutated Notch 3 receptor exhibits impaired intracellular trafficking and maturation but normal ligand-induced signaling. *Proc Natl Acad Sci U S A* **99**(26): 17119–17124.

Kast, J., Hanecker, P., *et al.* (2014). Sequestration of latent TGF-beta binding protein 1 into CADASIL-related Notch3-ECD deposits. *Acta Neuropathol Commun* **2**: 96.

Lewandowska, E., Dziewulska, D., *et al.* (2011). Ultrastructure of granular osmiophilic material deposits (GOM) in arterioles of CADASIL patients. *Folia Neuropathol* **49**(3): 174–180.

Liao, Y. C., Hsiao, C.T., *et al.* (2015). Characterization of CADASIL among the Han Chinese in Taiwan: Distinct Genotypic and Phenotypic Profiles. *PLoS One* **10**(8): e0136501.

Liu, X.Y., Gonzalez-Toledo, M.E., *et al.* (2015). Stem cell factor and granulocyte colony-stimulating factor exhibit therapeutic effects in a mouse model of CADASIL. *Neurobiol Dis* **73**: 189–203.

Liu, H., Kennard, S., *et al.* (2009). NOTCH3 expression is induced in mural cells through an autoregulatory loop that requires endothelial-expressed JAGGED1. *Circ Res* **104**(4): 466–475.

Logeat, F., Bessia, C., *et al.* (1998). The Notch1 receptor is cleaved constitutively by a furin-like convertase. *Proc Natl Acad Sci USA* **95**(14): 8108–8112.

Low, W.C., Junna, M., *et al.* (2007). Hereditary multi-infarct dementia of the Swedish type is a novel disorder different from NOTCH3 causing CADASIL. *Brain* **130**(Pt 2): 357–367.

Low, W.C., Santa, Y., *et al.* (2006). CADASIL-causing mutations do not alter Notch3 receptor processing and activation. *Neuroreport* **17**(10): 945–949.

Markus, H.S., Martin, R.J., *et al.* (2002). Diagnostic strategies in CADASIL. *Neurology* **59**(8): 1134–1138.

McGill, M., Dho, S.E., *et al.* (2009). Numb regulates post-endocytic trafficking and degradation of Notch1. *J Biol Chem* **284**(39): 26427–26438.

Miao, Q., Paloneva, T., *et al.* (2006). Arterioles of the lenticular nucleus in CADASIL. *Stroke* **37**(9): 2242–2247.

Monet-Lepretre, M., Haddad, I., *et al.* (2013). Abnormal recruitment of extracellular matrix proteins by excess Notch3 ECD: A new pathomechanism in CADASIL. *Brain* **136**(Pt 6): 1830–1845.

Mozaffarian, D., Benjamin, E.J., *et al.* (2016). Heart Disease and Stroke Statistics-2016 Update: A report From the American Heart Association. *Circulation* **133**(4): e38–360.

Mykkanen, K., Savontaus, M.L., *et al.* (2004). Detection of the founder effect in Finnish CADASIL families. *Eur J Hum Genet* **12**(10): 813–819.

Nagase, H., Visse, R., *et al.* (2006). Structure and function of matrix metalloproteinases and TIMPs. *Cardiovasc Res* **69**(3): 562–573.

Olzscha, H., Schermann, S.M., *et al.* (2011). Amyloid-like aggregates sequester numerous metastable proteins with essential cellular functions. *Cell* **144**(1): 67–78.

Opherk, C., Peters, N., *et al.* (2004). Long-term prognosis and causes of death in CADASIL: A retrospective study in 411 patients. *Brain* **127**(Pt 11): 2533–2539.

Park, S., Park, B., *et al.* (2014). Case report: Bipolar disorder as the first manifestation of CADASIL. *BMC Psychiatry* **14**: 175.

Pescini, F., Donnini, I., *et al.* (2016). Circulating biomarkers in cerebral autosomal dominant arteriopathy with subcortical infarcts and leukoencephalopathy patients. *J Stroke Cerebrovasc Dis.* **26**(4): 823–833.

Peters, N., Freilinger, T., *et al.* (2008). Enhanced L-arginine-induced vasoreactivity suggests endothelial dysfunction in CADASIL. *J Neurol* **255**(8): 1203–1208.

Peters, N., Herzog, J., *et al.* (2004a). A two-year clinical follow-up study in 80 CADASIL subjects: Progression patterns and implications for clinical trials. *Stroke* **35**(7): 1603–1608.

Peters, N., Opherk, C., *et al.* (2004b). CADASIL-associated Notch3 mutations have differential effects both on ligand binding and ligand-induced Notch3 receptor signaling through RBP-Jk. *Exp Cell Res* **299**(2): 454–464.

Pfefferkorn, T., von Stuckrad-Barre, S., *et al.* (2001). Reduced cerebrovascular CO(2) reactivity in CADASIL: A transcranial Doppler sonography study. *Stroke* **32**(1): 17–21.

Preissner, K.T., de Boer, H., *et al.* (1996). Thrombin regulation by physiological inhibitors: The role of vitronectin. *Semin Thromb Hemost* **22**(2): 165–172.

Preissner, K.T., Reuning, U. (2011). Vitronectin in vascular context: Facets of a multitalented matricellular protein. *Semin Thromb Hemost* **37**(4): 408–424.

Qi, J.H., Ebrahem, Q., *et al.* (2003). A novel function for tissue inhibitor of metalloproteinases-3 (TIMP3): Inhibition of angiogenesis by blockage of VEGF binding to VEGF receptor-2. *Nat Med* **9**(4): 407–415.

Reyes, S., Kurtz, A., *et al.* (2012). Presymptomatic genetic testing in CADASIL. *J Neurol* **259**(10): 2131–2136.

Ruchoux, M.M., Domenga, V., *et al.* (2003). Transgenic mice expressing mutant Notch3 develop vascular alterations characteristic of cerebral autosomal dominant arteriopathy with subcortical infarcts and leukoencephalopathy. *Am J Pathol* **162**(1): 329–342.

Ruchoux, M.M., Guerouaou, D., *et al.* (1995). Systemic vascular smooth muscle cell impairment in cerebral autosomal dominant arteriopathy with subcortical infarcts and leukoencephalopathy. *Acta Neuropathol* **89**(6): 500–512.

Ruchoux, M.M., Maurage, C.A. (1997). CADASIL: Cerebral autosomal dominant arteriopathy with subcortical infarcts and leukoencephalopathy. *J Neuropathol Exp Neurol* **56**(9): 947–964.

Rutten, J.W., Dauwerse, H.G., *et al.* (2016). Therapeutic NOTCH3 cysteine correction in CADASIL using exon skipping: *In vitro* proof of concept. *Brain* **139**(Pt 4): 1123–1135.

Salamon, N. (2014). Neuroimaging of cerebral small vessel disease. *Brain Pathol* **24**(5): 519–524.

Schon, F., Martin, R.J., *et al.* (2003). "CADASIL coma": An underdiagnosed acute encephalopathy. *J Neurol Neurosurg Psychiatry* **74**(2): 249–252.

Shahien, R., Bianchi, S., *et al.* (2011). Cerebral autosomal dominant arteriopathy with subcortical infarcts and leukoencephalopathy in an Israeli family. *Neuropsychiatr Dis Treat* **7**: 383–390.

Singhal, S., Bevan, S., *et al.* (2004). The influence of genetic and cardiovascular risk factors on the CADASIL phenotype. *Brain* **127**(Pt 9): 2031–2038.

Sonninen, V., Savontaus, M.L. (1987). Hereditary multi-infarct dementia. *Eur Neurol* **27**(4): 209–215.

Sourander, P., Walinder, J. (1977). Hereditary multi-infarct dementia. Morphological and clinical studies of a new disease. *Acta Neuropathol* **39**(3): 247–254.

Steiner, T., Al-Shahi Salman, R., *et al.* (2014). European Stroke Organisation (ESO) guidelines for the management of spontaneous intracerebral hemorrhage. *Int J Stroke* **9**(7): 840–855.

Stevens, D.L., Hewlett, R.H., *et al.* (1977). Chronic familial vascular encephalopathy. *Lancet* **1**(8026): 1364–1365.

Stratman, A.N., Schwindt, A.E., *et al.* (2010). Endothelial-derived PDGF-BB and HB-EGF coordinately regulate pericyte recruitment during vasculogenic tube assembly and stabilization. *Blood* **116**(22): 4720–4730.

Tikka, S., Baumann, M., *et al.* (2014). CADASIL and CARASIL. *Brain Pathol* **24**(5): 525–544.

Tikka, S., Ng, Y.P., *et al.* (2012). CADASIL mutations and shRNA silencing of NOTCH3 affect actin organization in cultured vascular smooth muscle cells. *J Cereb Blood Flow Metab* **32**(12): 2171–2180.

Tournier-Lasserve, E., Joutel, A., *et al.* (1993). Cerebral autosomal dominant arteriopathy with subcortical infarcts and leukoencephalopathy maps to chromosome 19q12. *Nat Genet* **3**(3): 256–259.

Trikamji, B., Thomas, M., *et al.* (2016). An unusual case of cerebral autosomal-dominant arteriopathy with subcortical infarcts and leukoencephalopathy with occipital lobe involvement. *Ann Indian Acad Neurol* **19**(2): 272–274.

Tsai, C.F., Anderson, N., *et al.* (2016). Comparing risk factor rofiles between intracerebral hemorrhage and ischemic stroke in Chinese and white opulations: Systematic review and meta-analysis. *PLoS One* **11**(3): e0151743.

Tsai, C.F., Thomas, B., *et al.* (2013). Epidemiology of stroke and its subtypes in Chinese vs white populations: A systematic review. *Neurology* **81**(3): 264–272.

Tuominen, S., Miao, Q., *et al.* (2004). Positron emission tomography examination of cerebral blood flow and glucose metabolism in young CADASIL patients. *Stroke* **35**(5): 1063–1067.

Vaccari, T., Lu, H., *et al.* (2008). Endosomal entry regulates Notch receptor activation in Drosophila melanogaster. *J Cell Biol* **180**(4): 755–762.

Valenti, R., Pescini, F., *et al.* (2011). Major depression and bipolar disorders in CADASIL: A study using the DSM-IV semi-structured interview. *Acta Neurol Scand* **124**(6): 390–395.

van Bogaert, L. (1955). Encephalopathie sous-corticale progressive (Binswanger) a evolution rapide sous deux soeurs. *Med Hellen* **24**: 961–972.

Villa, N., Walker, L., *et al.* (2001). Vascular expression of Notch pathway receptors and ligands is restricted to arterial vessels. *Mech Dev* **108**(1–2): 161–164.

Vishnevetsky, A., Inca-Martinez, M., *et al.* (2016). The first report of CADASIL in Peru: Olfactory dysfunction on initial presentation. *eNeurologicalSci* 5: 15–19.

Viswanathan, A., Gray. F., *et al.* (2006). Cortical neuronal apoptosis in CADASIL. *Stroke* **37**(11): 2690–2695.

Watanabe, M., Adachi, Y., *et al.* (2012). An unusual case of elderly-onset cerebral autosomal dominant arteriopathy with subcortical infarcts and leukoencephalopathy (CADASIL) with multiple cerebrovascular risk factors. *J Stroke Cerebrovasc Dis* **21**(2): 143–145.

Watanabe-Hosomi, A., Watanabe, Y., *et al.* (2012). Transendocytosis is impaired in CADASIL-mutant NOTCH3. *Exp Neurol* **233**(1): 303–311.

Weller, M., Dichgans, J., *et al.* (1998). Acetazolamide-responsive migraine in CADASIL. *Neurology* **50**(5): 1505.

Yamamoto, Y., Craggs, L.J., *et al.* (2013). Brain microvascular accumulation and distribution of the NOTCH3 ectodomain and granular osmiophilic material in CADASIL. *J Neuropathol Exp Neurol* **72**(5): 416–431.

Dural Fistula 9

Daniel L. Cooke, Matthew R. Amans, and Van V. Halbach

Department of Radiology and Biomedical Imaging
University of California, San Francisco, San Francisco, California, USA

Vascular malformation is a generic term used to describe pathological and anatomical variant arterial, capillary, and venous lesions. Vascular malformations are of interest largely because of their potential to cause stroke, particularly intracerebral hemorrhage (ICH), as well as their amenability to definitive surgical therapies. Dural arteriovenous fistulae (dAVF) are a subset of the broader vascular malformation category and, though they have been referred to as malformations, the term *fistula* is the preferred nomenclature. Like so many of the other malformations, they are relatively rare, an epidemiological constraint that limits their study largely to case series. Additionally, dAVF are a putative non-inheritable disorder limiting genetic knockout animal models and the expected isolation of particular molecular pathway(s) implicated in their formation and natural history. Despite such limitations, there has been voluminous clinical and basic science research into their pathophysiology that has shaped our current understanding of the disease.

9.1 Definition

Simply put, a dAVF is a direct connection between artery and vein typically within the dura. They have been recognized as a distinct vascular malformation for almost a century. Ernest Sachs has been credited for his description of the disease in 1931 (Sachs, 1931), and his definition as an AV shunt has largely remained intact over that time (Aminoff, 1973; Dandy, 1946; Fincher, 1951) They differ from arteriovenous malformations (AVM) not only in location, but also in their absence of a distinct vascular nidus intervening the arterial and venous components. The fistulous connection may comprise a single or multiple arterial afferents that coalesce within the wall of sinus (Nishijima *et al.*, 1992; Sakaki *et al.*, 1991) or bridging vein (Baltsavias *et al.*, 2015a) that then drains into the sinus itself and/or an adjacent bridging or emissary vein. They are most often found as solitary lesions, and multiple in some cases. Barnwell *et al.* noted 7 of 105 cases (Barnwell *et al.*, 1991), van Dijk *et al.*, 19 of 235 cases (van Dijk *et al.*, 2002b), and Ha *et al.* 14 of 179 cases (Ha *et al.*, 2012) of dAVF patients manifesting with multiple lesions. These series also noted higher rates of leptomeningeal venous drainage and associated aggressive clinical presentation. In general, dAVFs can occur anywhere throughout the dura, but they most often appear at the cavernous sinus (Figure 1) or transverse-sigmoid (Figure 2) sinus junction.

The clinical significance given to dAVF location is directly related to the anatomic venous constraints the site defines, particularly as it relates to natural history (Lasjaunias *et al.*, 1986). There are differences that each anatomic site carries as it relates to treatment as well, and both issues will be discussed in subsequent sections.

9.2 Epidemiology

Of brain vascular malformations, dAVF are a small fraction, with Newton and Cronqvist noting 9 pure dural malformations (7%) in their review of 129 vascular malformations (Newton & Cronqvist, 1969) and Piippo *et al.* noting 32% in their population-based analysis of 251 patients (Piippo *et al.*, 2013a). The incidence of dAVF varies depending on location, with a European series noting 0.16 per 100,000 (Al-Shahi *et al.*, 2003) adults per year, while Finnish and Japanese series noting 0.51 per 100,000 (Piippo *et al.*, 2013a) and 0.29 per 100,000 (Kuwayama, 2016), respectively. The higher prevalence of dAVF in the latter two populations mirrors that of aneurysms in those communities. Despite this similarity, as well as evidence presented by Gross *et al.* that dAVF patients may manifest aneurysms in fractions larger than the general population (Gross *et al.*, 2012) (8 of 56 patients, 14%), there is no evidence that patients with aneurysms more often manifest dAVFs. Overall, women are more often affected with dAVF than men, although when categorized by anatomic location, this association may increase (e.g., cavernous) or invert (e.g., tentorial) (Celik *et al.*, 2010; Chung *et al.*, 2002). While the anatomic locations most often encountered are the cavernous sinus and the transverse-sigmoid sinus junction, the relative fraction of these two types varies

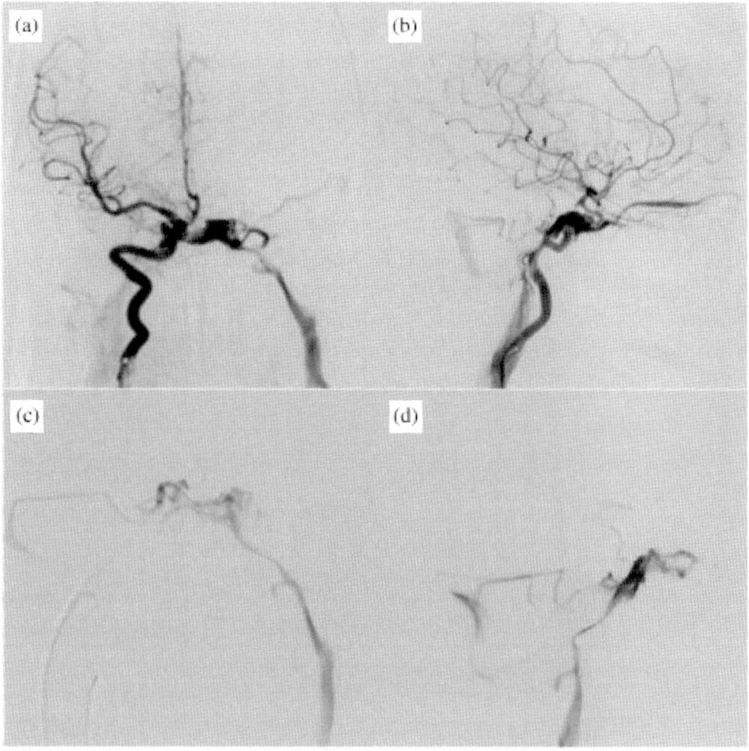

Figure 1. Composite image of a high-risk right cavernous carotid fistula. AP (a) and lateral (b) right internal carotid artery DSA images demonstrating AV shunting along the dorsum of the right cavernous sinus with outflow to the contralateral cavernous and inferior petrosal sinuses, left superior ophthalmic veins, and retrograde leptomeningeal venous drainage via lateral ponto-mesencephalic veins to the superior petrosal sinus. AP (c) and lateral (d) right cavernous sinus DSA images better demonstrating retrograde leptomeningeal venous drainage. Note, cross-sectional images from this case are depicted in Figure 8.

depending on continent (Cho *et al.*, 2013; Chung *et al.*, 2002; Hiramatsu *et al.*, 2014; Kim *et al.*, 2002; Li *et al.*, 2015). The disease most often manifests in the 6th decade of life, marking it later than AVMs (5th decade) and earlier than aneurysms (7th decade) (Greving *et al.*, 2014; Kim *et al.*, 2014; Kobayashi & Al-Shahi Salman, 2014). Children may be affected as well, with clear subgroups presenting as neonates and infants, although these cases are considered congenital and have different clinical presentation and management goals (Hetts *et al.*, 2016). From a stroke standpoint, dAVFs are a minority of all causes of ICH (van Asch *et al.*, 2015).

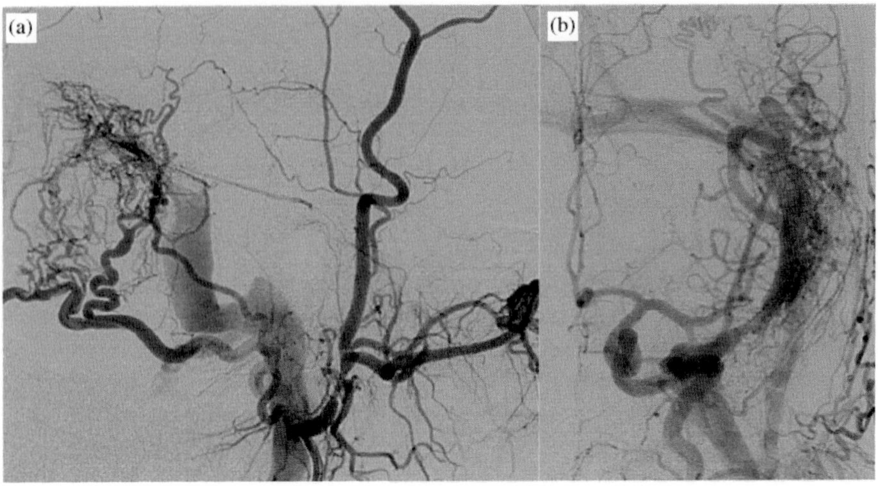

Figure 2. Lateral external (a) and AP common (b) left carotid artery DSA images demonstrating a low-risk transverse-sigmoid sinus junction dAVF with anterograde venous outflow. Note the collection of enlarged transosseous afferents from the occipital artery. Note, cross-sectional images from this case are depicted in Figure 10.

Their impact on the collective presentation of other neurological symptoms, such as headache, seizure, double vision, or dementia, is unknown but expected to be minimal. They do represent a measurable cause of tinnitus, or more specifically, pulse-synchronous tinnitus. The varied clinical presentations of dAVF will be discussed in more detail in subsequent sections.

9.3 Anatomy

The fine anatomic details of the dura and its vasculature are outside of the scope of this chapter, though some general information regarding their arrangement is important to understand disease origins and progression. The meninges originate from the meninx primitiva, a mesenchymal sheath inclusive of cells from the neural crest (Adeeb *et al.*, 2012). The outermost pachymeningeal layer, or dura mater, is affixed to the periostium and continues with the epineuria and adventitia when accompanying a nerve or artery at a boney foramina (Adeeb *et al.*, 2012). This interwoven quality with the skull base and calvarium, along with its vascular contents, is evidenced by the oft involvement of transosseus afferents to dAVFs. The dural sinuses pass between the endosteal and meningeal layers of the dura mater.

They are valveless and lack the muscular and adventitial layers of other veins. The classically named sinuses are accurately described as such, though the cavernous sinus is more a septated plexus than a venous sink. It gives passage to the internal carotid artery, ophthalmic and maxillary divisions of the trigeminal nerve, and oculomotor and abducens nerves. It serves as a conduit for venous flow to the ophthalmic veins, pterygoid plexus, superior and inferior petrosal sinuses, and sphenobasal and sphenoparietal sinuses (Barrow *et al.*, 1985; Harris and Rhoton, 1975, 1976; Joo *et al.*, 2012, Lasjaunias *et al.*, 1984; McKenzie *et al.*, 1996; Rhoton *et al.*, 1977, 1979; Seoane *et al.*, 1998, 2000; Yasuda *et al.*, 2004, 2005, 2008). The rich vascular anatomy it represents and its dural housing makes it a common location for fistulous formation. Overall, the general dural arterial anatomy is composed of primary and secondary anastomotic vessels that are seated in the outermost portions of the dura, with finer branches within the deeper aspects extending to the subdural space (Hardy *et al.*, 1980; Kerber & Newton, 1973; Martins *et al.*, 2005). The dural veins run in parallel with the primary dural arteries connecting with sinuses, emissary veins, or diploic lakes. Bridging or transdural veins, though originating from the pial surface, are also essential in dAVF pathogenesis given their monikered connection between the cortical surface and the dural sinuses (Baltsavias *et al.*, 2015b; Miyachi *et al.*, 2011). As for sensory mechanisms, the dura is largely innervated by trigeminal nerve branches, stimulation of which may cause a trigeminocardiac reflex characterized by hypotension and bradycardia (Amiridze & Darwish, 2009; LV *et al.*, 2009, 2014; Nyberg *et al.*, 2013).

9.4 Pathophysiology

9.4.1 *Anatomy*

There are varying theories regarding the events that cause fistulas and the location from which they form; there is evidence as well regarding which patients in general may be at greater risk for fistulas. In a histopathological review from dAVF patients, Kerber and Newton noted "crack-like" vessels within the dura surrounding an affected sinus (Kerber & Newton, 1973). They characterized these channels as physiological shunts between the arterial and venous anatomy of the dura. Ishikawa *et al.* also noted these channels, categorizing them into direct and indirect shunts (Ishikawa *et al.*, 2007). They posited that such channels might be the substrate for dAVF formation if other genetic and environmental cues permitted. Portnoy *et al.*, in their work on cerebrospinal fluid regulation, noted the role of para-sinus venous lacunae in the autoregulatory maintenance of the leptomeningeal veins (Portnoy *et al.*, 1983, 1994). In melding these two evidentiary streams,

Mironov suggested that the dural AV shunts contribute to this autoregulatory function and may be altered due to inflammation and/or thrombosis, and that through thrombotic recanalization, a sump effect might evolve causing positive feedback on the shunt (Mironov, 1994). He further posited that the channels described by Kerber and Newton permitted for decompressive shunting when pressures were elevated. Other non-dural veins have also been theorized as the anatomic sites of dAVF origins. Miyachi *et al.* considered the role of the emissary vein as the location of dAVF origin (Miyachi *et al.*, 2011). They similarly theorized an inciting traumatic, ischemic, inflammatory, and/or thrombotic event as the initial step, but instead focused on the emissary vein as the nidus of the disease rather than the dural AV channels. This latter theory is, however, somewhat diminished by its lack of explanation for dAVF sites traditionally remote from emissary channels. Another potential non-dural dAVF site is the bridging vein, the distal aspects of which in its connection to sinuses take on similar sinusoidal histological organization (Baltsavias *et al.*, 2015a). In support of this hypothesis, Baltsavias *et al.* reviewed 211 dAVF patients with emphasis on those with leptomeningeal venous drainage (n = 107), noting subtypes based on bridging vein or sinus wall shunt location (Baltsavias *et al.*, 2015a). Their review noted the role of sino-venous outflow stenosis and/or thrombosis manifesting with 94.3% of cases demonstrating leptomeningeal venous outflow, a finding that leads into a discussion of the flow parameters involved in dAVF formation.

9.4.2 *Hemodynamic*

The position that venous thrombosis and/or stenosis is central to dAVF progression is one that has evolved from earlier observational studies noting the relative increase in these findings as well as aberrant sinus anatomy (Chaudhary *et al.*, 1982; Djindjian *et al.*, 1978). Houser *et al.* noted this in their review of 28 patients, and it has been consistently documented in subsequent series (Houser *et al.*, 1972, 1979), with a number of case reports noting *de novo* dAVF formation in formerly thrombosed or surgically altered sinuses (Bai *et al.*, 2012; Goddard & Khangure, 2002; Gupta *et al.*, 2005; Ha *et al.*, 2012; Hamada *et al.*, 1997; Hetts *et al.*, 2015; Kiyosue *et al.*, 2002; Kubo *et al.*, 2002; Kurata *et al.*, 2006; Kusaka *et al.*, 2001; Tsai *et al.*, 2004). In support of the role of thrombosis in dAVF formation, there are a number of studies that have documented the disproportionate number of sinus thromboses in dAVFs, as well as a collection of studies implicating procoagulant disorders in dAVF patients (Goddard & Khangure, 2002; Izumi *et al.*, 2007; Safavi-Abbasi *et al.*, 2008). Kraus *et al.* (Kraus *et al.*, 2000, 2002) and Singh *et al.* (Singh *et al.*, 2001) demonstrated the respective increases in Factor V Leiden (FVL) and the factor II G20210A allele variant in dAVF patients relative to the general population. Gerlach *et al.* in a review of 25 dAVF cases, replicated the increased prevalence of the G20210A variant but did not find an association with FVL (Gerlach *et al.*, 2008). A 2007 meta-analysis of 121 dAVF cases (178 controls) by van Dijk *et al.* noted an increase in FVL (OR 4.69, CI 1.24–17.69) as well as the prothrombin G20210A (OR 10.87, CI 1.32–89.51) in dAVF patients, but given how infrequent the mutations were stilled identified, they did not recommend screening of patients (van Dijk *et al.*, 2007). Others, however, contend that sinus thrombosis, as important as it may be to fistula formation, may only be a step toward, or even a manifestation of, elevated venous pressures that are the primary driver of the disease.

9.4.3 *Basic science*

The genetic and environmental origins of dAVFs remain incomplete, though there is ample evidence describing the molecular and cellular events that characterize disease progression. The primary animal model used to study dAVFs utilizes the surgical anastomosis of the external carotid artery-to-jugular vein (Terada *et al.*, 1994). This model, predominantly using rats (though mice, dogs, and rabbits have also been described), has been modified through the addition of contralateral jugular ligation and sinus thrombosis to better simulate the clinical conditions encountered. Studies by Lawton *et al.* revealed the importance of sinus thrombosis, but more importantly, venous hypertension and its associated local-regional tissue ischemia as the environmental cues that initiate angiogenesis and fistula formation (Lawton *et al.*, 1997) . Evidenced animal data have confirmed the speculated role of venous hypertension in the angiogenesis of dAVF formation, documenting increases in vascular endothelial growth factor (VEGF) as well as other genes (Terada *et al.*, 1996) implicated in neovascularity (Kojima *et al.*, 2007; Shin *et al.*, 2007). Zhu *et al.*, using the carotid-jugular anastomotic rat model (Terada *et al.*, 1994), noted significant increases in hypoxia-induced factor-1 (Hif-1) and VEGF in the venule endothelium and para-sinus astrocytes within 1 and 7 days of post-venous hypertensive induction (Zhu *et al.*, 2006). Using the same model, Chen *et al.* noted that those animals with venous hypertension, with or without sinus thrombosis, more often had lasting (though local) changes in cerebral perfusion pressure as opposed to those with sinus thrombosis alone. They also noted persistent angiogenesis and increases in VEGF and matrix metalloproteinase-9 expression in the occipital lobe and dura (Chen *et al.*, 2009). Reciprocally, Li *et al.*

demonstrated that blocking the VEGF pathway actually impeded dAVF formation (Li *et al.*, 2014).

In parallel with the efforts from the neurovascular research community, the external carotid artery–jugular vein model has also served vascular surgery research groups in their understanding of dialysis arteriovenous fistula. Whereas dAVF groups focused on the intracranial process, the dialysis groups focused on the recipient vein of the surgical anastomosis. Though different in anatomic location, much may be learned about the changes that occur to an arterialized vein whether in the neck, arm, or dura. Gene microarray studies on both animals and human surgical AVF demonstrate significant changes in the endothelium particularly as it relates to extracellular matrix (ECM). Using the mouse carotid-jugular anastomotic model, Abeles *et al.* defined three major categories of differential gene expression in the ipsilateral jugular vein relative to sham and contralateral jugular veins: chemokines, platelet-derived growth factor (PDGF), and ECM organization (Abeles *et al.*, 2006). These collective gene groups manage multiple aspects of cell movement, in particular, the alterations to the vascular smooth muscles cells noted to progressively infiltrate into the venous media during the early phases of AV shunt formation (Abeles *et al.*, 2006). The higher pressures that the recipient vein experiences induce remodeling of the vessel wall, necessitating reorganization of the ECM and its support cells.

These animal studies have been supported by work in humans. A review by Tirakotai *et al.* and Uranishi *et al.* of dural samples from surgically treated dAVF patients ($n = 8$ and $n = 9$) noted similar patterns of immunohistochemical staining positive for VEGF and Hif-1, as well as a majority of samples also positive for transforming growth factor alpha (TGFα), basic fibroblastic growth factor (bFGF), and the vascular endothelial growth factor receptor 1 (Flt-1), genes involved in hypoxia and angiogenesis (Tirakotai *et al.*, 2005; Uranishi *et al.*, 1999; Vallon *et al.*, 2014). In a microarray analysis of 6 human dialysis fistulas compared to 4 normal cephalic veins, Hashimoto *et al.*, noted increased expression in cytokine, inflammatory, cell adhesion, and TGFβ signaling mechanisms (Hashimoto *et al.*, 2013). Additionally, human serum analyses have demonstrated the increase and subsequent decrease in circulating VEGF levels before and after treatment, respectively (Klisch *et al.*, 2005), while radiological studies in humans have documented similar perfusion abnormalities using radionuclide imaging, noting relative decline in oxygenation in brain-adjacent dAVFs (Kai *et al.*, 2003; Kanemaru *et al.*, 2015; Kawaguchi *et al.*, 2000; Nencini *et al.*, 1993; Shidoh *et al.*, 2014; Tominaga *et al.*, 2003; Waragai *et al.*, 2006). While the evidence for venous hypertension and the associated hypoxia and angiogenesis it invokes is impressive, there are other mechanisms that have been implicated in dAVF formation.

9.4.4 *Environmental factors*

Many theorize that the initiating event in dAVF formation may be traumatic, ischemic, inflammatory, and/or thrombotic. There are a number of individual reports and case series documenting this phenomenon in the setting of head injury, surgical or endovascular instrumentation (Gupta *et al.*, 2005; Gutierrez *et al.*, 2004; Pabaney *et al.*, 2016; Vadivelu *et al.*, 2012; Yassari *et al.*, 2002), and even acupuncture (Chung *et al.*, 2002). Of note, however, is the higher frequency of spontaneous thrombosis of many of these dAVFs, suggesting that the persistence of the AV shunt requires more factors than its presence alone represents (Luciani *et al.*, 2001; Piippo *et al.*, 2013a). Despite the evidence for trauma in instigating dAVF formation, the number of dAVF patients annually is a fraction of the overall trauma events (Lauw *et al.*, 2013). Similarly, there is significant discord between the numbers of cases of meningitis annually and dAVF reports. There a few case reports documenting the phenomenon, but not with the frequency of association of trauma and sinus thrombosis (Hai *et al.*, 2011; Itokawa *et al.*, 2010; Tsutsumi *et al.*, 2008) and dAVF presence. As for ischemia, there are few case reports noting dAVF formation in the setting of remote ischemic stroke or diminished cerebrovascular reserve (Cooke *et al.*, 2011; Hanaoka *et al.*, 2011; Killory *et al.*, 2008) and prolonged AV shunting. As it relates to diminished cerebrovascular reserve, Parammasivam *et al.* noted secondary dAVF formation within the dura surrounding the vein of Galen in the post-treatment vein of Galen malformation (VOGM) patients as well as along the course of venous outflow from pial AVFs (Paramasivam *et al.*, 2014; Paramasivam *et al.*, 2013a). They posited that chronic dural ischemia created angiogenic demand leading to dAVF formation. Finally, there are also reports of higher rates of dAVFs within patients with dural-based and other brain tumors (Vellimana *et al.*, 2014), though there is heterogeneity in the relationship between the tumor and its effect on the dAVF-involved sinus; again, the overall association is infrequent.

9.4.5 *Summary: Pathophysiology*

The etiology of dAVF is multifactorial, a combination of genetic and environmental conditions that lead to AV shunt formation and persistence. The microvasculature of the dura and its associated leptomeningeal veins is the site of dAVF formation, in part triggered by mechanical and/or inflammatory stimuli. Though not uniformly noted, the role of venous hypertension and/or thrombosis looms large in these incendiary processes, and the implications of pro-thrombotic disorders in contributing to dAVF formation is supported by the literature. Both of these hypertensive and thrombotic venous disorders manifest local–regional tissue hypoxia and associated angiogenesis, molecular mechanisms thought to drive dAVF formation and progression over time. The longstanding effects of such hypertension may lead to further venous thrombosis and/or diversion of venous flow to leptomeningeal veins, the latter of which increases the likelihood of aggressive clinical presentation.

9.5 Natural History

9.5.1 *Classification*

Much has been written about dAVFs and their natural history, with each series building on its predecessor, but all suffering from a limited number of patients, particularly those undergoing observational therapy. The hemorrhagic potential of the disease has been the focus of many retrospective reviews and alternative clinical presentations have also been studied. Houser *et al.*, and later Djindian *et al.*, noted the importance of leptomeningeal venous involvement in the hemorrhagic presentation of dAVF. This phenomenon was codified by Djindjian and Merland (Djindjian *et al.*, 1973; Merland *et al.*, 1977a, 1977b) and later revised by Cognard *et al.* (Cognard *et al.*, 1995) in a widely used five-point classification system for dAVFs. This system centers on the venous architecture of the dAVF, ignoring age, clinical presentation, other medical co-morbidities, and even largely anatomic location (e.g., Cognard V). Cognard *et al.* in their review of 258 cases (103 located at the transverse sinus) championed the concept that impedance of venous flow is the root cause of more aggressive clinical presentations, and that flow resistance is incrementally increased as flow changes from sinus anterograde (I, $n = 84$, 0 ICH) to sinus retrograde (IIa, $n = 27$, 0 ICH), to mixed leptomeningeal and sinus anterograde (IIb, $n = 10$, 2 ICH) or a combination thereof (IIa + b, $n = 18$, 1 ICH), to leptomeningeal only (III, $n = 25$, 10 ICH), and to leptomeningeal with evidence of venous ectasia (IV, $n = 29$, 19 ICH) ($p = 0.0001$) (Cognard *et al.*, 1995). The fifth subgroup they specified takes into account the anatomic location of the venous drainage (retrograde spinal) and, by definition, somewhat involves leptomeningeal veins. This methodology was simplified by Borden and Shucart into a three-tiered system condensing the Cognard II subgroups (Borden II) and the Cognard III and IV (Borden III) into single groups (Borden *et al.*, 1995). Davies *et al.* validated both systems in an analysis of a series of 98 patients (102 lesions, 44 transverse and 14 cavernous sinus lesions), in which 16 presented with ICH (Davies *et al.*, 1996). Of 55 Borden I lesions, 26 patients were observed with mean follow-up of 33 months and clinical improvement in 21 (81%) with no hemorrhagic events noted (Davies *et al.*, 1997a). Of 46 Borden II (n = 18, 2 ICH) and III (n = 29, 14 ICH) lesions, 14 patients were observed, 22 underwent surgery, and 20 underwent embolization. The Borden classification accurately described ICH and aggressive clinical presentation ($p < 0.0001$). In the observation cohort, 4 patients died (29%) during the 249 lesion-months follow-up interval, while of those patients undergoing embolization ($n = 11$) and/or surgery ($n = 14$) where residual leptomeningeal drainage persisted, 4 had subsequent events (1 death and 3 new non-hemorrhagic neurological deficits) (Davies *et al.*, 1997b). Similar results as it relates to ICH and aggressive clinical presentation were noted for stratification using the Cognard classification ($p < 0.0001$) (Davies *et al.*, 1996). Both systems were accurate in their association with aggressive clinical presentation (inclusive of ICH) (Chung *et al.*, 2002; Davies *et al.*, 1996; Cohen *et al.*, 2009) and showcase the importance of identifying leptomeningeal venous drainage as the singular most important feature of any dAVF.

9.5.2 *Prognostic risk stratification; angioarchitecture*

These classification systems and the series from which they derive were not, however, designed to be prognostic, either as it relates to future risk of hemorrhage or change in categorization over time. A collection of case series has been published, attempting to characterize those clinical and imaging parameters that portend stroke risk. For those patients without evidence of leptomeningeal drainage, future events are rare. Satomi *et al.* in their review of 68 such patients with long-term follow-up, noted benign or tolerable symptoms in 98.5% of patients (Satomi *et al.*, 2002). Conversely, others have demonstrated that the presence of leptomeningeal venous drainage does increase the risk of future ICH or other aggressive clinical presentation (Satomi *et al.*, 2015). Brown *et al.* in a review of 52 patients followed for mean of 6.6 years noted new ICH in 5 patients (1.6% crude ICH per year) and 1 death related to the disease (Brown *et al.*, 1994). They observed that the presence of venous ectasia was significantly ($p < 0.05$) associated with subsequent ICH as well as those with leptomeningeal venous drainage, though it did not meet statistical significance. In Butlers *et al.*'s review of 75 dAVF patients with leptomeningeal venous drainage, a 13% rate of new ICH per year with distinct differences in annual ICH event rates for those with (27%, n = 31) and those without (3.5%, n = 44) venous ectasia ($p = 0.03$) was noted (Bulters *et al.*, 2012). It is noteworthy that they found no statistically significant association between ICH presentation or the more general aggressive clinical presentation category and future ICH risk. Shin *et al.* reviewed 41 patients with untreated (n = 7) or partially treated leptomeningeal drainage dAVFs (median follow up 1.5 years) and noted an annual new ICH rate of 4.5% (n = 5) and morbidity and mortality rates of 11.6% (n =13) and

2.6% (n = 3) (Shin *et al.*, 2013). Van Dijk *et al.* noted, in a review of 20 patients with untreated (n = 14) or partially treated leptomeningeal drainage dAVFs (median follow up 1.5 years), an annual new ICH rate of 8.1% (n = 7) and mortality rate of 10.4% (n = 9) (van Dijk *et al.*, 2002a). Dutzmann *et al.* reviewed 53 cases and found 3 (15%) new hemorrhagic events from time of diagnosis to definitive treatment among 19 Borden II and III patients presenting with aggressive symptoms (Dützmann *et al.*, 2011). In the subacute setting, Duffau *et al.*, in a review of 20 dAVF patients presenting with ICH, noted repeat hemorrhage in 7 (35%) within a mean 20 days from ictus (Duffau *et al.*, 1999). More recently, Gross *et al.*, reviewing 465 cases, noted no subsequent ICH for Borden I lesions, 6% annual ICH for Borden II lesions, and 10% annual ICH for all Borden III lesions and 21% for Borden III lesions with ectasia (Gross & Du, 2012).

Beyond ICH risk, changes in angioarchitecture have also been described in observational series. In the aforementioned Gross *et al.* study, there was a 1.4% risk of increase in venous drainage risk stratification for Borden I lesions. Cognard *et al.* reviewed 7 sinus-only drainage (5 type I and 2 type IIa) cases (6 of which underwent partial embolization) that presented with new clinical symptoms in a delayed fashion and were noted to have increased Cognard angioarchitectural scaling (2 with new retrograde sinus drainage and 5 with leptomeningeal venous drainage) (Cognard *et al.*, 1997). They remarked on interval-associated venous stenosis or thrombosis in 4, increase in arterial afferents in 2, and a *de novo* fistula in a single case. Kim *et al.* reviewed 112 Borden I cases foregoing treatment and noted 4 cases with interval higher risk venous drainage (Kim *et al.*, 2010), while Satomi *et al.* noted 2 conversions to leptomeningeal drainage in symptom-driven angiographic follow-up in 50 patients (Satomi *et al.*, 2002).

9.5.3 *Prognostic risk stratification; clinical presentation*

More recently, Zipfel *et al.* looked at the utility of incorporating clinical presentation into natural history prediction (Zipfel *et al.*, 2009). Based on reviews by Soderman *et al.* (Söderman *et al.*, 2008) (n = 85) and Strom *et al.* (Strom *et al.*, 2009) (n = 28) that found elevated rates in subsequent ICH in those originally presenting with hemorrhage (7.4% or 7.6%) versus those without hemorrhage (1.5% or 1.4%), they suggested altering the Borden scheme by designating type III lesions as having a symptomatic versus asymptomatic presentation (Zipfel *et al.*, 2009). Gross *et al.* replicated this result in their review, noting a 2% annual ICH event rates for Borden II and III lesions with non-aggressive symptoms that increased to 10% for those presenting with aggressive (non-hemorrhagic) symptoms, and ultimately to 46% for those presenting with ICH (Gross & Du, 2012). However, these analyses have been criticized for incorporating incidental dAVF presentation as being a non-aggressive presentation as well as the short-term follow up intervals (van Dijk *et al.*, 2009). A scenario of incidental detection is more often seen, given the frequency of imaging performed today, and this consequently creates lead-time bias as it relates to dAVF. With this in mind, Baltsavias *et al.* instead proposed a system that takes into account involvement of the leptomeningeal veins in reference to how *Direct* the flow is to them, the *Exclusivity* of their involvement, and any evidence of *Strain* (DES) defined as ectasia or generalized pseudophlebectatisia. Their system, based on the review of 107 patients with leptomeningeal venous drainage, yielded 8 categories with incremental risk of aggressive clinical presentation ($p < 0.001$). The presence of each named component contributed more to risk of aggressive presentation than its absence (note this stratum does not account for lesions with exclusively sinus drainage) (Baltsavias *et al.*, 2015c, 2015d). They noted 0 of 23 patients with aggressive presentation within the non-direct, non-exclusive, non-strain group, and 33 of 37 within the direct, exclusive, strain group presenting as such (Figure 3).

This work emphasizes the need for careful evaluation of leptomeningeal venous drainage, particularly the evidence of venous decompensation where of all subtypes manifesting strain (45 of 59 or 76.2%) had an aggressive presentation. Of those subgroups not manifesting strain, only those with direct and exclusive drainage had significantly elevated rates (54%, 7 of 16) of aggressive presentation (Baltsavias *et al.*, 2015d).

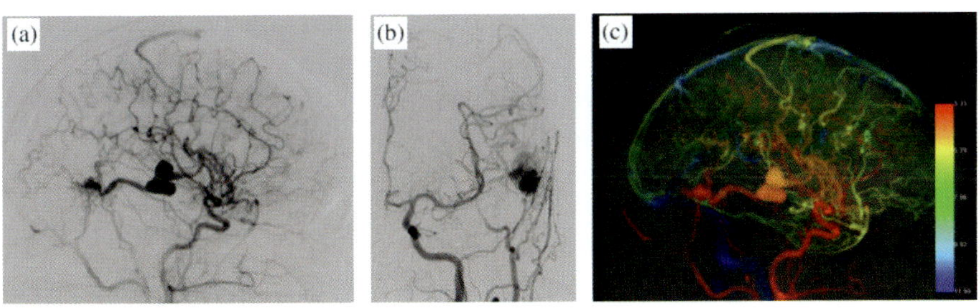

Figure 3. Lateral (a) and AP (b) left common carotid artery DSA images depicting a Labbe-type dAVF supplied by middle meningeal and occipital artery afferents. Note the venous ectasia as the leptomeningeal vein courses along the temporal lobe. Lateral iFlow left common carotid artery DSA image depicting parametric color-coding of the dAVF. Note, cross-sectional images from this case are depicted in Figure 9.

9.5.4 *Prognostic risk stratification: clinical outcomes*

Beyond just designation of those lesions that are at higher risk for hemorrhage, hemorrhagic types and the clinical outcomes they represent bear discussion. ICH related to dAVF most often presents as intraparenchymal hemorrhage, though it may also appear as subarachnoid or subdural hemorrhage. Cordonnier *et al.*, in review of 141 adults presenting with ICH in the setting of vascular malformation, noted 8 dAVF related cases, 4 of which were purely parenchymal and 4 with SAH, SDH, and/or IVH components (Cordonnier *et al.*, 2008). Compared to AVMs, dAVFs had similar hemorrhagic volumes (median volumes 16 ml versus 14 ml). There were no angioarchitectural features that predicted hemorrhagic subtype, including the presence of flow-related aneurysms, a phenomenon seen more frequently in AVMs. Also important is the tendency for more benign clinical outcomes related to the dAVF ICH spectrum than compared to other secondary vascular etiologies. Jolink *et al.* in the literature and case series review of 326 dAVFs presenting with ICH (58% IPH, 25% SAH, 11% IVH, 6% SDH), documented a median 1-year follow-up mortality of 4.7% and poor clinical outcome (mRS>2 or GOS<4) of 8.3% (Jolink *et al.*, 2015). These values are lower than reports for AVMs and primary ICH and may reflect a higher fraction of non-deep anatomic hemorrhages in the dAVF group. Importantly, despite a higher fraction (73%) of men within the cohort, gender, age, or presence of IPH did not predict case fatality rate. This is important when counseling patients and their families about the risks of natural history relative to surgical treatment. A smaller subset of patients will manifest progressive disease despite treatment. Hetts *et al.* described 19 of these "runaway" type lesions in a review of 579 dAVFs and noted that: (1) these lesions more often occurred in younger patients (36 years versus 53 years), including 6 children; and (2) the anatomy reflected longstanding venous hypertension, including sinus ectasia and a pseudophlebectasia (Hetts *et al.*, 2015). Ultimately, the clinical long view of dAVFs is limited due to the scarcity of series with delayed follow-up. Our best estimate comes from the Finnish study by Piippo *et al.* who noted ($n = 227$, median 10 years follow-up) that patients presenting with hemorrhage had a tendency to have excess mortality only up to 7 years after admission, but that patients who had dAVF with leptomeningeal drainage, whether presenting with hemorrhage or not, continued to have excess mortality even thereafter (Piippo *et al.*, 2013b).

9.5.5 *Summary: Natural history*

There are a number of classification systems used for dAVF categorization, the most common being the Cognard and Borden methods. Both of these, as well as others, emphasize the pattern of venous drainage as their central and guiding feature. The evidence supports a generally benign course for those patients with exclusively sinus drainage and risk of stroke or other neurological disorders in patients with any leptomeningeal venous drainage. More recent series have validated these methods emphasizing the importance of identifying venous decompensation as well as pointing out the importance of a patient's clinical presentation in predicting future events. As it relates to dAVF-related stroke outcomes, they tend to be more benign than other neurovascular disorders, though the supporting data is limited.

9.6 Clinical Presentation

As reviewed above, the venous drainage pattern more often than not determines the natural history risk. It follows then that the clinical presentation is also referenced by the venous angioarchtecture (Lasjaunias *et al.*, 1986). The majority of dAVF patients present with pulse synchronous tinnitus, headache, or ocular signs, or symptoms reflecting the higher overall prevalence of transverse-sigmoid and cavernous sinus lesions and their proximity to the petrous bone. The last location is associated with the fistulous effects on the cranial nerves that pass through the sinus as well as venous congestion created by egress to the ophthalmic veins, though such a clinical presentation does not always predict a cavernous dAVF location (Robert *et al.*, 2016a). Though these signs and symptoms convey the majority of dAVFs, there is greater concern for the more clinically insidious lesions that too often do not manifest until they cause significant cerebral venous congestion. In these cases, clinical presentation is more often hemorrhagic but there may also be cognitive change (Chahbazian *et al.*, 2014; Gonçalves *et al.*, 2008; Hasumi *et al.*, 2007; Holekamp *et al.*, 2016; Kai *et al.*, 2009; Lantz *et al.*, 2009; Matsuda *et al.*, 1999; Nencini *et al.*, 1993; Tanaka *et al.*, 1999), seizure (Le Guennec *et al.*, 2015), myelopathy (Haryu *et al.*, 2014), or neuropathy (Du *et al.*, 2003). For these types of clinical presentations, leptomeningeal venous drainage is almost assured and often times involves drainage along the temporal lobe (Dong *et al.*, 2014), deep nuclei (Gonçalves *et al.*, 2008; Kim *et al.*, 2015; Morparia *et al.*, 2012) (Figure 4), and/or brainstem (Kulwin *et al.*, 2012; Murphy & Van Gompel, 2012) (Figures 5 and 6), particularly veins bridging lateral and deep sinuses such as the lateral mesencephilc vein of the lateral recess of the fourth ventricle, reducing cerebral blood flow to the neighboring parenchymal anatomy (Cannizzaro *et al.*, 2016; Kanemaru *et al.*, 2015; Satoh *et al.*, 2001, 2005).

For the small subset of patients with perimedullary and spinal venous drainage, an aggressive clinical presentation more often occurs (Khan *et al.*, 2009;

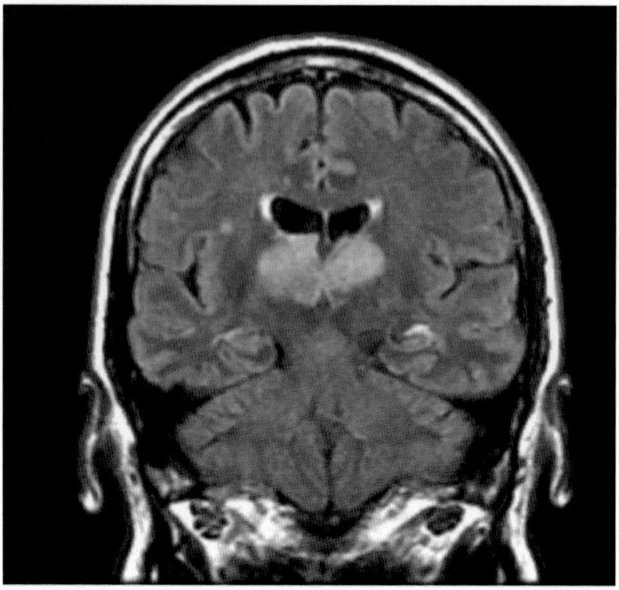

Figure 4. Coronal T2-weighted MR image demonstrating high signal within the thalami related to a tentorial dAVF (not pictured).

Ricolfi *et al.*, 1999; van Rooij *et al.*, 2007). Brunereau *et al.*, in a review of 12 patients with spinal venous drainage, found that those with eventual cervical radiculo-medullary venous egress pathways did not manifest myelopathy (Brunereau *et al.*, 1996).

Many series, including formal classifications systems (Barrow *et al.*, 1985), have focused on the anatomic location of the dAVF to explain their symptomatology, though ultimately these relationships are artifactual as they relate to natural history. Hemorrhagic presentation occurs in 7.5–35% of patients depending on the cohort (Davies *et al.*, 1996; Dützmann *et al.*, 2011; Li *et al.*, 2015; Malik *et al.*, 1984, Peng *et al.*, 2014; Piippo *et al.*, 2013a; Singh *et al.*, 2008), though a 2014 meta-analysis of 2,278 patients noted a 17% (n = 386) overall ICH presentation (Kobayashi & Al-Shahi Salman, 2014). Across series the most predictive variable associated with ICH presentation is leptomeningeal venous drainage, with Kobayahsi and Salman noting a pooled (n = 855) odds ratio of 23.2 (CI 13.8 – 39) in these cases (Kobayashi & Al-Shahi Salman, 2014). Other features have also been found to be associated with ICH and aggressive presentation. Awad *et al.*, in a review of the literature (n = 377) and their own case series (n = 17), compared

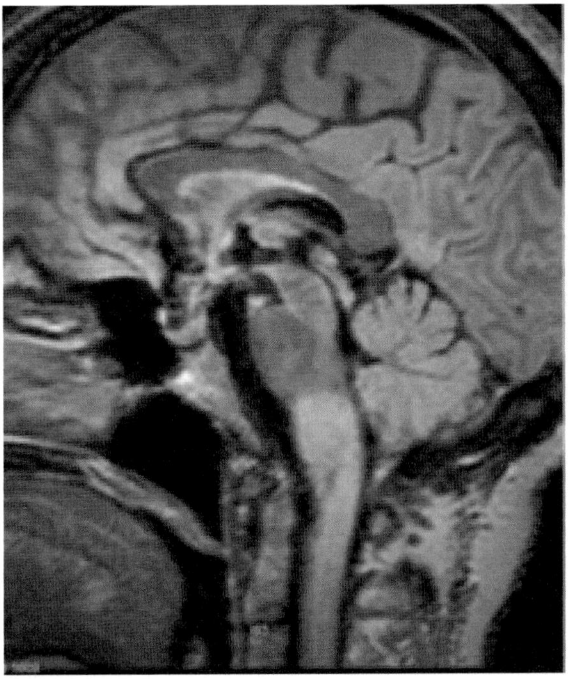

Figure 5. Sagittal T2-weighted MR image demonstrating high signal within the medulla and upper cervical cord.

benign and aggressive clinical presentations noting significant association of the latter with leptomeningeal venous drainage ($p < 0.001$), venous ectasia ($p < 0.001$), and Galenic drainage ($p < 0.05$) (Awad *et al.*, 1990). Geibprasert *et al.* examined dAVF with emphasis on the embryological contributions to the epidural anatomy designating ventral, dorsal, and lateral locations (Geibprasert *et al.*, 2008). Based on a review of 300 cases, they noted, of the 150 ventral, osteocartilaginous, type patients (cavernous, anterior marginal, and sigmoid sinuses and basilar plexus), a female predominance (2.3:1) and general absence of leptomeningeal venous drainage and its associated aggressive clinical presentation (8%), unless concomitant extensive thrombosis was noted ($p < 0.001$). Of the 67 dorsal, osteomembranous, type patients (torcula, transverse, superior, occipital, and posterior marginal sinuses), there was no gendered division, the age at presentation was younger (28 years versus 51 or 56 years) driven by more pediatric cases (31 of 67), there was a higher number of multiple lesions (36%) and more frequent leptomeningeal venous drainage and its associated aggressive clinical presentation (63%), again when concomitant extensive thrombosis was noted ($p < 0.001$). Of the 63 lateral

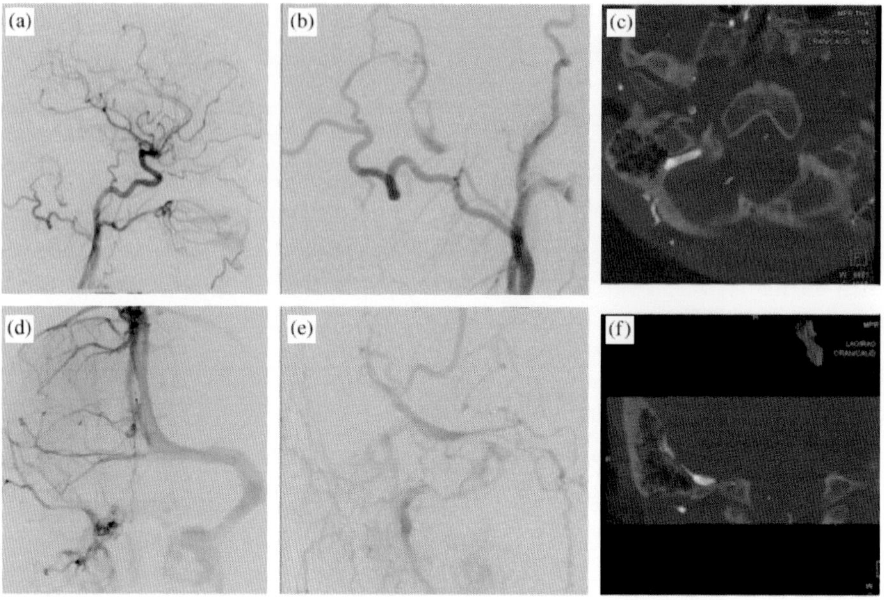

Figure 6. Composite image of a high-risk sigmoid sinus dAVF. Lateral projection right common carotid (a) and external carotid artery (b) DSA images demonstrating fistulous connection to an isolated sigmoid sinus. AP Low (d) and high magnification (e) DSA images demonstrating leptomeningeal venous drainage along the anterior medullary vein. Axial (c) and coronal (f) flat-panel CT images demonstrating the isolated sigmoid sinus and anterior medullary drainage. Note, cross-sectional images from this case are depicted in Figure 5.

type patients (falco- or basitentorial, petrosal, lateral marginal sinus, and ethmoidal), they were more often men (81%), and uniformly had leptomeningeal venous drainage, more frequent aggressive clinical presentation (86%), and less common venous outflow restriction ($p < 0.001$). As to how this embryonically based organization relates to Borden and Cognard classifications, the DE group was equivocal to Borden III and Cognard III and IV types, while LE and VE groups, given their sinus involvement, largely landed in Borden and Cognard I and II categories (Geibprasert *et al.*, 2008). They also point out the gendered weight given to anterior (female) and lateral (male) lesions, the latter of which has been documented by others as being an overall risk factor for ICH or other aggressive clinical symptoms (Chung *et al.*, 2002; Peng *et al.*, 2014; Singh *et al.*, 2008), originates in dural morphogenesis.

Geibprasert *et al.* applied this methodology to a review of 446 dAVF patients from Asian, European, and South American centers, examining the epidemiological variables across ethnicities (Geibprasert *et al.*, 2009). They noted no difference in ventral, dorsal, and lateral locations between continental groups, though they did note higher prevalence of venous outflow restriction and associated leptomeningeal venous drainage in the Asian dorsal epidural populations (Geibprasert *et al.*, 2009). Overall, anatomy dictates venous drainage and it, in turn, typically dictates clinical presentation (e.g., ICH). Beyond the associated treatment risks each anatomic location represents, the site of the fistula does not have additive bearing on natural history beyond which its venous anatomy denotes.

9.6.1 *Summary: Clinical presentation*

The clinical signs and symptoms of dAVFs are closely related to the pattern of venous outflow, with those with leptomeningeal drainage more often presenting with hemorrhage or other neurological deficits. The presence of such drainage is primarily a function of the physiological venous anatomy at any given anatomic location. Consequently, there is more commonality as it relates to the clinical presentation and natural history between two lesions anatomically remote from one another if venous drainage pattern is similar (e.g., anterior cranial fossa and tentorium) than neighboring anatomic sites if venous drainage pattern is different (e.g., anterior cranial fossa and cavernous sinus). Given that most dAVFs are either located at the cavernous sinus or transverse-sigmoid sinus junction, the majority of patients will not have an aggressive presentation. Nonetheless, identifying those patients with aggressive, non-hemorrhagic signs and symptoms is important, as the risk of stroke is considerable.

9.7 Diagnosis

9.7.1 *Overview*

Terminology is essential to consistently characterizing risk for many disorders and is no different in the case of dAVFs. There are no established clinical criteria that reliably screen for a dAVF, though dAVF is the most common diagnosis in patients with pulsatile tinnitus that can be appreciated by the examiner upon auscultation (Dietz *et al.*, 1994). Beyond that, the history and physical exam must be supplemented by cross-sectional imaging and digital subtraction angiography (DSA). The modalities are widely available at most institutions, though expertise in dAVFs is also essential in identifying them.

9.7.2 *Computed tomography*

In the setting of acute neurological changes, CT and CT angiography are excellent methods to screen for hemorrhage (Chung *et al.*, 2002). In a review of 298 patients presenting with non-traumatic ICH, van Asch *et al.* noted a dAVF etiology in 13 (4.3%) patients, of whom 9 were detected with CT alone (van Asch *et al.*, 2015). In addition to ICH features, non-contrast CT provides excellent information regarding bony anatomy, as dAVFs may manifest changes in the bony foramina (e.g., foramen spinosum) from flow-related enlargement of the dural vasculature as well as more permeative or wormian to portions of the calvarium (e.g., mastoid) from transosseous afferents. CT will also clearly demonstrate parenchymal calcifications (Yang *et al.*, 2005), of which the subcortical white matter type may be present in dAVFs. These findings, though non-specific, may represent longstanding venous hypertension, classically seen in the delayed presentation of VOGMs typical of hydrovenous disorders.

These features aside, the key to diagnosing a dAVF is demonstration of the AV shunt, a finding best made with the assistance of temporal angiographic information. Conventional CTA is obtained at a single moment in time and therefore usually lacks the temporal resolution to distinguish early filling of veins, but asymmetrical enhancement of venous and/or sinus structures (Meckel *et al.*, 2005) and dilated or ectatic leptomeningeal veins can be suggestive of dAVF (Lekkhong *et al.*, 2011). Other features are an increased number of vessels, arteries or veins, in and along a sinus or dural reflection, enlargement of the external carotid branches (and the foramena through which they course), and/or parenchymal enhancement (Figure 7).

These findings are sensitive to the presence of dAVF, though their absence does not rule out the disease (Cohen *et al.*, 2009; Lekkhong *et al.*, 2011). With larger (e.g., 128 or 256) multidetector CT systems, temporal encoding is now possible and with it the ability to better resolve the presence and precise location of dAVF shunts. There are a number of small case series suggesting 4D CTA may lead to improved detection, anatomic delineation, and classification of dAVFs (Beijer *et al.*, 2013; Biswas *et al.*, 2015; Willems *et al.*, 2011). The performance of cross-sectional imaging varies and is in part a function of the experience of the radiologist reading the study as much as the technique itself (Bink *et al.*, 2012).

9.7.3 *Magnetic resonance*

MR, like CT, provides excellent spatial resolution and can perform time-encoded vascular imaging; however, it also offers superior tissue characterization and more

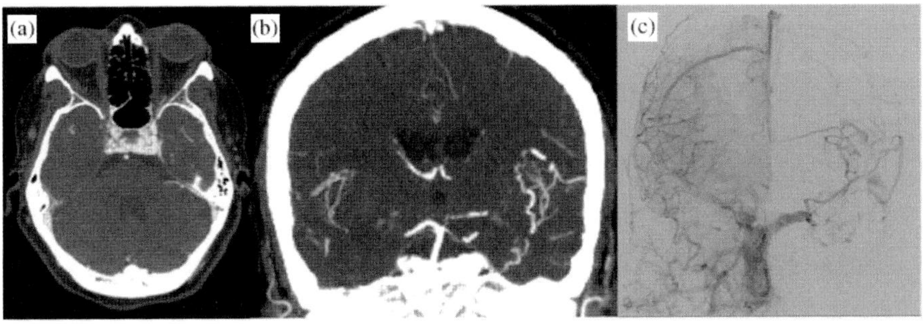

Figure 7. Axial (a) and coronal (b) CTA images demonstrating early, robust enhancement of the cavernous sinus as well as increased conspicuity of leptomeningeal veins within the left Sylvian triangle. (c) AP right internal carotid artery DSA image demonstrating delayed phase from a right cavernous carotid fistula. Note the prominence of the retrograde leptomeningeal venous drainage from the left cavernous in and along the temporal and frontal lobes.

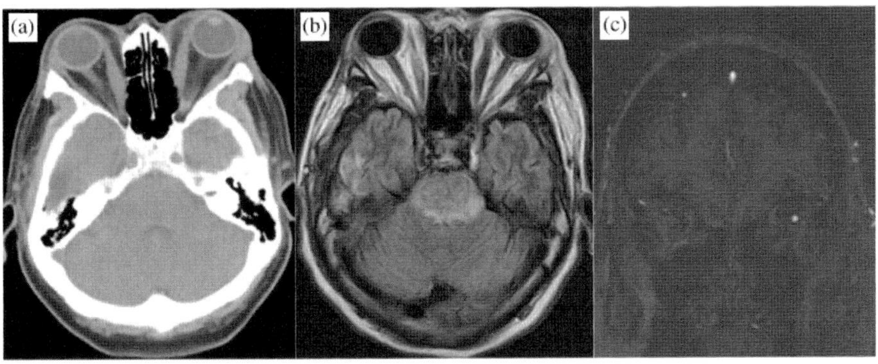

Figure 8. Composite image of a high-risk cavernous carotid fistula. Axial non-contrast CT (a) and T2 weighted MR (b) images demonstrating pontine edema. Coronal MRV (c) demonstrating relative enlargement of the left superior ophthalmic vein.

sensitive flow- and susceptibility-sensitive techniques to help locate dAVFs. The presence of edema, as evidenced by T2 prolongation within the parenchyma (Figure 8), is non-specific but is much easier to detect on MR rather than CT (Chung *et al.*, 2002; Zeidman *et al.*, 1995).

From an angiographic standpoint, as with CTA, MR-based assays are useful and demonstrate high degrees of reproducibility between readers for dAVF

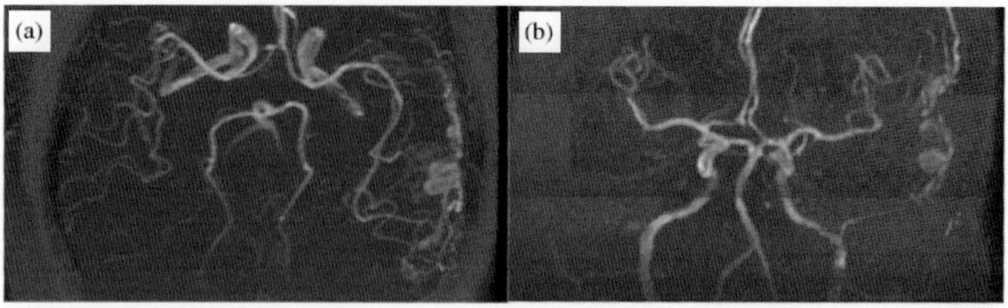

Figure 9. Axial (a) and coronal (b) time-of-flight (TOF) MRA MIP images demonstrating a dAVF draining into an ipsilateral Labbe-type leptomeningeal vein. Note the venous ectasia.

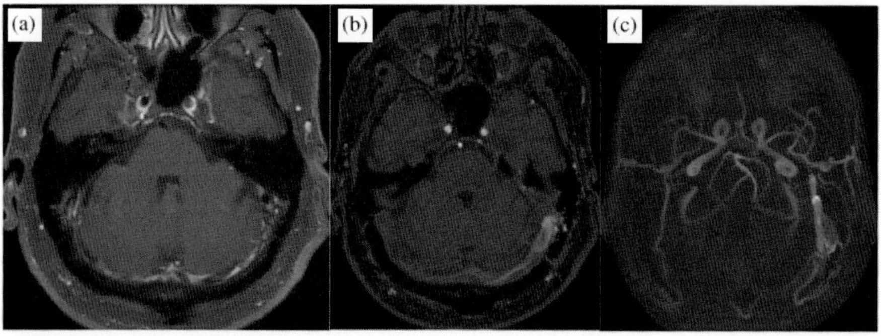

Figure 10. Composite image of low-risk left transverse-sigmoid sinus dAVF. Axial post-gadolinium (a) and TOF MRA (b) images demonstrating a relative increase in the number of flow voids within the and around the left transverse-sigmoid sinus junction as well as asymmetrically high signal within the left transverse sinus. (c) Axial TOF MRA MIP image demonstrating multiple enlarged branches of the occipital artery coalescing on the left transverse-sigmoid sinus junction.

location, presence of leptomeningeal involvement, arterial afferents, and classification (Figures 9 and 10).

There are a number of small case series comparing TOF (Azuma *et al.*, 2015; Hori *et al.*, 2011) and gadolinium-enhanced MR angiographic techniques (Cohen *et al.*, 2009; Nishimura *et al.*, 2010; Schanker *et al.*, 2011), using conventional and temporally encoded methods, noting the sensitivity and specificity relative DSA. Azuma *et al.*, in a review of 26 DSA-proven dAVFs compared to 3T TOF MR angiography, noted excellent ($k = 0.97$) agreement with DSA for lesion location while only good agreement for the main arterial afferents ($k = 0.81$) and venous efferents ($k = 0.84$) (Azuma *et al.*, 2015). Bink *et al.*, in a review of 38 patients (19 with DSA confirmed dAVFs) undergoing 3T TOF and time-resolved gadolinium enhanced MR angiography, demonstrated significant heterogeneity in inter-observer performance particularly for lesions outside the cavernous sinus, and largely fair and moderate agreement in reference to Borden classification and identification of arterial afferents (Bink *et al.*, 2012). For the one more experienced reader in this study, agreement with DSA imagining was "good" to "very good" for site location, Borden classification, and arterial afferent identity.

For those temporally encoded angiographic techniques, there is similar diagnostic performance. Edjlali *et al.* described the use of selective flow-tracking cartography in concert with 4D flow MR techniques in 8 DSA-proven dVAF patients with excellent intermodality agreement for main arterial ($k = 0.92$) and venous ($k = 1$) anatomy as well as Cognard classification ($k = 1$) (Edjlali *et al.*, 2014). More recently, Ertl *et al.* reviewed 24 DSA-proven transverse-sigmoid dAVFs cases using contrast-enhanced MRA and temporally resolved MRA (i.e., TRICKS) (Figure 11), noting excellent intermodality agreement on Cognard classification ($k = 1$), though overall moderate ($k = 0.4$) agreement for major feeding arteries (Ertl *et al.*, 2016a).

Farb *et al.*, in a review of 42 patients with dAVFs (pre- and post-treatment), also noted excellent agreement (39/42) between DSA and 3T temporally resolved MR angiography in identifying fistulas as well as between readers ($k = 0.94$) (Farb *et al.*, 2009).

9.7.4 *Adjuvant MR techniques*

MR also has the ability to perform arterial spin labeling (ASL), a technique that designates a cross-section of blood as a function of time and directionality relative the cranial–caudal axis of the patient (Amukotuwa *et al.*, 2016). If there is asymmetry in this flow through the head (i.e., from an AV shunt), ASL can identify even subtle differences seen as high-intensity vessels relative to the expected low-intensity vasculature (Noguchi *et al.*, 2010). It still lacks high-spatial resolution, but can direct radiologists to areas of interest for more close interrogation. In review of 34 DSA-proven dAVF patients, Amukotuwa *et al.* noted that ASL had high (91% and 96%) sensitivity and specificity for the detection of leptomeningeal venous drainage as well as very good ($k = 0.82$) agreement with DSA in defining Borden classification (Amukotuwa *et al.*, 2016). The use of temporally encoded ASL has also demonstrated promise in supplementing DSA in dAVF work up

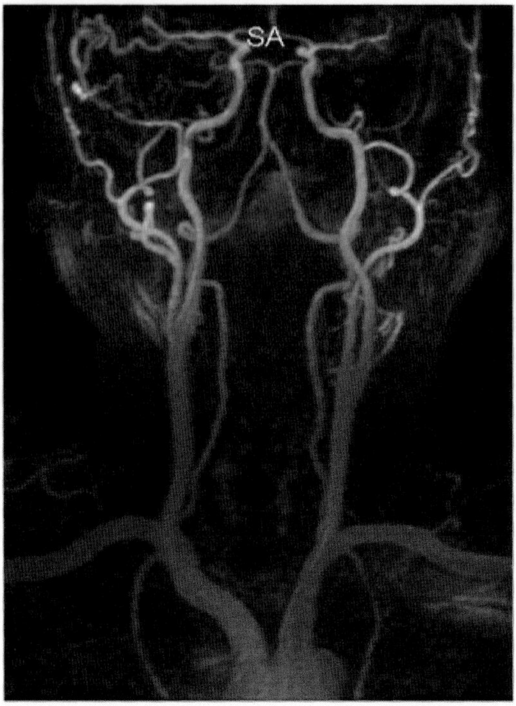

Figure 11. Single frame coronal image from a TRICKS MRA demonstrating enlarged occipital and meningeal branches coalescing at the right transverse sinus.

(Jang *et al.*, 2014). Iryo *et al.* studied 9 DSA-proven dAVF patients using 3T 4D ASL MR angiography, noting excellent ($k = 1.00$) agreement with DSA regarding fistula location and venous drainage, though only good ($k = 0.8$) for the main arterial afferents (Iryo *et al.*, 2014). Just as ASL is a helpful method in dAVF characterization, susceptibility-weighted imaging (SWI) is a technique used to identify the leptomeningeal veins via emphasizing the decreased oxygen content within their lumen (Figure 12).

These vessels are often very small and run in parallel to cortical arteries that may confound their identification. In the setting of an AV shunt, the relative increase in oxygenated blood in the leptomeningeal veins can help identify them and in turn the presence of a dAVF (Hodel *et al.*, 2011). Jain *et al.* in a review of 26 DSA-proven dAVF patients, noted that the addition of SWI sequence increased fistula localization to 75% from 27% using conventional MR sequences alone

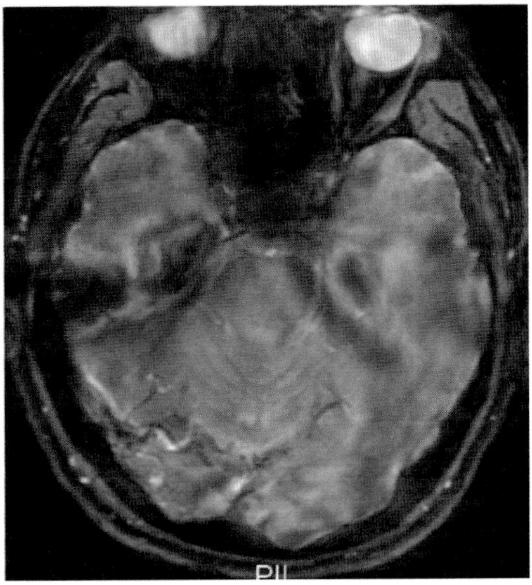

Figure 12. Axial SWI MR image demonstrating high-intensity leptomeningeal veins along the right tentorium consistent with AV shunting.

(Jain *et al.*, 2016). They also noted uniform detection of leptomeningeal drainage compared to DSA. Both MRA and CTA provide for perfusion imaging, a technique not largely used for dAVF identification, but has been demonstrated in certain cases (Kim & Krings, 2011; Noguchi *et al.*, 2006, 2010, 2011; Shidoh *et al.*, 2014). Overall, MR and CT are valuable in the diagnosis of dAVFs. In a 2016 meta-analysis by Lin *et al.* reviewing 13 studies representing 179 (MR) and 65 (CT) dAVFs with accompanying 225 and 53 controls, they noted that MR had a sensitivity of 0.90 and specificity of 0.94 while CT had a sensitivity of 0.80 and specificity of 0.87 (p = 0.02). Contrast medium use and time-resolved MR angiography did not improve MR diagnostic performance (p = 0.31 and 0.44, respectively) (Lin *et al.*, 2016).

9.7.5 *Digital subtraction angiography*

Despite the merits of MR and CT, DSA remains the standard for any patient for whom dAVF is clinically considered. DSA provides the most sensitive temporal and spatial resolution and the ability to select individual vessels of interest. This

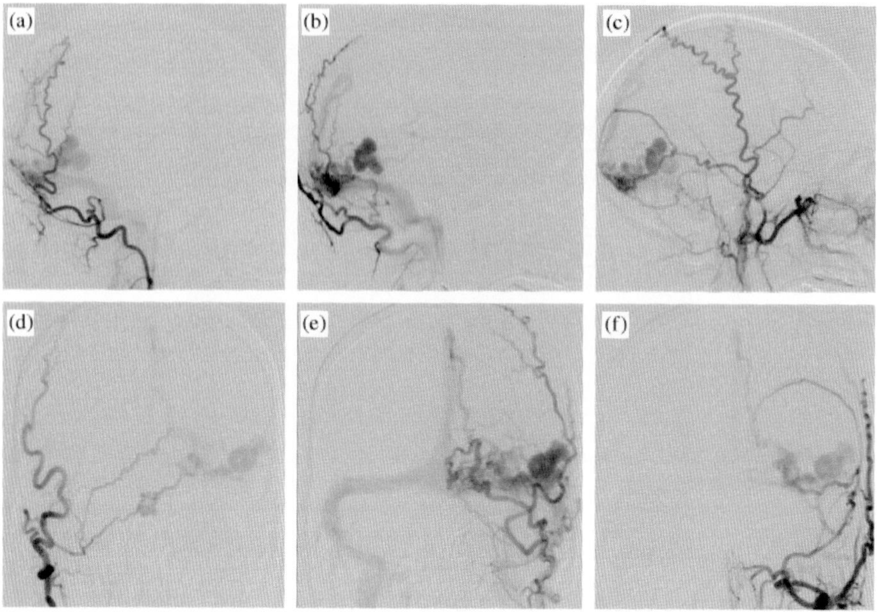

Figure 13. Composite image of a high-risk left transverse sinus dAVF. Lateral (a) and AP (d) left occipital artery DSA images demonstrating transosseous branches coalescing at the dorsum of left transverse sinus with mixed sinus and leptomenigeal venous drainage. dAVF depiction from left occipital (lateral, b; AP, e) and distal external carotid (lateral, c; AP, f). Note the venous varix or ectasia of the posterior temporal leptomeningeal vein.

provides the best method to identify the fistula location as well its afferent and efferent anatomy (Lasjaunias *et al.*, 1986). Contemporary units also have flat-panel CT (Figure 6) capability enabling high-resolution cross-sectional imaging that can further help identify AV shunt locations when 2D fluoroscopic images are inconclusive (Hiu *et al.*, 2009). As to reporting standards, a dAVF is conventionally named by its dural location, though this method is often coarse and misses many of the nuances required in treatment. Ideally, a dAVF should be named by its primary recipient venous structure with reporting on all arterial afferents. Additionally, directionality of venous drainage (anterograde or retrograde), involvement of dural sinuses, presence of venous ectasia (Figure 13) or pseudophlebecatsia (Figure 14), and dural sinus stenosis and/or thrombosis should be recorded.

Willinsky *et al.*, in a review of 122 dAVF patients undergoing MR, noted pseudophlebecatsia in 42% ($n = 51$) of cases, and its presence was more often associated

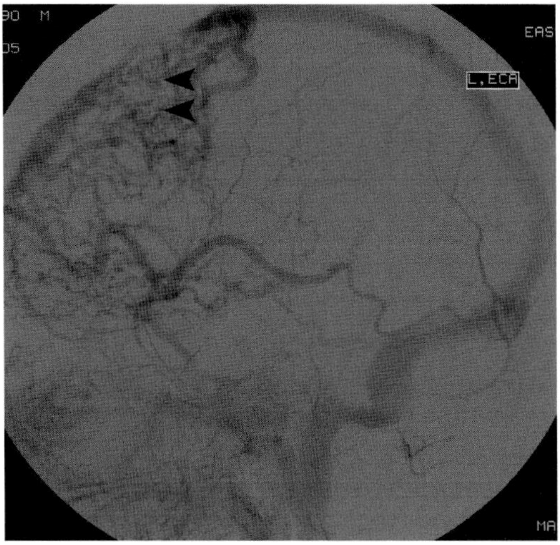

Figure 14. Lateral projection left external carotid artery DSA image in venous phase demonstrating a pseudophlebectatic pattern (arrows) around the frontal lobe.

with an aggressive presentation (73%), leptomeningeal drainage (81%), and increased parenchymal T2 signal (88%) (Willinsky *et al.*, 1999). The patterns of physiological drainage are also essential to treatment planning as those venous pathways utilized by the dAVF may also serve the brain proper and thus their sacrifice may cause injury. DSA evaluation requires a 6-vessel angiogram with close interrogation of films prior to embarking on treatment. In observation or post-treatment settings, DSA also remains the standard for evaluating the patency of dAVF.

9.7.6 *Summary: Diagnosis*

Dural AVFs are rare and often difficult to diagnose either clinically or through conventional CT and MR imaging. The dependency on temporally resolved angiographic information is essential to confirming the diagnosis, and the relative absence of this technique in most screening CT and MR studies makes dAVF identification challenging. There are, however, non-specific secondary imaging features that both modalities elucidate, MR more than CT, and the use of time-sensitive techniques is becoming more common. Ultimately, DSA remains the reference standard in dAVF evaluation and, as with any modality, identifying leptomeningeal venous drainage is an important and major goal.

9.8 Treatment

9.8.1 *Overview*

The treatment of dAVFs has evolved over the past 50 years but the core concept remains the same: focused disconnection of the arteriovenous shunt (Djindjian *et al.*, 1973; Kiyosue *et al.*, 2004; Miller & Gandhi, 2015; Piippo *et al.*, 2013a; Sarma & ter Brugge, 2003). Despite substantial literature on the topic, there is no prospective study that has examined the course of any dAVF subtype randomized to any treatment modality relative to observation, or studies that pit treatment modalities relative to one another. The literature represents the clinical experience in case series and their operators' surgical preferences. For a subset of lesions, conservative management, including observation and even ipsilateral carotid compression therapy, may be sufficient. Invasive treatments may not be necessary in patients with benign symptoms and no leptomeningeal venous drainage. A few of these "low-risk" lesions could even undergo spontaneous regression (Al-Afif *et al.*, 2015).

Conservative management aside, the two primary means of dAVF treatment are endovascular and microsurgical (Gandhi *et al.*, 2012). Both techniques have evolved over the years, and, on occasion, the same fistula may be best treated using a combination of these techniques; both techniques have also found their respective niches depending on anatomic location (Lucas *et al.*, 1997). Currently, dAVFs are approached with an endovascular method, in part because most lesions involve either the cavernous or transverse-sigmoid sinus regions. The former is difficult to access from a microsurgical standpoint, while the latter requires extensive skeletalization to sufficiently disconnect arterial inputs. A clip occluding the venous outflow is more problematic as the fistula is embedded within the sinus wall, and trapping the sinus is itself a higher risk. Microsurgical techniques are now more often suited to lesions not easily accessible from either arterial or venous approaches. In a 2014 meta-analysis, Kobayashi and Salman reviewed 24 studies representing 2,513 patients, 2,329 of whom underwent treatment, noting 1.2% mortality, 0.5% or 0.7% stroke risk (ICH or infarction), and an overall complication risk of 2.5% (Kobayashi & Al-Shahi Salman, 2014). The overall risk was slightly lower (2.2%) in those patients undergoing embolization alone rather than multimodality treatment (2.8%).

9.8.2 *Endovascular*

The performance of endovascular interventions varies depending on the materials used, operator experience, and the anatomic location and presence of sinus drainage. Endovascular techniques may be divided into arterial and venous approaches. These are largely undertaken using a transfemoral approach, though direct surgical access has been used with success when conventional endovascular routes do not permit (Cowie *et al.*, 2014; Rivet *et al.*, 2006). As general rule, the transarterial approach is preferred for most lesions. Although a transvenous method is more often first-line therapy for lesions at the cavernous sinus and those centered at and around the skull base (e.g., marginal sinus, anterior or posterior condyle, hypoglossal canal), in part because the closer proximity to the jugular veins and paravertebral plexus, as well as risks accompanying embolization of the arterial afferents that supply these skull base lesions due to their physiological supply to the lower cranial nerves (Arai *et al.*, 2004; Halbach *et al.*, 1989).

9.8.3 *Transarterial*

The transarterial approach most often employs the anterior division of the middle meningeal artery, a branch not without potentially dangerous collateral connections to the orbit or facial nerve; in large part, it provides the safest and most direct access to most lesions. Other routinely encountered external carotid branches include: (1) the occipital artery, often via transmastoid perforating branches; (2) the posterior auricular artery, often via the stylomastoid artery; and (3) the ascending pharyngeal artery, via branches of the neuromeningeal trunk (NMT) and superior pharyngeal artery. Embolization through these branches is possible, but again care must be taken particularly for any branch that may potentially supply the cranial nerves. Within the posterior circulation, transarterial embolization is routinely performed via the posterior meningeal artery as well as the dural branches of the superior cerebellar arteries. Of note, dAVFs may also recruit pial supply, and identification of this supply is essential especially during infusion of the ethylene vinyl alcohol co-polymer, Onyx (Wu *et al.*, 2016).

9.8.4 *Transvenous*

Efforts to access the venous system, particularly when thrombosed, may be challenging, and a number of methods have been described to enable such access (Hanaoka *et al.*, 2006; Lekkhong *et al.*, 2011; Weber *et al.*, 2003). Despite such challenges, there are multiple series documenting the success of transvenous treatment. Meyers *et al.*, in a review of 135 cavernous sinus dAVF patients undergoing transvenous embolization, noted 90% cure at a mean follow-up of 56 months with 131 (97%) patients showing good recovery (mRS < 3), and permanent complications occurred in 2.3% (Meyers *et al.*, 2002). Subsequent series have demonstrated similar efficacy and safety (Bink *et al.*, 2010; Ducruet *et al.*, 2013; Ellis *et al.*, 2012; He *et al.*, 2007; Kim *et al.*, 2006; Kirsch *et al.*, 2006; Klisch *et al.*, 2003; Korkmazer *et al.*, 2013; Lekkhong *et al.*, 2011; Marques *et al.*, 2010; Miller, 2012; Ng *et al.*, 2003; Pashapour *et al.*, 2014; Rhim *et al.*, 2015; Riina & Zumofen, 2014; Rodrigues *et al.*, 2014; Théaudin *et al.*, 2007; Thiex *et al.*, 2014; Yoshida *et al.*, 2010; Zanaty *et al.*, 2014), with some noting variable success based on sinus coil localization (Takahashi *et al.*, 2004) as well as higher rates of post-procedural cranial nerve palsies related to coil location and packing density (Nishino *et al.*, 2008). It is important to note leptomeningeal drainage patterns, often via connections through the sphenoparietal and superior petrosal sinues (Shimada *et al.*, 2013), in cavernous dAVF treatment as their occlusion should be undertaken prior to sinus embolization (Figure 1).

9.9 Embolic Agents

Embolization may utilize coils, particles, and/or liquid agents (e.g., ETOH, EVOH, nBCA) in isolation or combination. Ethanol is a potent sclerotherapy agent and is best suited for delivery into smaller, low-flow structures to maximize contact time with the endothelium (Alexander *et al.*, 2016; Caragine *et al.*, 2003; Phatouros *et al.*, 1999). Because of this potency and its potential to diffuse, non-target embolization may cause skin necrosis, cranial neuropathy, blindness, stroke, cardiopulmonary arrest, as well as other complications (Alexander *et al.*, 2016; Behnia, 1995; Berenguer *et al.*, 1999; Chi &Wu, 1991; Do *et al.*, 2005; Hammer *et al.*, 2001; Mitchell *et al.*, 2006; Phatouros *et al.*, 1999; Wong *et al.*, 2006; Yakes *et al.*, 1997). A more commonly used agent is *n*-Butyl cyanoacrylate (nBCA). It has been used for decades, and there are many series noting its success for various anatomical dAVF locations. Agid *et al.*, in a review of 24 ethmoidal dAVFs (Figure 15), noted an nBCA-based endovascular cure of 7 of 11 lesions and no major complications (Agid *et al.*, 2009).

Cannizzaro *et al.* demonstrated complete occlusion of 13 ethmoidal lesions using either endovascular (46.1%) or surgical (53.9%) methods with 2 perioperative, non-permanent complications (1 from each treatment cohort) (Cannizzaro *et al.*, 2016). The authors noted greater technical success while using nBCA concentration of 50%. Conversely, Kim *et al.*, in a review of 115 dAVF cases undergoing nBCA embolization, noted improved occlusion rates (55% versus 10%) in Borden III lesions when using more dilute nBCA (23% versus 37%) (Kim *et al.*, 2011), though these lesions were largely not ethmoidal. The need for more dilute nBCA in part reflects the goal of more distal penetration of embolic material, an effect that may be enabled through the use of simultaneous guide-catheter

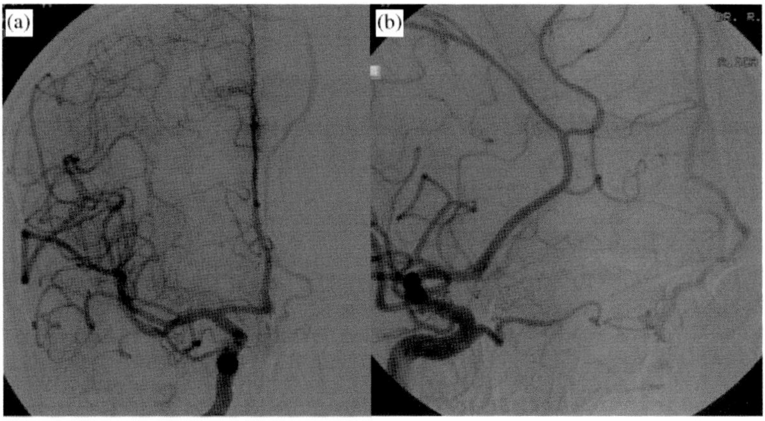

Figure 15. AP (a) and oblique (b) right internal carotid DSA images demonstrating an ethmoidal dAVF supplied by branches of the ophthalmic artery and with leptomeningeal venous drainage.

5% dextrose solution infusion (Moore *et al.*, 2006) or wedge catheter position (Nelson *et al.*, 2003); both techniques delay polymerization through reduction in nBCA contact with blood.

Since its introduction around 2005, Onyx has grown to be the primary means of embolization, and with it, an increase in single-time angiographic cures have been noted (Maimon *et al.*, 2011). In a series of 44 dAVFs undergoing Onyx embolization, Abud *et al.* noted angiographic cure in 35 cases, 27 of which were cured from a single arterial pedicle. Complications occurred in 6 patients, all transient, and 4 of which were neuropathic in nature (Abud *et al.*, 2011). In their series of 30 dAVF patients undergoing Onyx embolization, Cognard *et al.* noted occlusion in 24 (20 in a single session) and there were no permanent complications (Cognard *et al.*, 2008). Additional series have been published noting similar efficacy and safety (Cha *et al.*, 2013; Cognard *et al.*, 2008; Guedin *et al.*, 2010; Hu *et al.*, 2011; Lv *et al.*, 2009; Maimon *et al.*, 2011; Rabinov *et al.*, 2013; Wu *et al.*, 2016). This is in large part related to its non-adhesive quality, permitting prolonged embolization sessions that allow for retrograde casting of alternative arterial pedicles as well as efferent veins. This quality, however, also allows for embolization of non-target arteries and veins and places a premium on a complete understanding of not only those vessels involved in the fistula, but those branches occult on DSA imaging that supply cranial nerves and/or the brain itself that may be involved. New detachable tip catheters and the use of balloons to assist in catheter (Zhao *et al.*, 2012) and embolic agent flow directionality have also emerged. The balloons may be used for transarterial purposes to help accelerate anterograde embolization and prevent retrograde filling of collateral branches (Chiu *et al.*, 2013; Deng *et al.*, 2013; Gonzalez *et al.*, 2012; Jittapiromsak *et al.*, 2013; Kim, J.W. *et al.*, 2016; Liang *et al.*, 2013; Paramasivam *et al.*, 2013b; Shi *et al.*, 2013a; Spiotta *et al.*, 2011; Yu *et al.*, 2012). They may also be used in the arterial anatomy to help prevent retrograde filling of the internal carotid artery when transvenous Onyx delivery is used for cavernous-carotid fistulas (Gonzalez *et al.*, 2012; Yu *et al.*, 2012). They may be used within the venous anatomy for similar accelerated embolization and vein/sinus protection purposes, as sinus preservation may prove difficult using Onyx alone from a transarterial position (Chiu *et al.*, 2013; Jittapiromsak *et al.*; 2013, Shi *et al.*, 2013a; Zhang *et al.*, 2015). For example, Abud *et al.* noted that in 13 of 20 dAVFs the affected sinus (Borden I and II) was occluded as a result of Onyx penetration (Abud *et al.*, 2011). Generally, a more recent series from Gross *et al.* reported on the chronological improvement in the technical success of endovascular intervention over a 20-year interval. In a review of 251 patients, they noted an overall occlusion rate of 70% with a 3% recurrence and 3% permanent complication rate. Following the introduction of Onyx, occlusion rates rose from 60% to 76% ($p = 0.01$) as well as use of a single arterial pedicle (11% versus 29%, $p = 0.002$) and exclusively transarterial (23% versus 43%, $p = 0.002$) treatment (Gross *et al.*, 2016a). Although there is limited data, the success of Onyx has also led to alternative DMSO-based embolic agents (Akmangit *et al.*, 2014; Leyon *et al.*, 2016). In comparing nBCA and Onyx agents, Rabinov *et al.*, in a review of 53 dAVF, compared those treated with nBCA ($n = 21$) with those treated with Onyx ($n = 35$), noting higher rates of occlusion with the latter (33% versus 83%, $p < 0.001$) with comparable rates of permanent complications (Rabinov *et al.*, 2013).

9.10 Anatomic Considerations

As noted above, fistulas involving the sinus may prove more difficult to treat via a transarterial approach (even with the use of Onyx) if sinus preservation is desired (Lv *et al.*, 2011). In instances where sinus sacrifice is required, liquids, coils, or even surgery may be safely employed, though it is essential that such treatments isolate those occlusive methods to only those sinus segments no longer involved in physiological drainage (Carlson *et al.*, 2013; Tanoue *et al.*, 2012). In a review of 21 patients with benign (Borden I) lesions, Bink *et al.* noted comparable rates of pulsatile tinnitus resolution in the treated (44.4%) and non-treated (41.6%) groups, though of the 9 patients undergoing treatment only two cases demonstrated complete occlusion (both transvenous coil embolization of a parallel venous channel) and both had resolution of symptoms (Bink *et al.*, 2011). Kim *et al.*, in a review of 68 dAVFs treated with Onyx, noted occlusion in 92% of non-sinus fistulae and only 63% in those involving a sinus (Kim, *et al.*, 2016). Cho *et al.* in a review of 34 patients (27 Borden II or III) with transverse-sigmoid dAVFs undergoing endovascular treatment, found initial occlusion in 50% with increase to 64.5% in longer term follow-up with 4 cases demonstrating interval occlusion and 5 cases of angiographic recurrence (Cho *et al.*, 2013). Lasting occlusion more often occurred in patients undergoing some form of transvenous embolization (14 of 19, 74%) and permanent complications were noted in 3 patients. Similarly, Ertl *et al.*, reviewing 36 transverse-sigmoid dAVFs undergoing endovascular treatment, noted sacrifice of the sinus had higher rates of occlusion than sinus-sparing techniques (93% versus 71%), though complications were also higher (20% versus 0%) (Ertl *et al.*, 2016b). Kirsch *et al.* also noted excellent (81%) occlusion rates in their review of 150 transverse-sigmoid dAVFS using transvenous embolization alone, though permanent complications were lower (2%) (Kirsch *et al.*, 2009). Given this trade-off between potentially higher complication rates for higher complete occlusive results, some advocate for embolization targeting high-risk features. Van Dijk *et al.* in a review of 23 dAVFs with combined sinus and leptomeningeal drainage, described selective occlusion and microsurgical ligation of the high-risk components noting comparable clinical outcomes and lower overall complication rates over a mean 27-month follow-up period to those with complete occlusion (van Dijk *et al.*, 2004).

Reciprocally, higher rates of transarterial occlusion have been found in those patients with exclusively leptomeningeal drainage (Figure 16). Baltsavias *et al.* in a review of 170 dAVF cases, largely treated in a single session (85%), transarterial manner (85%) using nBCA (73%), noted that the long-term occlusion rate was 85.5%, with 78% success in those without leptomeningeal drainage and 89.5% with leptomeningeal drainage (Baltsavias & Valavanis, 2014). There were no deaths and 2.3% permanent complications, all occurring in the leptomeningeal drainage group. Guedin *et al.* in a series of 38 dAVFs with leptomeningeal drainage, noted complete occlusion using transarterial nBCA in 34 patients and embolization of high-risk components only in the remainder of cases with no complications (Guedin *et al.*, 2010). Chandra *et al.* in a review of 40 dAVF patients (26 Borden III) treated with Onyx, noted 95% initial occlusion (85% in a single session) and 2% permanent complications (Chandra *et al.*, 2014). They did note 2 of 35 (6%) cases with short-term follow-up demonstrating angiographic recurrence, though no clinical symptoms manifested-and none of the long-term follow-up cases had recurrence. In a meta-analysis of 274 tentorial dAVF cases with predominantly exclusively leptomeningeal drainage, Cannizzaro *et al.*

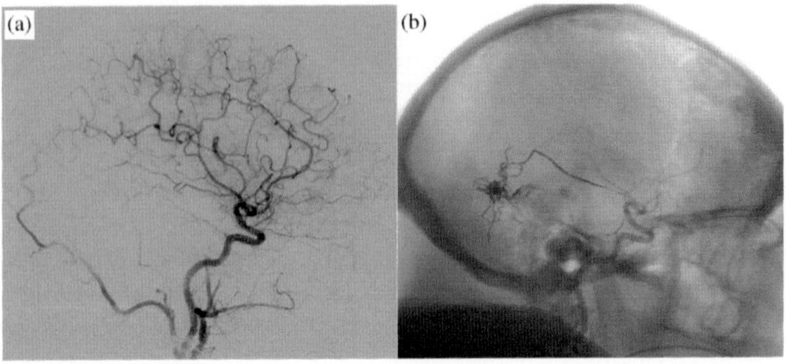

Figure 16. Lateral subtracted (a) and unsubtracted (b) left common carotid artery DSA images demonstrating complete occlusion of the Labbe-type dAVF with Onyx casting the fistula and the middle meningeal artery delivery pedicle. Note, this case is also depicted in Figures 3 and 9.

documented significant differences among treatment modalities in their respective occlusion rates with endovascular at 71%, surgery at 81%, and combination therapy at 84% ($p < 0.01$). There were significantly lower rates of perioperative morbidity in the endovascular cohort (6%) relative to surgical (18%) or combined groups (14%, $p < 0.01$), though this effect did not persist as it relates to long-term morbidity. Daniels *et al.* in a review of a largely (75%) tentorial series of 28 dAVFs presenting with hemorrhage, again noted the disproportionate number of males (86%) and that despite ICH presentation most (71%) had an excellent clinical outcome (mRS < 2) (Daniels *et al.*, 2013). There are additional, smaller series and case reports looking at treatment at various other anatomic locations such as the inferior petrosal sinus (Barnwell *et al.*, 1990), marginal sinus (Choi *et al.*, 2012; Wang *et al.*, 2015), posterior condylar canal (Kiyosue *et al.*, 2007; Maus *et al.*, 2016), hypoglossal canal (Manabe *et al.*, 2008; Spittau *et al.*, 2015), and middle cranial fossa (Shi *et al.*, 2013b).

9.11 Follow-up

Despite increases in first-time angiographic cure, whether by Onyx or any combination of improvements in endovascular imaging and delivery systems, a fair number of cases will have some residual AV shunt at case conclusion. Depending on the venous anatomy (e.g., high risk versus low risk) and the patient's symptoms, angiographic follow-up may be short term (< 3 months), mid-range (6–12 months), or longer. For example, Fok *et al.* demonstrated interval thrombosis in 3 patients treated using transarterial nBCA during short-term (<30 day) follow-up (Fok *et al.*, 2004). Nonetheless, adjuvant use of pro-thrombotic agents has been employed to assist in this process. In a series of 7 dAVFs treated with epsilon-aminocaproic acid (Amicar), Kallmes *et al.* noted progressive thrombosis of 2 cases and reduction in AV shunt velocity in 2 other cases (Kallmes *et al.*, 2000), though this management technique is not widely used. Additionally, even in cases where occlusion is noted during initial treatment, angiographic follow-up is important. Ambekar *et al.* in a review of 21 dAVF cases deemed successfully occluded, noted 3 (14.1%) with angiographic recurrence at a mean follow-up of 14 months (Ambekar *et al.*, 2016). In a review of 104 dAVF patients effectively treated with clinical and imaging follow-up, Baltsavias *et al.* similarly noted 9 with recurrence (Ramalingaiah *et al.*, 2013).

9.12 Treatment

9.12.1 *Surgery*

Microsurgery is not typically the preferred initial treatment method for the vast majority of dAVF. However, it plays an important adjunct role in association with embolization, and can play a more primary role in cases where embolization is either not as effective or not as safe due to anatomic considerations.

Surgery for dAVFs has evolved over time to reflect those anatomic locations for which endovascular therapies are less effective. (Kosnik *et al.*, 1974; Sundt and Piepgras, 1983; Youssef *et al.*, 2014). In a literature review by Cannizzaro *et al.* of 274 tentorial dAVF patients, the proportion of patients undergoing surgery alone declined from 38.7% from 1980–1995 to 20.4% from 2006–2014, while the same intervals yielded an increase from 16.1 to 48% for those undergoing endovascular care alone (Cannizzaro *et al.*, 2015). Despite this trend, tentorial lesions often have dural afferents arising from internal carotid and superior cerebellar arteries, sometimes limiting microcatheter access, and rarely involve sinuses larges enough to permit transvenous passage. Microsurgery is also used more frequently in settings where definitive occlusion of the AV shunt is required to eliminate leptomeningeal venous drainage, often in combination with endovascular embolization and often with higher rates of occlusion than when either modality is used in isolation (Lucas *et al.*, 1997). For example, anterior cranial fossa or ethmoidal dAVFs, though at times accessible to endovascular techniques (Robert *et al.*, 2016b), are more often than not best treated with open surgery. In a review of 24 lesions, Agid *et al.* noted obliteration of all 15 cases undergoing microsurgery and without complications

(Agid *et al.*, 2009). Similarly, Gross *et al.* in a review of 24 cases, noted occlusion in all without permanent complications (Gross *et al.*, 2016b). There are additional series documenting the excellent safety and efficacy of microsurgery for these lesions (Al-Mahfoudh *et al.*, 2015; Martin *et al.*, 1990). Beyond these two locations, there are a number of case series documenting general surgical success for other, non-cavernous dAVFs. In a general review of 53 cases, Dutzmann *et al.* reported on outcomes of 24 undergoing surgical treatment, all with occlusive outcomes and 8% complication rate (Dützmann *et al.*, 2011). Eftekhar *et al.* reported on 42 transverse-sigmoid dAVFs undergoing surgery noting occlusion in all cases and without complications (Eftekhar and Morgan, 2013).

9.12.2 *Radiosurgery*

Radiosurgery is a longstanding method for the treatment of vascular malformations and has been used with success for dAVFs as well (Dalyai *et al.*, 2013; Loumiotis *et al.*, 2011; Söderman *et al.*, 2006). Similar to its use in AVMs, radiosurgery is most effective when the lesion is well defined. To optimize that identification, Link *et al.* recommended performing targeting and treatment prior to embolization, as the latter may obscure anatomic details (Link *et al.*, 1996). In a literature review of 142 cases culled from 4 case series, Dalyai reported angiographic occlusion rates of 68–93% at follow-up intervals of 2 years or longer and with 13 complications (5 hemorrhagic) (Dalyai *et al.*, 2013). Similar results were seen in a 2012 meta-analysis of 558 cases by Gross *et al.* that noted an overall occlusion rate of 71% with permanent complications in 2.4% (1 death) and post-radiation hemorrhage occurring in 1.6% of cases (4.8% of those with leptomeningeal venous drainage). More specifically, they noted an 84% occlusion rate for cavernous lesions, though lower rates for transverse-sigmoid (58%) and for tentorial (59%) lesions ($p = 0.032$). Additionally, occlusion rates were higher for lesions without leptomeningeal drainage (80% versus 60%, $p < 0.001$). There are a number of other series documenting radiosurgical success not only as a standalone therapy (Park *et al.*, 2016; Söderman *et al.*, 2013), but also as an adjuvant to endovascular and microsurgical techniques (Guo *et al.*, 1998; Koebbe *et al.*, 2005; Yang *et al.*, 2013).

9.12.3 *Pediatric*

AV fistulae in children are rare, and the dural subtype even more so. Garcia-Monaco *et al.* reported 11 dAVF cases of 112 total AV shunting lesions noting 4 that presented with multiple lesions (Garcia-Monaco *et al.*, 1991). Similarly, Hetts *et al.* noted 22 cases out of 423 pediatric shunting lesions (Hetts *et al.*, 2016). As briefly touched on above, pediatric dAVFs, as described by Lasjaunias *et al.* and others (Kincaid *et al.*, 2001), may be divided into dural sinus malformations (neonatal) and infantile and adolescent types, the last of which is similar in natural history and clinical presentation to adult dAVFs (Lasjaunias *et al.*, 1996). In their review of 29 pediatric cases, Lasjaunias *et al.* described 19 neonatal lesions typified by involvement of the posterior sinus, with or without torcular involvement, and the presence of venous lakes and slow flow or those with normal sinus anatomy and high flow (this latter type manifesting with congestive heart failure). Of 7 infantile type dAVFs, higher flow was encountered with absence of venous lakes. For those with more venous outflow restriction, hydrovenous disorders emerged characterized by disproportionate increase in head circumference, developmental delay, hydrocephalus, white matter calcification, seizures, hemorrhage, and parenchymal volume loss (i.e., melting brain syndrome) (Lasjaunias *et al.*, 1996). Pediatric dAVFs may also represent an element of a disease spectrum that include other vascular malformations and morphological abnormalities (Bun *et al.*, 2009). As to treatment, in the Hetts *et al.* series, endovascular treatment achieved complete occlusion in 8 of 21 cases. Around 77% of patients presenting after 1 year, as opposed to 57% patients presenting before 1 year, had a good (mRS < 3) outcome; there were 6 deaths (Hetts *et al.*, 2016).

9.12.4 *Alternative therapies and research*

The study of dAVFs, like so many other rare disorders, is largely limited to retrospective case series from centers of excellence around the world. These collective efforts are invaluable, but ultimately cannot best answer questions of natural history, standardized approaches to treatment, or methods to assess innovative therapies. Nonetheless, prospective study is also difficult. AVMs are considerably more common than dAVFs, and the ARUBA trial, the only prospective trial to assess natural history relative to best surgical management, was plagued by slow enrollment at treating centers. As discussed above, in addition to the issue related to prevalence, dAVFs also lack a strong inheritable component, making animal modeling difficult. Lastly, with endovascular therapies being the primary method of treatment, surgical tissue collection is rare, and thus our ability to better develop radio-genomic data is slowed.

To solve these issues, investigators must work collaboratively in and among centers of excellence. Organizations such as the NIH-supported Brain Vascular

Malformation Consortium help establish reporting standards for systematized data collection and centralized analytic groups for optimization of review. From such organizations, research has grown using alternative biomarkers for diseases, including advanced imaging measures (e.g., computational fluid dynamic analysis and vessel wall enhancement patterns) to better understand the physical and cellular events mirroring traditional angioarchitectural details. Additionally, methods for endovascular-based endothelial cell collection continue to evolve, permitting single cell gene expression analysis (Sun *et al.*, 2014) and better animal models for genetic study (Yang *et al.*, 2015). Such methods would enable genome-wide association studies to help identify corollaries with other disease pathways, e.g., diabetes (Lee *et al.*, 2013). It is through these efforts, as well as other anatomic discovery(Louveau *et al.*, 2015), that we hope to expand our understanding of dAVFs and the repertoire of surgical and, potentially, medical or gene therapies (Gross *et al.*, 2013; Janardhanan *et al.*, 2013; Yang *et al.*, 2013) available to patients and providers.

9.13 Conclusion

dAVF is a rare cause of stroke and other neurological disorders, though one that has excellent and lasting therapeutic options. Consequently, it is essential to any provider dealing with neurovascular disease to consider dAVF as a cause of a patient's symptoms when appropriate. Over the past 50 years, much has been learned, through both human and animal study, about the origins of dAVFs, particularly the roles of venous thrombosis and hypertension in the formation and progression of the disease. Additionally, advances in angiography have permitted for more refined assessment of the dAVF angioarchitecture consistently noting the association between more aggressive clinical presentations, inclusive of stroke, and leptomeningeal venous drainage. As it relates to natural history, those dAVFs with exclusively sinus drainage have an almost uniformly benign clinical course, though those with any leptomeningeal venous drainage have some risk of stroke. Diagnostic imaging, namely, CT and MR, have advanced considerably over this time and the techniques of each have improved in their identification of secondary dAVF features as well as the essential localization of the AV shunt itself. MR and its flow-sensitive techniques in particular have proven most useful in the work-up of a presumed dAVF. In parallel to imaging progression, endovascular devices and techniques have also changed. Endovascular therapy is the first-line method for dAVFs, regardless of location, with demonstrable improvements in overall treatment success with accompanying declines in peri-procedural morbidity over the past 30 years. Surgery still plays an important role as well, particularly in those cases where endovascular therapy is incomplete and definitive treatment is required. Despite all these advances, much is still unknown about dAVF given their scarcity and seeming lack of germline genetic mechanism, as well as the continued need for safer and more definitive therapies. Dural hypoxia and its subsequent angiogenesis have demonstrated essential features of dAVF pathophysiology and are logical targets for future therapies.

References

Abeles, D., Kwei, S., *et al.* (2006). Gene expression changes evoked in a venous segment exposed to arterial flow. *J Vasc Surg* **44**(4): 863–870.

Abud, T.G., Nguyen, A., *et al.* (2011). The use of Onyx in different types of intracranial dural arteriovenous fistula. *AJNR Am J Neuroradiol* **32**(11): 2185–2191.

Adeeb, N., Mortazavi, M.M., *et al.* (2012). The cranial dura mater: A review of its history, embryology, and anatomy. *Childs Nerv Syst* **28**(6): 827–837.

Agid, R., Terbrugge, K., *et al.* (2009). Management strategies for anterior cranial fossa (ethmoidal) dural arteriovenous fistulas with an emphasis on endovascular treatment. *J Neurosurg* **110**(1): 79–84.

Akmangit, I., Daglioglu, E., *et al.* (2014). Preliminary experience with squid: A new liquid embolizing agent for AVM, AV fistulas and tumors. *Turk Neurosurg* **24**(4): 565–570.

Al-Afif, S., Nakamura, M., *et al.* (2015). Spontaneous closure of a dural arteriovenous fistula. *J Neurointerv Surg* **7**(8): e28.

Al-Mahfoudh, R., Kirollos, R., *et al.* (2015). Surgical disconnection of the cortical venous reflux for high-grade intracranial dural arteriovenous fistulas. *World Neurosurg* **83**(4): 652–656.

Al-Shahi, R., Bhattacharya, J., *et al.* (2003). Prospective, population-based detection of intracranial vascular malformations in adults: The Scottish Intracranial Vascular Malformation Study (SIVMS). *Stroke* **34**(5): 1163–1169.

Alexander, M.D., Halbach, V., *et al.* (2016). Transvenous ethanol sclerotherapy of feeding arteries for treatment of a dural arteriovenous fistula. *J Neurointerv Surg*. **9**(3): e12

Ambekar, S., Gaynor, B.G., *et al.* (2016). Long-term angiographic results of endovascularly "cured" intracranial dural arteriovenous fistulas. *J Neurosurg* **124**(4): 1123–1127.

Aminoff, M.J. (1973). Vascular anomalies in the intracranial dura mater. *Brain* **96**(3): 601–612.

Amiridze, N., Darwish, R. (2009). Hemodynamic instability during treatment of intracranial dural arteriovenous fistula and carotid cavernous fistula with Onyx: Preliminary results and anesthesia considerations. *J Neurointerv Surg* **1**(2): 146–150.

Amukotuwa, S.A., Heit, J.J., *et al.* (2016). Detection of cortical venous drainage and determination of the borden type of dural arteriovenous fistula by means of 3D pseudocontinuous arterial spin-labeling MRI. *AJR Am J Roentgenol* **207**(1): 163–169.

Arai, Y., Ishii, H., *et al.* (2004). Dural arteriovenous fistula within the hypoglossal canal successfully treated by transvenous embolization. Case report. *Interv Neuroradiol* **10**(1): 59–62.

Awad, I.A., Little, J.R., *et al.* (1990). Intracranial dural arteriovenous malformations: Factors predisposing to an aggressive neurological course. *J Neurosurg* **72**(6): 839–850.

Azuma, M., Hirai, T., *et al.* (2015). Evaluation of intracranial dural arteriovenous fistulas: Comparison of unenhanced 3T 3D time-of-flight MR angiography with digital subtraction angiography. *Magn Reson Med Sci* **14**(4): 285–293.

Bai, Y., He, C., *et al.* (2012). *De novo* multiple dural arteriovenous fistulas and arteriovenous malformation after embolization of cerebral arteriovenous fistula: Case report. *Childs Nerv Syst* **28**(11): 1981–1983.

Baltsavias, G., Kumar, R., *et al.* (2015a). Cranial dural arteriovenous shunts, Part 2. The shunts of the bridging veins and leptomeningeal venous drainage. *Neurosurg Rev* **38**(2): 265–271; discussion 272.

Baltsavias, G., Parthasarathi, V., *et al.* (2015b). Cranial dural arteriovenous shunts. Part 1. Anatomy and embryology of the bridging and emissary veins. *Neurosurg Rev* **38**(2): 253–263; discussion 263–254.

Baltsavias, G., Roth, P., *et al.* (2015c). Cranial dural arteriovenous shunts. Part 3. Classification based on the leptomeningeal venous drainage. *Neurosurg Rev* **38**(2): 273–281; discussion 281.

Baltsavias, G., Spiessberger, A., *et al.* (2015d). Cranial dural arteriovenous shunts. Part 4. Clinical presentation of the shunts with leptomeningeal venous drainage. *Neurosurg Rev* **38**(2): 283–291; discussion 291.

Baltsavias, G., Valavanis, A. (2014). Endovascular treatment of 170 consecutive cranial dural arteriovenous fistulae: Results and complications. *Neurosurg Rev* **37**(1): 63–71.

Barnwell, S.L., Halbach, V.V., *et al.* (1990). Dural arteriovenous fistulas involving the inferior petrosal sinus: Angiographic findings in six patients. *AJNR Am J Neuroradiol* **11**(3): 511–516.

Barnwell, S.L., Halbach, V.V., *et al.* (1991). Multiple dural arteriovenous fistulas of the cranium and spine. *AJNR Am J Neuroradiol* **12**(3): 441–445.

Barrow, D.L., Spector, R.H., *et al.* (1985). Classification and treatment of spontaneous carotid-cavernous sinus fistulas. *J Neurosurg* **62**(2): 248–256.

Behnia, R. (1995). Systemic effects of absolute alcohol embolization in a patient with a congenital arteriovenous malformation of the lower extremity. *Anesth Analg* **80**(2): 415–417.

Beijer, T.R., van Dijk, E.J., *et al.* (2013). 4D-CT angiography differentiating arteriovenous fistula subtypes. *Clin Neurol Neurosurg* **115**: 1313–1316.

Berenguer, B., Burrows, P., *et al.* (1999). Sclerotherapy of craniofacial venous malformations: Complications and results. *Plast Reconstr Surg* **104**(1): 1–11; discussion 12–15.

Bink, A., Berkefeld, J., *et al.* (2011). Long-term outcome in patients treated for benign dural arteriovenous fistulas of the posterior fossa. *Neuroradiology* **53**(7): 493–500.

Bink, A., Berkefeld, J., *et al.* (2012). Detection and grading of dAVF: Prospects and limitations of 3T MRI. *Eur Radiol* **22**(2): 429–438.

Bink, A., Goller, K., *et al.* (2010). Long-term outcome after coil embolization of cavernous sinus arteriovenous fistulas. *AJNR Am J Neuroradiol* **31**(7): 1216–1221.

Biswas, S., Chandran, A., *et al.* (2015). Accuracy of four-dimensional CT angiography in detection and characterisation of arteriovenous malformations and dural arteriovenous fistulas. *Neuroradiol J* **28**(4): 376–384.

Borden, J.A., Wu, J.K., *et al.* (1995). A proposed classification for spinal and cranial dural arteriovenous fistulous malformations and implications for treatment. *J Neurosurg* **82**(2): 166–179.

Brown, R.D., Wiebers, D.O., *et al.* (1994). Intracranial dural arteriovenous fistulae: Angiographic predictors of intracranial hemorrhage and clinical outcome in nonsurgical patients. *J Neurosurg* **81**(4): 531–538.

Brunereau, L., Gobin, Y.P., *et al.* (1996). Intracranial dural arteriovenous fistulas with spinal venous drainage: Relation between clinical presentation and angiographic findings. *AJNR Am J Neuroradiol* **17**(8): 1549–1554.

Bulters, D.O., Mathad, N., *et al.* (2012). The natural history of cranial dural arteriovenous fistulae with cortical venous reflux — the significance of venous ectasia. *Neurosurgery* **70**(2): 312–318; discussion 318–319.

Bun, Y.Y., Ming, C.K., *et al.* (2009). Endovascular treatment of a neonate with dural arteriovenous fistula and other features suggestive of cerebrofacial arteriovenous metameric syndromes. *Childs Nerv Syst* **25**(3): 383–387.

Cannizzaro, D., Brinjikji, W., *et al.* (2015). Changing Clinical and Therapeutic Trends in Tentorial Dural Arteriovenous Fistulas: A Systematic Review. *AJNR Am J Neuroradiol* **36**(10): 1905–1911.

Cannizzaro, D., Peschillo, S., *et al.* (2016). Endovascular and surgical approaches of ethmoidal dural fistulas: A multicenter experience and a literature review. *Neurosurg Rev.*

Cannizzaro, D., Rammos, S.K., *et al.* (2016). The lateral mesencephalic vein: Surgical anatomy and its role in the drainage of tentorial dural arteriovenous fistulae. *World Neurosurg* **85**: 163–168.

Caragine, L., Halbach, V., *et al.* (2003). Parallel venous channel as the recipient pouch in transverse/sigmoid sinus dural fistulae. *Neurosurgery* **53**(6): 1261–1266; discussion 1266–1267.

Carlson, A.P., Alaraj, A., *et al.* (2013). Endovascular approach and technique for treatment of transverse-sigmoid dural arteriovenous fistula with cortical reflux: The importance of venous sinus sacrifice. *J Neurointerv Surg* **5**(6): 566–572.

Celik, O., Piippo, A., *et al.* (2010). Management of dural arteriovenous fistulas - Helsinki and Kuopio experience. *Acta Neurochir Suppl* **107**: 77–82.

Cha, K.C., Yeon, J.Y., *et al.* (2013). Clinical and angiographic results of patients with dural arteriovenous fistula. *J Clin Neurosci* **20**(4): 536–542.

Chahbazian, K., Théaudin, M., *et al.* (2014). Reversible pseudo-Creutzfeldt–Jakob syndrome related to cerebral dural arteriovenous fistula. *J Am Geriatr Soc* **62**(10): 2024–2026.

Chandra, R.V., Leslie-Mazwi, T.M., *et al.* (2014). Transarterial onyx embolization of cranial dural arteriovenous fistulas: Long-term follow-up. *AJNR Am J Neuroradiol* **35**(9): 1793–1797.

Chaudhary, M.Y., Sachdev, V.P., *et al.* (1982). Dural arteriovenous malformation of the major venous sinuses: An acquired lesion. *AJNR Am J Neuroradiol* **3**(1): 13–19.

Chen, L., Mao, Y., *et al.* (2009). Local chronic hypoperfusion secondary to sinus high pressure seems to be mainly responsible for the formation of intracranial dural arteriovenous fistula. *Neurosurgery* **64**(5): 973–983; discussion 983.

Chi, L., Wu, W. (1991). Mechanism of hemolysis of red blood cell mediated by ethanol. *Biochim Biophys Acta* **1062**(1): 46–50.

Chiu, A.H., Aw, G., *et al.* (2013). Double-lumen arterial balloon catheter technique for Onyx embolization of dural arteriovenous fistulas: Initial experience. *J Neurointerv Surg* **6**(5): 400–403.

Cho, W.S., Han, J.H., *et al.* (2013). Treatment outcomes of intracranial dural arteriovenous fistulas of the transverse and sigmoid sinuses from a single institute in Asia. *J Clin Neurosci* **20**(7): 1007–1012.

Choi, H.S., Kim, D.I., *et al.* (2012). Endovascular treatment of dural arteriovenous fistula involving marginal sinus with emphasis on the routes of transvenous embolization. *Neuroradiology* **54**(2): 163–169.

Chung, S.J., Kim, J.S., *et al.* (2002). Intracranial dural arteriovenous fistulas: Analysis of 60 patients. *Cerebrovasc Dis* **13**(2): 79–88.

Cognard, C., Gobin, Y., *et al.* (1995). Cerebral dural arteriovenous fistulas: Clinical and angiographic correlation with a revised classification of venous drainage. *Radiology* **194**(3): 671–680.

Cognard, C., Houdart, E., *et al.* (1997). Long-term changes in intracranial dural arteriovenous fistulae leading to worsening in the type of venous drainage. *Neuroradiology* **39**(1): 59–66.

Cognard, C., Januel, A., *et al.* (2008). Endovascular treatment of intracranial dural arteriovenous fistulas with cortical venous drainage: New management using Onyx. *AJNR Am J Neuroradiol* **29**(2): 235–241.

Cohen, S.D., Goins, J.L., *et al.* (2009). Dural arteriovenous fistula: Diagnosis, treatment, and outcomes. *Laryngoscope* **119**(2): 293–297.

Cooke, D.L., Johnson, G., *et al.* (2011). Synchronous chronic middle cerebral artery occlusion and ipsilateral dural arteriovenous fistula. *Clin Nucl Med* **36**(7): 570–573.

Cordonnier, C., Al-Shahi Salman, R., *et al.* (2008). Differences between intracranial vascular malformation types in the characteristics of their presenting haemorrhages: Prospective, population-based study. *J Neurol Neurosurg Psychiatry* **79**(1): 47–51.

Cowie, C.J., Pešić-Smith, J.D., *et al.* (2014). Minimally invasive neurosurgical access techniques to facilitate endovascular treatment of dural arteriovenous fistulae. *Br J Neurosurg*: 1–3.

Dalyai, R.T., Ghobrial, G., *et al.* (2013). Radiosurgery for dural arterio-venous fistulas: A review. *Clin Neurol Neurosurg* **115**(5): 512–516.

Dandy, W.E. (1946). Arteriovenous aneurysms of the scalp and face. *Arch Surg* **52**: 1–32.

Daniels, D.J., Vellimana, A.K., *et al.* (2013). Intracranial hemorrhage from dural arteriovenous fistulas: Clinical features and outcome. *Neurosurg Focus* **34**(5): E15.

Davies, M., TerBrugge, K., *et al.* (1996). The validity of classification for the clinical presentation of intracranial dural arteriovenous fistulas. *J Neurosurg* **85**(5): 830–837.

Davies, M.A., Saleh, J., *et al.* (1997a). The natural history and management of intracranial dural arteriovenous fistulae. Part 1: Benign lesions. *Interv Neuroradiol* **3**(4): 295–302.

Davies, M.A., Ter Brugge, K., *et al.* (1997b). The natural history and management of intracranial dural arteriovenous fistulae. Part 2: Aggressive lesions. *Interv Neuroradiol* **3**(4): 303–311.

Deng, J.P., Zhang, T., *et al.* (2013). Treatment of dural arteriovenous fistula by balloon-assisted transarterial embolization with Onyx. *Clin Neurol Neurosurg* **115**(10): 1992–1997.

Dietz, R.R., Davis, W.L., *et al.* (1994). MR imaging and MR angiography in the evaluation of pulsatile tinnitus. *AJNR Am J Neuroradiol* **15**(5): 879–889.

Djindjian, R., Cophignon, J., *et al.* (1973). Superselective arteriographic embolization by the femoral route in neuroradiology. Study of 50 cases. II. Embolization in vertebromedullary pathology. *Neuroradiology.* **6**(3): 132–142.

Do, Y.S., Yakes, W.F., *et al.* (2005). Ethanol embolization of arteriovenous malformations: Interim results. *Radiology* **235**(2): 674–682.

Dong, Y., Cao, W., *et al.* (2014). Ipsilateral intracranial edema associated with drainage patterns of dural arteriovenous fistula. *J Stroke Cerebrovasc Dis* **23**(5): 1094–1098.

Du, R., Binder, D.K., *et al.* (2003). Trigeminal neuralgia in a patient with a dural arteriovenous fistula in Meckel's cave: Case report. *Neurosurgery* **53**(1): 216–221; discussion 221.

Ducruet, A.F., Albuquerque, F.C., *et al.* (2013). The evolution of endovascular treatment of carotid cavernous fistulas: A single-center experience. *World Neurosurg* **80**(5): 538–548.

Duffau, H., Lopes, M., *et al.* (1999). Early rebleeding from intracranial dural arteriovenous fistulas: Report of 20 cases and review of the literature. *J Neurosurg* **90**(1): 78–84.

Dützmann, S., Beck, J., *et al.* (2011). Management, risk factors and outcome of cranial dural arteriovenous fistulae: A single-center experience. *Acta Neurochir* **153**(6): 1273–1281.

Edjlali, M., Roca, P., *et al.* (2014). MR selective flow-tracking cartography: A postprocessing procedure applied to four-dimensional flow MR imaging for complete characterization of cranial dural arteriovenous fistulas. *Radiology* **270**(1): 261–268.

Eftekhar, B., Morgan, M.K. (2013). Surgical management of dural arteriovenous fistulas of the transverse-sigmoid sinus in 42 patients. *J Clin Neurosci* **20**(4): 532–535.

Ellis, J.A., Goldstein, H., *et al.* (2012). Carotid-cavernous fistulas. *Neurosurg Focus* **32**(5): E9.

Ertl, L., Brückmann, H., *et al.* (2016a). Assessment and treatment planning of lateral intracranial dural arteriovenous fistulas in 3 T MRI and DSA: A detailed analysis under consideration of time-resolved imaging of contrast kinetics (TRICKS) and CE-MRA sequences. *Eur Radiol* **26**(12): 4284–4292.

Ertl, L., Brückmann, H., *et al.* (2016b). Endovascular therapy of low- and intermediate-grade intracranial lateral dural arteriovenous fistulas: A detailed analysis of primary success rates, complication rates, and long-term follow-up of different technical approaches. *J Neurosurg* **126**(2): 360–367.

Farb, R.I., Agid, R., *et al.* (2009). Cranial dural arteriovenous fistula: Diagnosis and classification with time-resolved MR angiography at 3T. *AJNR Am J Neuroradiol* **30**(8): 1546–1551.

Fincher, E.F. (1951). Arteriovenous fistula between the middle meningeal artery and the greater petrosal sinus; case report. *Ann Surg* **133**(6): 886–888.

Fok, K.F., Agid, R., *et al.* (2004). Thrombosis of aggressive dural arteriovenous fistula after incomplete embolization. *Neuroradiology* **46**(12): 1016–1021.

Gandhi, D., Chen, J., *et al.* (2012). Intracranial dural arteriovenous fistulas: Classification, imaging findings, and treatment. *AJNR Am J Neuroradiol* **33**(6): 1007–1013.

Garcia-Monaco, R., Rodesch, G., *et al.* (1991). Multifocal dural arteriovenous shunts in children. *Childs Nerv Syst* **7**(8): 425–431.

Geibprasert, S., Krings, T., *et al.* (2009). Clinical characteristics of dural arteriovenous shunts in 446 patients of three different ethnicities. *Interv Neuroradiol* **15**(4): 395–400.

Geibprasert, S., Pereira, V., *et al.* (2008). Dural arteriovenous shunts: A new classification of craniospinal epidural venous anatomical bases and clinical correlations. *Stroke* **39**(10): 2783–2794.

Gerlach, R., Boehm-Weigert, M., *et al.* (2008). Thrombophilic risk factors in patients with cranial and spinal dural arteriovenous fistulae. *Neurosurgery* **63**(4): 693–698; discussion 698–699.

Goddard, A.J., Khangure, M.S. (2002). Multiple dural arteriovenous fistulas. Radiologic progression and endovascular cure. Case report. *Interv Neuroradiol* **8**(2): 183–191.

Gonçalves, M.B., Maia Jr, O., *et al.* (2008). Dural arteriovenous fistula presenting as thalamic dementia. *Arq Neuropsiquiatr* **66**(2A): 264–267.

Gonzalez, L.F., Chalouhi, N., *et al.* (2012). Treatment of carotid-cavernous fistulas using intraarterial balloon assistance: Case series and technical note. *Neurosurg Focus* **32**(5): E14.

Greving, J.P., Wermer, M.J., *et al.* (2014). Development of the PHASES score for prediction of risk of rupture of intracranial aneurysms: A pooled analysis of six prospective cohort studies. *Lancet Neurol* **13**(1): 59–66.

Gross, B.A., Albuquerque, F.C., *et al.* (2016a). Evolution of treatment and a detailed analysis of occlusion, recurrence, and clinical outcomes in an endovascular library of 260 dural arteriovenous fistulas. *J Neurosurg* **126**(6): 1884–1893.

Gross, B.A., Du, R. (2012). The natural history of cerebral dural arteriovenous fistulae. *Neurosurgery* **71**(3): 594–602; discussion 602–593.

Gross, B.A., Lai, P.M., *et al.* (2013). Angiotensin-converting enzyme-inhibitors, statins and the risk of hemorrhage from cerebral dural arteriovenous fistulae. *J Clin Neurosci* **20**(9): 1228–1231.

Gross, B.A., Moon, K., *et al.* (2016b). Clinical and anatomic insights from a series of ethmoidal dural arteriovenous fistulas at Barrow Neurological Institute. *World Neurosurg* **93**: 94–99.

Gross, B.A., Ropper, A.E., *et al.* (2012). Cerebral dural arteriovenous fistulas and aneurysms. *Neurosurg Focus* **32**(5): E2.

Guedin, P., Gaillard, S., *et al.* (2010). Therapeutic management of intracranial dural arteriovenous shunts with leptomeningeal venous drainage: Report of 53 consecutive patients with emphasis on transarterial embolization with acrylic glue. *J Neurosurg* **112**(3): 603–610.

Guo, W.Y., Pan, D.H., *et al.* (1998). Radiosurgery as a treatment alternative for dural arteriovenous fistulas of the cavernous sinus. *AJNR Am J Neuroradiol* **19**(6): 1081–1087.

Gupta, R., Horowitz, M., *et al.* (2005). *De novo* development of a remote arteriovenous fistula following transarterial embolization of a carotid cavernous fistula: Case report and review of the literature. *AJNR Am J Neuroradiol* **26**(10): 2587–2590.

Gutierrez, A., Do, H.M., *et al.* (2004). Alteration in the venous drainage of a dural arteriovenous fistula following angioplasty. *AJNR Am J Neuroradiol* **25**(6): 1086–1088.

Ha, S.Y., Kwon, Y.S., *et al.* (2012). Clinical and angiographic characteristics of multiple dural arteriovenous shunts. *AJNR Am J Neuroradiol* **33**(9): 1691–1695.

Hai, J., Zhang, L., *et al.* (2011). A dural arteriovenous fistula in cavernous sinus developed from viral meningitis. *Acta Neurol Belg* **111**(2): 146–148.

Halbach, V.V., Higashida, R.T., *et al.* (1989). Transvenous embolization of dural fistulas involving the cavernous sinus. *AJNR Am J Neuroradiol* **10**(2): 377–383.

Hamada, Y., Goto, K., *et al.* (1997). Histopathological aspects of dural arteriovenous fistulas in the transverse-sigmoid sinus region in nine patients. *Neurosurgery* **40**(3): 452–456; discussion 456–458.

Hammer, F., Boon, L., *et al.* (2001). Ethanol sclerotherapy of venous malformations: Evaluation of systemic ethanol contamination. *J Vasc Interv Radiol* **12**(5): 595–600.

Hanaoka, M., Matsubara, S., *et al.* (2011). Dural arteriovenous fistulae after cerebral infarction: Report of two cases. *Neurosurgery* **68**(2): E575–579; discussion E580.

Hanaoka, M., Satoh, K., *et al.* (2006). Microcatheter pull-up technique in the transvenous embolization of an isolated sinus dural arteriovenous fistula. Technical note. *J Neurosurg* **104**(6): 974–977.

Hardy, D.G., Peace, D.A., *et al.* (1980). Microsurgical anatomy of the superior cerebellar artery. *Neurosurgery* **6**(1): 10–28.

Harris, F., Rhoton, A.L. (1975). Microsurgical anatomy of the cavernous sinus. *Surg Forum* **26**: 462–463.

Harris, F. S., Rhoton, A.L. (1976). Anatomy of the cavernous sinus. A microsurgical study. *J Neurosurg* **45**(2): 169–180.

Haryu, S., Endo, T., *et al.* (2014). Cognard type V intracranial dural arteriovenous shunt: Case reports and literature review with special consideration of the pattern of spinal venous drainage. *Neurosurgery* **74**(1): E135–142; discussion E142.

Hashimoto, Y., Okamoto, A., *et al.* (2013). Gene expression changes in venous segment of overflow arteriovenous fistula. *Int J Nephrol* **2013**: 980923.

Hasumi, T., Fukushima, T., *et al.* (2007). Focal dural arteriovenous fistula (DAVF) presenting with progressive cognitive impairment including amnesia and alexia. *Intern Med* **46**(16): 1317–1320.

He, H.W., Jiang, C.H., *et al.* (2007). Transvenous embolization of cavernous dural arteriovenous fistula: Report of 28 cases. *Chin Med J* **120**(24): 2229–2232.

Hetts, S.W., Moftakhar, P., *et al.* (2016). Pediatric intracranial dural arteriovenous fistulas: Age-related differences in clinical features, angioarchitecture, and treatment outcomes. *J Neurosurg Pediatr*: 1–9.

Hetts, S.W., Tsai, T., *et al.* (2015). Progressive versus nonprogressive intracranial dural arteriovenous fistulas: Characteristics and outcomes. *AJNR Am J Neuroradiol* **36**(10): 1912–1919.

Hiramatsu, M., Sugiu, K., *et al.* (2014). Epidemiology of dural arteriovenous fistula in Japan: Analysis of Japanese Registry of Neuroendovascular Therapy (JR-NET2). *Neurol Med Chir* **54**(1): 63–71.

Hiu, T., Kitagawa, N., *et al.* (2009). Efficacy of DynaCT digital angiography in the detection of the fistulous point of dural arteriovenous fistulas. *AJNR Am J Neuroradiol* **30**(3): 487–491.

Hodel, J., Gerber, S., *et al.* (2011). MR imaging findings in intracranial dural arteriovenous fistula shunt with retrograde cortical venous drainage using susceptibility-weighted angiography. *AJNR Am J Neuroradiol* **32**(10): E196–E197.

Holekamp, T.F., Mollman, M.E., *et al.* (2016). Dural arteriovenous fistula-induced thalamic dementia: Report of 4 cases. *J Neurosurg* **124**(6): 1752–1765.

Hori, M., Aoki, S., *et al.* (2011). Utility of time-resolved three-dimensional magnetic resonance digital subtraction angiography without contrast material for assessment of intracranial dural arterio-venous fistula. *Acta Radiol* **52**(7): 808–812.

Houser, O.W., Baker, H.L., *et al.* (1972). Intracranial dural arteriovenous malformations. *Radiology* **105**(1): 55–64.

Houser, O.W., Campbell, J.K., *et al.* (1979). Arteriovenous malformation affecting the transverse dural venous sinus--an acquired lesion. *Mayo Clin Proc* **54**(10): 651–661.

Hu, Y.C., Newman, C.B., *et al.* (2011). Cranial dural arteriovenous fistula: Transarterial Onyx embolization experience and technical nuances. *J Neurointerv Surg* **3**(1): 5–13.

Iryo, Y., Hirai, T., *et al.* (2014). Intracranial dural arteriovenous fistulas: Evaluation with 3-T four-dimensional MR angiography using arterial spin labeling. *Radiology* **271**(1): 193–199.

Ishikawa, T., Sato, S., *et al.* (2007). Histologic study of arteriovenous shunts in the normal dura mater adjacent to the transverse sinus. *Surg Neurol* **68**(3): 272–276; discussion 276.

Itokawa, K., Fukui, M., *et al.* (2010). Dural arteriovenous fistula as a possible cause of Tolosa-Hunt syndrome: a case report. *J Neurol* **257**(5): 846–847.

Izumi, T., Miyachi, S., *et al.* (2007). Thrombophilic abnormalities among patients with cranial dural arteriovenous fistulas. *Neurosurgery* **61**(2): 262–268; discussion 268–269.

Jain, N.K., Kannath, S.K., *et al.* (2016). The application of susceptibility-weighted MRI in pre-interventional evaluation of intracranial dural arteriovenous fistulas. *J Neurointerv Surg* **9**(5): 502–507.

Janardhanan, R., Yang, B., *et al.* (2013). Simvastatin reduces venous stenosis formation in a murine hemodialysis vascular access model. *Kidney Int* **84**(2): 338–352.

Jang, J., Schmitt, P., *et al.* (2014). Non-contrast-enhanced 4D MR angiography with STAR spin labeling and variable flip angle sampling: A feasibility study for the assessment of Dural Arteriovenous Fistula. *Neuroradiology* **56**(4): 305–314.

Jittapiromsak, P., Ikka, L., *et al.* (2013). Transvenous balloon-assisted transarterial Onyx embolization of transverse-sigmoid dural arteriovenous malformation. *Neuroradiology* **55**(3): 345–350.

Jolink, W.M., van Dijk, J.M., *et al.* (2015). Outcome after intracranial haemorrhage from dural arteriovenous fistulae; a systematic review and case-series. *J Neurol* **262**(12): 2678–2683.

Joo, W., Yoshioka, F., *et al.* (2012). Microsurgical anatomy of the abducens nerve. *Clin Anat* **25**(8) 1030–1042.

Kai, Y., Hamada, J., *et al.* (2003). Pre- and post-treatment MR imaging and single photon emission CT in patients with dural arteriovenous fistulas and retrograde leptomeningeal venous drainage. *AJNR Am J Neuroradiol* **24**(4): 619–625.

Kai, Y., Ito, K., *et al.* (2009). Reversibility of cognitive disorder after treatment of dural arteriovenous fistulae. *Neuroradiology* **51**(11): 731–739.

Kallmes, D. F., Marx, W. F., *et al.* (2000). Adjuvant use of epsilon-aminocaproic acid (Amicar) in the endovascular treatment of cranial arteriovenous fistulae. *Neuroradiology* **42**(4): 302–308.

Kanemaru, K., Kinouchi, H., *et al.* (2015). Cerebral hemodynamic disturbance in dural arteriovenous fistula with retrograde leptomeningeal venous drainage: A prospective study using (123)I-iodoamphetamine single photon emission computed tomography. *J Neurosurg* **123**(1): 110–117.

Kawaguchi, T., Kawano, T., *et al.* (2000). [rCBF study with 123I-IMP SPECT of dural arteriovenous fistula]. *No To Shinkei* **52**(11): 991–996.

Kerber, C.W., Newton, T.H. (1973). The macro and microvasculature of the dura mater. *Neuroradiology* **6**(4): 175–179.

Khan, S., Polston, D.W., *et al.* (2009). Tentorial dural arteriovenous fistula presenting with quadriparesis: Case report and review of the literature. *J Stroke Cerebrovasc Dis* **18**(6): 428–434.

Killory, B.D., Gonzalez, L.F., *et al.* (2008). Simultaneous unilateral moyamoya disease and ipsilateral dural arteriovenous fistula: Case report. *Neurosurgery* **62**(6): E1375–1376; discussion E1376.

Kim, B., Jeon, P., *et al.* (2016). Predictive factors for response of intracranial dural arteriovenous fistulas to transarterial onyx embolization: Angiographic subgroup analysis of treatment outcomes. *World Neurosurg* **88**: 609–618.

Kim, D.J., Kim, D.I., *et al.* (2006). Results of transvenous embolization of cavernous dural arteriovenous fistula: A single-center experience with emphasis on complications and management. *AJNR Am J Neuroradiol* **27**(10): 2078–2082.

Kim, D.J., Krings, T. (2011). Whole-brain perfusion CT patterns of brain arteriovenous malformations: A pilot study in 18 patients. *AJNR Am J Neuroradiol* **32**(11): 2061–2066.

Kim, D.J., terBrugge, K., *et al.* (2010). Spontaneous angiographic conversion of intracranial dural arteriovenous shunt: Long-term follow-up in nontreated patients. *Stroke* **41**(7): 1489–1494.

Kim, D.J., Willinsky, R.A., *et al.* (2011). Intracranial dural arteriovenous shunts: Transarterial glue embolization — experience in 115 consecutive patients. *Radiology* **258**(2): 554–561.

Kim, H., Al-Shahi Salman, R., *et al.* (2014). Untreated brain arteriovenous malformation: Patient-level meta-analysis of hemorrhage predictors. *Neurology* **81**(7): 590–597.

Kim, H.R., Lee, J.Y., *et al.* (2015). Dural arteriovenous fistula-associated reversible Parkinsonism with presynaptic dopaminergic loss. *J Mov Disord* **8**(3): 141–143.

Kim, J.W., Kim, B.M., *et al.* (2016). Onyx embolization for isolated type dural arteriovenous fistula using a dual-lumen balloon catheter. *Neurosurgery* **78**(5): 627–636.

Kim, M.S., Han, D.H., *et al.* (2002). Clinical characteristics of dural arteriovenous fistula. *J Clin Neurosci* **9**(2): 147–155.

Kincaid, P.K., Duckwiler, G.R., *et al.* (2001). Dural arteriovenous fistula in children: Endovascular treatment and outcomes in seven cases. *AJNR Am J Neuroradiol* **22**(6): 1217–1225.

Kirsch, M., Henkes, H., *et al.* (2006). Endovascular management of dural carotid–cavernous sinus fistulas in 141 patients. *Neuroradiology* **48**(7): 486–490.

Kirsch, M., Liebig, T., *et al.* (2009). Endovascular management of dural arteriovenous fistulas of the transverse and sigmoid sinus in 150 patients. *Neuroradiology* **51**(7): 477–483.

Kiyosue, H., Hori, Y., *et al.* (2004). Treatment of intracranial dural arteriovenous fistulas: Current strategies based on location and hemodynamics, and alternative techniques of transcatheter embolization. *Radiographics* **24**(6): 1637–1653.

Kiyosue, H., Okahara, M., *et al.* (2007). Dural arteriovenous fistula involving the posterior condylar canal. *AJNR Am J Neuroradiol* **28**(8): 1599–1601.

Kiyosue, H., Tanoue, S., *et al.* (2002). Recurrence of dural arteriovenous fistula in another location after selective transvenous coil embolization: Report of two cases. *AJNR Am J Neuroradiol* **23**(4): 689–692.

Klisch, J., Huppertz, H.J., *et al.* (2003). Transvenous treatment of carotid cavernous and dural arteriovenous fistulae: Results for 31 patients and review of the literature. *Neurosurgery* **53**(4): 836–856; discussion 856–837.

Klisch, J., Kubalek, R., *et al.* (2005). Plasma vascular endothelial growth factor and serum soluble angiopoietin receptor sTIE-2 in patients with dural arteriovenous fistulas: A pilot study. *Neuroradiology* **47**(1): 10–17.

Kobayashi, A., Al-Shahi Salman, R. (2014). Prognosis and treatment of intracranial dural arteriovenous fistulae: A systematic review and meta-analysis. *Int J Stroke* **9**(6): 670–677.

Koebbe, C.J., Singhal, D., *et al.* (2005). Radiosurgery for dural arteriovenous fistulas. *Surg Neurol* **64**(5): 392–398; discussion 398–399.

Kojima, T., Miyachi, S., *et al.* (2007). The relationship between venous hypertension and expression of vascular endothelial growth factor: Hemodynamic and immunohistochemical examinations in a rat venous hypertension model. *Surg Neurol* **68**(3): 277–284; discussion 284.

Korkmazer, B., Kocak, B., *et al.* (2013). Endovascular treatment of carotid cavernous sinus fistula: A systematic review. *World J Radiol* **5**(4): 143–155.

Kosnik, E.J., Hunt, W.E., *et al.* (1974). Dural arteriovenous malformations. *J Neurosurg* **40**(3): 322–329.

Kraus, J.A., Stüper, B.K., *et al.* (2002). Molecular analysis of thrombophilic risk factors in patients with dural arteriovenous fistulas. *J Neurol* **249**(6): 680–682.

Kraus, J.A., Stüper, B.K., *et al.* (2000). Significantly increased prevalence of factor V Leiden in patients with dural arteriovenous fistulas. *J Neurol* **247**(7): 521–523.

Kubo, M., Kuwayama, N., *et al.* (2002). Dural arteriovenous fistulae developing at different locations after resolution of previous fistulae: Report of three cases and review of the literature. *AJNR Am J Neuroradiol* **23**(5): 787–789.

Kulwin, C., Bohnstedt, B.N., *et al.* (2012). Dural arteriovenous fistulas presenting with brainstem dysfunction: Diagnosis and surgical treatment. *Neurosurg Focus* **32**(5): E10.

Kurata, A., Suzuki, S., *et al.* (2006). New development of a dural arteriovenous fistula (AVF) of the superior sagittal sinus after transvenous embolization of a left sigmoid sinus dural AVF. Case report and review of the literature. *Interv Neuroradiol* **12**(4): 363–368.

Kusaka, N., Sugiu, K., *et al.* (2001). The importance of venous hypertension in the formation of dural arteriovenous fistulas: A case report of multiple fistulas remote from sinus thrombosis. *Neuroradiology* **43**(11): 980–984.

Kuwayama, N. (2016). Epidemiologic survey of dural arteriovenous fistulas in Japan: Clinical frequency and present status of treatment. *Acta Neurochir Suppl* **123**: 185–188.

Lantz, E.R., Lavine, S.D., *et al.* (2009). Acute confusional syndrome from a dural arteriovenous fistula. *Neurosurgery* **65**(1): E208–209; discussion E209.

Lasjaunias, P., Chiu, M., *et al.* (1986). Neurological manifestations of intracranial dural arteriovenous malformations. *J Neurosurg* **64**(5): 724–730.

Lasjaunias, P., Lopez-Ibor, L., *et al.* (1984). Radiological anatomy of the vascularization of cranial dural arteriovenous malformations. *Anat Clin* **6**(2): 87–99.

Lasjaunias, P., Magufis, G., *et al.* (1996). Anatomoclinical aspects of dural arteriovenous shunts in children. Review of 29 cases. *Interv Neuroradiol* **2**(3): 179–191.

Lasjaunias, P., Manelfe, C., *et al.* (1986). Angiographic architecture of intracranial vascular malformations and fistulas--pretherapeutic aspects. *Neurosurg Rev* **9**(4): 253–263.

Lauw, M.N., Barco, S., *et al.* (2013). Cerebral venous thrombosis and thrombophilia: A systematic review and meta-analysis. *Semin Thromb Hemost* **39**(8): 913–927.

Lawton, M., Jacobowitz, R., *et al.* (1997). Redefined role of angiogenesis in the pathogenesis of dural arteriovenous malformations. *J Neurosurg* **87**(2): 267–274.

Le Guennec, L., Leclercq, D., *et al.* (2015). Dural arteriovenous fistula mimicking a brainstem glioma. *J Neuroimaging* **25**(6): 1053–1055.

Lee, D.G., Zhao, L.B., *et al.* (2013). Relationship between diabetes mellitus with dural arteriovenous fistula. *Neuroradiology*.

Lekkhong, E., Pongpech, S., *et al.* (2011). Transvenous embolization of intracranial dural arteriovenous shunts through occluded venous segments: Experience in 51 Patients. *AJNR Am J Neuroradiol* **32**(9): 1738–1744.

Leyon, J.J., Chavda, S., *et al.* (2016). Preliminary experience with the liquid embolic material agent PHIL (Precipitating Hydrophobic Injectable Liquid) in treating cranial and spinal dural arteriovenous fistulas: technical note. *J Neurointerv Surg* **8**(6): 596–602.

Li, C., Wang, Y., *et al.* (2015). Clinical and angioarchitectural risk factors associated with intracranial hemorrhage in dural arteriovenous fistulas: A single-center retrospective study. *PLoS One* **10**(6): e0131235.

Li, Q., Zhang, Q., *et al.* (2014). A pivotal role of the vascular endothelial growth factor signaling pathway in the formation of venous hypertension-induced dural arteriovenous fistulas. *Mol Med Rep* **9**(5): 1551–1558.

Liang, G., Gao, X., *et al.* (2013). Endovascular treatment for dural arteriovenous fistula at the foramen magnum: Report of five consecutive patients and experience with balloon-augmented transarterial Onyx injection. *J Neuroradiol* **40**(2): 134–139.

Lin, Y.H., Lin, H.H., *et al.* (2016). Diagnostic performance of CT and MRI on the detection of symptomatic intracranial dural arteriovenous fistula: A meta-analysis with indirect comparison. *Neuroradiology* **58**(8): 753–763.

Link, M.J., Coffey, R.J., *et al.* (1996). The role of radiosurgery and particulate embolization in the treatment of dural arteriovenous fistulas. *J Neurosurg* **84**(5): 804–809.

Loumiotis, I., Lanzino, G., *et al.* (2011). Radiosurgery for intracranial dural arteriovenous fistulas (DAVFs): A review. *Neurosurg Rev* **34**(3): 305–315; discussion 315.

Louveau, A., Smirnov, I., *et al.* (2015). Structural and functional features of central nervous system lymphatic vessels. *Nature* **523**(7560): 337–341.

Lucas, C.P., Zabramski, J.M., *et al.* (1997). Treatment for intracranial dural arteriovenous malformations: A meta-analysis from the English language literature. *Neurosurgery* **40**(6): 1119–1130; discussion 1130–1112.

Luciani, A., Houdart, E., *et al.* (2001). Spontaneous closure of dural arteriovenous fistulas: Report of three cases and review of the literature. *AJNR Am J Neuroradiol* **22**(5): 992–996.

Lv, X., Jiang, C., *et al.* (2011). The limitations and risks of transarterial Onyx injections in the treatment of grade I and II DAVFs. *Eur J Radiol* **80**(3): e385–388.

Lv, X., Jiang, C., *et al.* (2009). Complications related to percutaneous transarterial embolization of intracranial dural arteriovenous fistulas in 40 patients. *AJNR Am J Neuroradiol* **30**(3): 462–468.

Lv, X., Wu, Z., *et al.* (2014). Innervation of the cerebral dura mater. *Neuroradiol J* **27**(3): 293–298.

Maimon, S., Nossek, E., *et al.* (2011). Transarterial treatment with Onyx of intracranial dural arteriovenous fistula with cortical drainage in 17 patients. *AJNR Am J Neuroradiol* **32**(11): 2180–2184.

Malik, G.M., Pearce, J.E., *et al.* (1984). Dural arteriovenous malformations and intracranial hemorrhage. *Neurosurgery* **15**(3): 332–339.

Manabe, S., Satoh, K., *et al.* (2008). Characteristics, diagnosis and treatment of hypoglossal canal dural arteriovenous fistula: Report of nine cases. *Neuroradiology* **50**(8): 715–721.

Marques, M.C., Caldas, J.G., *et al.* (2010). Follow-up of endovascular treatment of direct carotid–cavernous fistulas. *Neuroradiology* **52**(12): 1127–1133.

Martin, N.A., King, W.A., *et al.* (1990). Management of dural arteriovenous malformations of the anterior cranial fossa. *J Neurosurg* **72**(5): 692–697.

Martins, C., Yasuda, A., *et al.* (2005). Microsurgical anatomy of the dural arteries. *Neurosurgery* **56**(2 Suppl): 211–251; discussion 211–251.

Matsuda, S., Waragai, M., *et al.* (1999). Intracranial dural arteriovenous fistula (DAVF) presenting progressive dementia and parkinsonism. *J Neurol Sci* **165**(1): 43–47.

Maus, V., Söderman, M., *et al.* (2016). Endovascular treatment of posterior condylar canal dural arteriovenous fistula. *J Neurointerv Surg* **9**(2): e7.

McKenzie, J.D., Dean, B.L., *et al.* (1996). Trigeminal-cavernous fistula: Saltzman anatomy revisited. *AJNR Am J Neuroradiol* **17**(2): 280–282.

Meckel, S., Lovblad, K.O., *et al.* (2005). Arterialization of cerebral veins on dynamic MDCT angiography: A possible sign of a dural arteriovenous fistula. *AJR Am J Roentgenol* **184**(4): 1313–1316.

Merland, J.J., Bories, J., *et al.* (1977a). The blood supply of the falx cerebri, the falx cerebelli and the tentorium cerebelli. *J Neuroradiol* **4**(2): 175–202.

Merland, J.J., Bories, J., *et al.* (1977b). Normal and pathological blood supply of the cranial vault. *J Neuroradiol* **4**(2): 95–127.

Meyers, P.M., Halbach, V.V., *et al.* (2002). Dural carotid cavernous fistula: Definitive endovascular management and long-term follow-up. *Am J Ophthalmol* **134**(1): 85–92.

Miller, N.R. (2012). Dural carotid-cavernous fistulas: Epidemiology, clinical presentation, and management. *Neurosurg Clin N Am* **23**(1): 179–192.

Miller, T.R. and Gandhi, D. (2015). Intracranial dural arteriovenous fistulae: Clinical presentation and management strategies. *Stroke* **46**(7): 2017–2025.

Mironov, A. (1994). Pathogenetical consideration of spontaneous dural arteriovenous fistulas (DAVFs). *Acta Neurochir (Wien)* **131**(1–2): 45–58.

Mitchell, S., Shah, A., *et al.* (2006). Pulmonary artery pressure changes during ethanol embolization procedures to treat vascular malformations: Can cardiovascular collapse be predicted? *J Vasc Interv Radiol* **17**(2 Pt 1): 253–262.

Miyachi, S.E., Izumi, T., *et al.* (2011). Mechanism of the formation of dural arteriovenous fistula: The role of the emissary vein. *Interv Neuroradiol* **17**(2): 195–202.

Moore, C., Murphy, K., *et al.* (2006). Improved distal distribution of n-butyl cyanoacrylate glue by simultaneous injection of dextrose 5% through the guiding catheter: Technical note. *Neuroradiology* **48**(5): 327–332.

Morparia, N., Miller, G., *et al.* (2012). Cognitive decline and hypersomnolence: Thalamic manifestations of a tentorial dural arteriovenous fistula (dAVF). *Neurocrit Care* **17**(3): 429–433.

Murphy, M., Van Gompel, J.J. (2012). Medullary venous hypertension secondary to a petrous apex dural arteriovenous fistula: A case report. *Case Rep Neurol* **4**(3): 187–193.

Nelson, P.K., Russell, S.M., *et al.* (2003). Use of a wedged microcatheter for curative transarterial embolization of complex intracranial dural arteriovenous fistulas: Indications, endovascular technique, and outcome in 21 patients. *J Neurosurg* **98**(3): 498–506.

Nencini, P., Inzitari, D., *et al.* (1993). Dementia with leucoaraiosis and dural arteriovenous malformation: Clinical and PET case study. *J Neurol Neurosurg Psychiatry* **56**(8): 929–931.

Newton, T.H., Cronqvist, S. (1969). Involvement of dural arteries in intracranial arteriovenous malformations. *Radiology* **93**(5): 1071–1078.

Ng, P., Higashida, R., *et al.* (2003). Endovascular strategies for carotid cavernous and intracerebral dural arteriovenous fistulas. *Neurosurg Focus* **15**(4): ECP1.

Nishijima, M., Takaku, A., *et al.* (1992). Etiological evaluation of dural arteriovenous malformations of the lateral and sigmoid sinuses based on histopathological examinations. *J Neurosurg* **76**(4): 600–606.

Nishimura, S., Hirai, T., *et al.* (2010). Evaluation of dural arteriovenous fistulas with 4D contrast-enhanced MR angiography at 3T. *AJNR Am J Neuroradiol* **31**(1): 80–85.

Nishino, K., Ito, Y., *et al.* (2008). Cranial nerve palsy following transvenous embolization for a cavernous sinus dural arteriovenous fistula: Association with the volume and location of detachable coils. *J Neurosurg* **109**(2): 208–214.

Noguchi, K., Kubo, M., *et al.* (2006). Intracranial dural arteriovenous fistulas with retrograde cortical venous drainage: assessment with cerebral blood volume by dynamic susceptibility contrast magnetic resonance imaging. *AJNR Am J Neuroradiol* **27**(6): 1252–1256.

Noguchi, K., Kuwayama, N., *et al.* (2011). Flow-sensitive alternating inversion recovery (fair) imaging for retrograde cortical venous drainage related to intracranial dural arteriovenous fistula. *Neuroradiology* **53**(3): 153–158.

Noguchi, K., Kuwayama, N., *et al.* (2010). Intracranial dural arteriovenous fistula with retrograde cortical venous drainage: Use of susceptibility-weighted imaging in combination with dynamic susceptibility contrast imaging. *AJNR Am J Neuroradiol* **31**(10): 1903–1910.

Noguchi, T., Irie, H., *et al.* (2010). Hemodynamic studies of intracranial dural arteriovenous fistulas using arterial spin-labeling MR imaging. *Interv Neuroradiol* **16**(4): 409–419.

Nyberg, E.M., Chaudry, M.I., *et al.* (2013). Transient cranial neuropathies as sequelae of Onyx embolization of arteriovenous shunt lesions near the skull base: Possible axonotmetic traction injuries. *J Neurointerv Surg* **5**(4): e21.

Pabaney, A.H., Robin, A.M., *et al.* (2016). Surgical management of dural arteriovenous fistula after craniotomy: Case report and review of literature. *World Neurosurg* **89**: 731.e737–731.e711.

Paramasivam, S., Toma, N., *et al.* (2013a). De novo development of dural arteriovenous fistula after endovascular embolization of pial arteriovenous fistula. *J Neurointerv Surg* **5**(4): 321–326

Paramasivam, S., Niimi, Y., *et al.* (2013b). Onyx embolization using dual-lumen balloon catheter: Initial experience and technical note. *J Neuroradiol* **40**(4): 294–302.

Paramasivam, S., Niimi, Y., *et al.* (2014). Dural arteriovenous shunt development in patients with vein of galen malformation. *Interv Neuroradiol* **20**(6): 781–790.

Park, K.S., Kang, D.H., *et al.* (2016). The efficacy of gamma knife radiosurgery alone as a primary treatment for intracranial dural arteriovenous fistulas. *Acta Neurochir* **158**(4): 821–828.

Pashapour, A., Mohammadian, R., *et al.* (2014). Long-term endovascular yreatment outcome of 46 patients with cavernous sinus dural arteriovenous fistulas presenting with ophthalmic symptoms. A non-controlled trial with clinical and angiographic follow-up. *Neuroradiol J* **27**(4): 461–470.

Peng, T., Liu, A., *et al.* (2014). Risk factors for dural arteriovenous fistula intracranial hemorrhage. *J Clin Neurosci* **21**(5): 769–772.

Phatouros, C.C., Higashida, R.T., *et al.* (1999). Embolization of the meningohypophyseal trunk as a cause of diabetes insipidus. *AJNR Am J Neuroradiol* **20**(6): 1115–1118.

Piippo, A., Niemelä, M., *et al.* (2013a). Characteristics and long-term outcome of 251 patients with dural arteriovenous fistulas in a defined population. *J Neurosurg* **118**(5): 923–934.

Piippo, A., Laakso, A., *et al.* (2013b). Early and long-term excess mortality in 227 patients with intracranial dural arteriovenous fistulas. *J Neurosurg* **119**(1): 164–171.

Portnoy, H.D., Branch, C., *et al.* (1994). The relationship of intracranial venous pressure to hydrocephalus. *Childs Nerv Syst* **10**(1): 29–35.

Portnoy, H.D., Chopp, M., *et al.* (1983). Hydraulic model of myogenic autoregulation and the cerebrovascular bed: The effects of altering systemic arterial pressure. *Neurosurgery* **13**(5): 482–498.

Rabinov, J.D., Yoo, A.J., *et al.* (2013). ONYX versus n-BCA for embolization of cranial dural arteriovenous fistulas. *J Neurointerv Surg* **5**(4): 306–310.

Ramalingaiah, A.H., Prasad, C., *et al.* (2013). Transarterial treatment of direct carotico–cavernous fistulas with coils and Onyx. *Neuroradiology* **55**(10): 1213–1220.

Rhim, J.K., Cho, Y.D., *et al.* (2015). Endovascular treatment of cavernous sinus dural arteriovenous fistula with ipsilateral inferior petrosal sinus occlusion: A single-center experience. *Neurosurgery* **77**(2): 192–199; discussion 199.

Rhoton, A.L., Hardy, D.G., *et al.* (1979). Microsurgical anatomy and dissection of the sphenoid bone, cavernous sinus and sellar region. *Surg Neurol* **12**(1): 63–104.

Rhoton, A.L., Harris, F.S., *et al.* (1977). Microsurgical anatomy of the sellar region and cavernous sinus. *Clin Neurosurg* **24**: 54–85.

Ricolfi, F., Manelfe, C., *et al.* (1999). Intracranial dural arteriovenous fistulae with perimedullary venous drainage. Anatomical, clinical and therapeutic considerations. *Neuroradiology* **41**(11): 803–812.

Riina, H., Zumofen, D. (2014). Perspectives on the evolution of endovascular treatment of carotid cavernous fistulas. *World Neurosurg* **82**(1–2): e169–170.

Rivet, D.J., Goddard, J.K., *et al.* (2006). Percutaneous transvenous embolization of a dural arteriovenous fistula through a mastoid emissary vein. Technical note. *J Neurosurg* **105**(4): 636–639.

Robert, T., Botta, D., *et al.* (2016a). Ocular signs caused by dural arteriovenous fistula without involvement of the cavernous sinus: A case series with review of the literature. *AJNR Am J Neuroradiol* **37**(10): 1870–1875.

Robert, T., Blanc, R., *et al.* (2016b). Endovascular treatment of cribriform plate dural arteriovenous fistulas: Technical difficulties and complications avoidance. *J Neurointerv Surg* **8**(9): 954–958.

Rodrigues, T., Willinsky, R., *et al.* (2014). Management of dural carotid cavernous fistulas: A single-centre experience. *Eur Radiol* **24**(12): 3051–3058.

Safavi-Abbasi, S., Di Rocco, F., *et al.* (2008). Thrombophilia due to Factor V and Factor II mutations and formation of a dural arteriovenous fistula: Case report and review of a rare entity. *Skull Base* **18**(2): 135–143.

Sakaki, S., Furuta, S., *et al.* (1991). Dural arteriovenous malformation of the transverse and sigmoid sinus with special reference to its pathological features. *Br J Neurosurg* **5**(1): 87–92.

Sarma, D., ter Brugge, K. (2003). Management of intracranial dural arteriovenous shunts in adults. *Eur J Radiol* **46**(3): 206–220.

Satoh, K., Satomi, J., *et al.* (2001). Cerebellar hemorrhage caused by dural arteriovenous fistula: A review of five cases. *J Neurosurg* **94**(3): 422–426.

Satoh, M., Kuriyama, M., *et al.* (2005). Brain stem ischemia from intracranial dural arteriovenous fistula: Case report. *Surg Neurol* **64**(4): 341–345; discussion 345.

Satomi, J., Ghaibeh, A.A., *et al.* (2015). Predictability of the future development of aggressive behavior of cranial dural arteriovenous fistulas based on decision tree analysis. *J Neurosurg* **123**(1): 86–90.

Satomi, J., van Dijk, J.M., *et al.* (2002). Benign cranial dural arteriovenous fistulas: Outcome of conservative management based on the natural history of the lesion. *J Neurosurg* **97**(4): 767–770.

Schanker, B.D., Walcott, B.P., *et al.* (2011). Time-resolved contrast-enhanced magnetic resonance angiography in the investigation of suspected intracranial dural arteriovenous fistula. *J Clin Neurosci* **18**(6): 837–839.

Seoane, E., Rhoton, A.L., *et al.* (1998). Microsurgical anatomy of the dural collar (carotid collar) and rings around the clinoid segment of the internal carotid artery. *Neurosurgery* **42**(4): 869–884; discussion 884–866.

Seoane, E., Tedeschi, H., *et al.* (2000). The pretemporal transcavernous approach to the interpeduncular and prepontine cisterns: Microsurgical anatomy and technique application. *Neurosurgery* **46**(4): 891–898; discussion 898–899.

Shi, Z.S., Loh, Y., *et al.* (2013a). Flow control techniques for Onyx embolization of intracranial dural arteriovenous fistulae. *J Neurointerv Surg* **5**(4): 311–316.

Shi, Z.S., Ziegler, J., *et al.* (2013b). Middle cranial fossa sphenoidal region dural arteriovenous fistulas: Anatomic and treatment considerations. *AJNR Am J Neuroradiol* **34**(2): 373–380.

Shidoh, S., Akiyama, T., *et al.* (2014). Cerebral perfusion change of venous hypertension on near-infrared spectroscopy signals after operation for dural arteriovenous fistula. *J Stroke Cerebrovasc Dis* **23**(5): 823–828.

Shimada, R., Kiyosue, H., *et al.* (2013). Superior petrosal sinus: Hemodynamic features in normal and cavernous sinus dural arteriovenous fistulas. *AJNR Am J Neuroradiol* **34**(3): 609–615.

Shin, N.Y., Kwon, Y.S., *et al.* (2013). Venous angioarchitectural features of intracranial dural arteriovenous shunt and its relation to the clinical course. *Neuroradiology* **55**(9): 1119–1127.

Shin, Y., Nakase, H., *et al.* (2007). Expression of angiogenic growth factor in the rat DAVF model. *Neurol Res* **29**(7): 727–733.

Singh, V., Meyers, P.M., *et al.* (2001). Dural arteriovenous fistula associated with prothrombin gene mutation. *J Neuroimaging* **11**(3): 319–321.

Singh, V., Smith, W.S., *et al.* (2008). Risk factors for hemorrhagic presentation in patients with dural arteriovenous fistulae. *Neurosurgery* **62**(3): 628–635; discussion 628–635.

Söderman, M., Dodoo, E., *et al.* (2013). Dural arteriovenous fistulas and the role of gamma knife stereotactic radiosurgery: The Stockholm experience. *Prog Neurol Surg* **27**: 205–217.

Söderman, M., Edner, G., *et al.* (2006). Gamma knife surgery for dural arteriovenous shunts: 25 years of experience. *J Neurosurg* **104**(6): 867–875.

Söderman, M., Pavic, L., *et al.* (2008). Natural history of dural arteriovenous shunts. *Stroke* **39**(6): 1735–1739.

Spiotta, A.M., Hughes, G., *et al.* (2011). Balloon-augmented Onyx embolization of a dural arteriovenous fistula arising from the neuromeningeal trunk of the ascending pharyngeal artery: Technical report. *J Neurointerv Surg* **3**(3): 300–303.

Spittau, B., Millán, D.S., *et al.* (2015). Dural arteriovenous fistulas of the hypoglossal canal: Systematic review on imaging anatomy, clinical findings, and endovascular management. *J Neurosurg* **122**(4): 883–903.

Strom, R.G., Botros, J.A., *et al.* (2009). Cranial dural arteriovenous fistulae: Asymptomatic cortical venous drainage portends less aggressive clinical course. *Neurosurgery* **64**(2): 241–247; discussion 247–248.

Sun, Z., Su, H., *et al.* (2014). Endothelial cell high-enrichment from endovascular biopsy sample by laser capture microdissection and fluorescence activated cell sorting. *J Biotechnol* **192PA**: 34–39.

Sundt, T.M., Piepgras, D.G. (1983). The surgical approach to arteriovenous malformations of the lateral and sigmoid dural sinuses. *J Neurosurg* **59**(1): 32–39.

Takahashi, S., Sakuma, I., *et al.* (2004). Transvenous embolization of dural arteriovenous fistula of the cavernous sinus. Fistulous points and route of catheterization. *Interv Neuroradiol* **10 Suppl 1**: 85–92.

Tanaka, K., Morooka, Y., *et al.* (1999). Dural arteriovenous malformation manifesting as dementia due to ischemia in bilateral thalami. A case report. *Surg Neurol* **51**(5): 489–493; discussion 493–484.

Tanoue, S., Kiyosue, H., *et al.* (2012). Turn-back embolization technique for effective transvenous embolization of dural arteriovenous fistulas. *AJNR Am J Neuroradiol* **33**(6): E88–91.

Terada, T., Higashida, R.T., *et al.* (1994). Development of acquired arteriovenous fistulas in rats due to venous hypertension. *J Neurosurg* **80**(5): 884–889.

Terada, T., Tsuura, M., *et al.* (1996). The role of angiogenic factor bFGF in the development of dural AVFs. *Acta Neurochir* **138**(7): 877–883.

Théaudin, M., Saint-Maurice, J.P., *et al.* (2007). Diagnosis and treatment of dural carotid-cavernous fistulas: A consecutive series of 27 patients. *J Neurol Neurosurg Psychiatry* **78**(2): 174–179.

Thiex, R., Gross, B.A., *et al.* (2014). Transvenous approach to carotid–cavernous fistula via facial vein cut down. *J Clin Neurosci* **21**(7): 1238–1240.

Tirakotai, W., Bertalanffy, H., *et al.* (2005). Immunohistochemical study in dural arteriovenous fistulas and possible role of local hypoxia for the de novo formation of dural arteriovenous fistulas. *Clin Neurol Neurosurg* **107**(6): 455–460.

Tominaga, T., Shamoto, H., *et al.* (2003). Selective loss of Purkinje cells in transverse and sigmoid dural arteriovenous fistulas. Report of two cases. *J Neurosurg* **98**(3): 617–620.

Tsai, L.K., Jeng, J.S., *et al.* (2004). Intracranial dural arteriovenous fistulas with or without cerebral sinus thrombosis: Analysis of 69 patients. *J Neurol Neurosurg Psychiatry* **75**(11): 1639–1641.

Tsutsumi, S., Yasumoto, Y., *et al.* (2008). Atypical dural arteriovenous fistula associated with meningitis. *Neurol Med Chir* **48**(2): 68–71.

Uranishi, R., Nakase, H., *et al.* (1999). Expression of angiogenic growth factors in dural arteriovenous fistula. *J Neurosurg* **91**(5): 781–786.

Vadivelu, S., Xin, X., *et al.* (2012). Iatrogenic dural arteriovenous fistula and aneurysmal subarachnoid hemorrhage. *Neurosurg Focus* **32**(5): E1.

Vallon, M., Chang, J., *et al.* (2014). Developmental and pathological angiogenesis in the central nervous system. *Cell Mol Life Sci* **71**(18): 3489–3506.

van Asch, C.J., Velthuis, B.K., *et al.* (2015). Diagnostic yield and accuracy of CT angiography, MR angiography, and digital subtraction angiography for detection of macrovascular causes of intracerebral haemorrhage: Prospective, multicentre cohort study. *BMJ* **351**: h5762.

van Dijk, J.M., TerBrugge, K.G., *et al.* (2007). Thrombophilic factors and the formation of dural arteriovenous fistulas. *J Neurosurg* **107**(1): 56–59.

van Dijk, J.M., terBrugge, K.G., *et al.* (2002a). Clinical course of cranial dural arteriovenous fistulas with long-term persistent cortical venous reflux. *Stroke* **33**(5): 1233–1236.

van Dijk, J.M., TerBrugge, K.G., *et al.* (2002b). Multiplicity of dural arteriovenous fistulas. *J Neurosurg* **96**(1): 76–78.

van Dijk, J.M., Terbrugge, K.G., *et al.* (2009). The natural history of dural arteriovenous shunts: The toronto experience. *Stroke* **40**(5): e412; author reply e413–414.

van Dijk, J.M., TerBrugge, K.G., *et al.* (2004). Selective disconnection of cortical venous reflux as treatment for cranial dural arteriovenous fistulas. *J Neurosurg* **101**(1): 31–35.

van Rooij, W.J., Sluzewski, M., *et al.* (2007). Intracranial dural fistulas with exclusive perimedullary drainage: The need for complete cerebral angiography for diagnosis and treatment planning. *AJNR Am J Neuroradiol* **28**(2): 348–351.

Vellimana, A.K., Daniels, D.J., *et al.* (2014). Dural arteriovenous fistulas associated with benign meningeal tumors. *Acta Neurochir* **156**(3): 535–544.

Wang, J.Y., Molenda, J., *et al.* (2015). Natural history and treatment of craniocervical junction dural arteriovenous fistulas. *J Clin Neurosci* **22**(11): 1701–1707.

Waragai, M., Takeuchi, H., *et al.* (2006). MRI and SPECT studies of dural arteriovenous fistulas presenting as pure progressive dementia with leukoencephalopathy: A cause of treatable dementia. *Eur J Neurol* **13**(7): 754–759.

Weber, W., Kis, B., *et al.* (2003). Endovascular treatment of a dural arteriovenous fistula of the transverse sinus by recanalisation, angioplasty and stent deployment. A case report and follow-up. *Interv Neuroradiol* **9**(1): 65–69.

Willems, P.W., Brouwer, P.A., *et al.* (2011). Detection and classification of cranial dural arteriovenous fistulas using 4D-CT angiography: Initial experience. *AJNR Am J Neuroradiol* **32**(1): 49–53.

Willinsky, R., Goyal, M., *et al.* (1999). Tortuous, engorged pial veins in intracranial dural arteriovenous fistulas: correlations with presentation, location, and MR findings in 122 patients. *AJNR Am J Neuroradiol* **20**(6): 1031–1036.

Wong, G., Armstrong, D., *et al.* (2006). Cardiovascular collapse during ethanol sclerotherapy in a pediatric patient. *Paediatr Anaesth* **16**(3): 343–346.

Wu, Q., Zhang, X. S., *et al.* (2016). Onyx embolization for tentorial dural arteriovenous fistula with pial arterial supply: Case series and analysis of complications. *World Neurosurg* **92**: 58–64.

Yakes, W.F., Krauth, L., *et al.* (1997). Ethanol endovascular management of brain arteriovenous malformations: Initial results. *Neurosurgery* **40**(6): 1145–1152; discussion 1152–1144.

Yang, B., Janardhanan, R., *et al.* (2013). Adventitial transduction of lentivirus-shRNA-VEGF-A in arteriovenous fistula reduces venous stenosis formation. *Kidney Int* **85**(2): 289–306.

Yang, H., Kano, H., *et al.* (2013). Stereotactic radiosurgery with or without embolization for intracranial dural arteriovenous fistulas. *Prog Neurol Surg* **27**: 195–204.

Yang, M.S., Chen, C.C., *et al.* (2005). Unilateral subcortical calcification: A manifestation of dural arteriovenous fistula. *AJNR Am J Neuroradiol* **26**(5): 1149–1151.

Yang, S.T., Rodriguez-Hernandez, A., *et al.* (2015). Adult mouse venous hypertension model: Common carotid artery to external jugular vein anastomosis. *J Vis Exp* (95): 50472.

Yassari, R., Jahromi, B., *et al.* (2002). Dural arteriovenous fistula after craniotomy for pilocytic astrocytoma in a patient with protein S deficiency. *Surg Neurol* **58**(1): 59–64; discussion 64.

Yasuda, A., Campero, A., *et al.* (2008). Microsurgical anatomy and approaches to the cavernous sinus. *Neurosurgery* **62**(6 Suppl 3): 1240–1263.

Yasuda, A., Campero, A., *et al.* (2005). Microsurgical anatomy and approaches to the cavernous sinus. *Neurosurgery* **56**(1 Suppl): 4–27; discussion 24–27.

Yasuda, A., Campero, A., *et al.* (2004). The medial wall of the cavernous sinus: Microsurgical anatomy. *Neurosurgery* **55**(1): 179–189; discussion 189–190.

Yoshida, K., Melake, M., *et al.* (2010). Transvenous embolization of dural carotid cavernous fistulas: A series of 44 consecutive patients. *AJNR Am J Neuroradiol* **31**(4): 651–655.

Youssef, P.P., Schuette, A.J., *et al.* (2014). Advances in surgical approaches to dural fistulas. *Neurosurgery* **74 Suppl 1**: S32–41.

Yu, Y., Huang, Q., *et al.* (2012). Use of onyx for transarterial balloon-assisted embolization of traumatic carotid cavernous fistulas: A report of 23 cases. *AJNR Am J Neuroradiol* **33**(7): 1305–1309.

Zanaty, M., Chalouhi, N., *et al.* (2014). Endovascular treatment of carotid–cavernous fistulas. *Neurosurg Clin N Am* **25**(3): 551–563.

Zeidman, S., Monsein, L., *et al.* (1995). Reversibility of white matter changes and dementia after treatment of dural fistulas. *AJNR Am J Neuroradiol* **16**(5): 1080–1083.

Zhang, Y., Li, Q., *et al.* (2015). Embolization of a superior sagittal sinus dural arteriovenous fistula under intrasinus balloon protection: A case report. *Interv Neuroradiol* **21**(1): 94–100.

Zhao, W. Y., Krings, T., *et al.* (2012). Balloon-assisted superselective microcatheterization for transarterial treatment of cranial dural arteriovenous fistulas: Technique and results. *Neurosurgery* **71**(2 Suppl Operative): ons269–273; discussion ons273.

Zhu, Y., Lawton, M. T., *et al.* (2006). Expression of hypoxia-inducible factor-1 and vascular endothelial growth factor in response to venous hypertension. *Neurosurgery* **59**(3): 687–696; discussion 687–696.

Zipfel, G., Shah, M., *et al.* (2009). Cranial dural arteriovenous fistulas: Modification of angiographic classification scales based on new natural history data. *Neurosurg Focus* **26**(5): E14.

Interventional Therapies for Cerebrovascular Diseases

10

Fabio Settecase and Steven W. Hetts

Department of Radiology and Biomedical Imaging,
University of California, San Francisco,
San Francisco, California, USA

The scope of diseases accessible to endovascular therapy is broad and includes not only primary disorder of the blood vessels themselves, but also such conditions as tumors that can be treated with embolic agents or chemotherapy, to achieve local tumor control before surgical resection or sometimes as stand-alone therapy. This chapter will focus on four primary disorders of blood vessels in the cervical and cerebral vasculature: ischemic stroke, atherosclerosis, arteriovenous malformations, and aneurysms.

10.1 Acute Ischemic Stroke

The goal of endovascular treatment of an acute ischemic stroke due to large vessel occlusion is rapid reperfusion (Khatri *et al.*, 2014; Rha & Saver, 2007). The use of intravenous tPA in acute ischemic stroke patients presenting within 3–4.5 hours of stroke onset (i.e., time last seen normal) is supported by multiple randomized controlled clinical trials (The National Institute of Neurological Disorders and Stroke rt-PA Stroke Study Group 1995b; del Zoppo *et al.*, 1992; Hacke *et al.*, 2008; Lees *et al.*, 2010). The success rate of IV tPA at lysis of larger and more proximal thromboocclusions (i.e., ICA and M1 segment middle cerebral artery), however, is low (Bhatia *et al.*, 2010; Riedel *et al.*, 2011). Approximately 35–40% of ischemic stroke cases are due to large vessel occlusions (ICA, MCA, or basilar arteries) and without recanalization, the prognosis for these types of stroke is poor (Furlan *et al.*, 1999; Jansen *et al.*, 1995; Lima *et al.*, 2014).

Efforts to increase the likelihood of recanalization of larger more proximal vessels initially focused on local intra-arterial delivery of thrombolytic agents into the clot via endovascular techniques. Despite an increased frequency of early symptomatic intracranial hemorrhage, PROACT-II, a multicenter randomized clinical trial of IA prourokinase and IV heparin within 6 hours of the onset of acute ischemic stroke caused by MCA occlusion significantly improved clinical outcome at 90 days, compared with IV heparin only (Furlan *et al.*, 1999). While this trial ushered in the era of endovascular therapy for acute ischemic stroke patients and extended the treatment window to 6 hours, the recanalization rates achieved using IA tPA were still relatively low.

Over the last few decades, several different devices have been developed to mechanically disrupt or remove clot causing an acute ischemic stroke. The effectiveness of earlier generations of these devices, such as the Merci device, or the EKOS catheter, was marginal. Stent-retrievers have become the mainstay of mechanical thrombectomy procedures, owing to their relatively high (approximately 70–80%) efficacy at clot removal and achievement of TICI 2B or 3 reperfusion compared with other devices (Nogueira *et al.*, 2012; Saver, *et al.*, 2012).

In late 2014 and early 2015, five different multicenter randomized clinical trials were published comparing outcomes in select patients undergoing mechanical thrombectomy without or without IV tPA to patients receiving IV tPA alone (Berkhemer *et al.*, 2015; Campbell *et al.*, 2015; Goyal *et al.*, 2015; Jovin *et al.*, 2015; Saver *et al.*, 2015). All trials demonstrated a significant improvement in clinical outcomes (defined as mRS 0–2 at 90 days) when using stent-retrievers for mechanical thrombectomy with or without IV tPA compared with IV tPA alone in patients presenting within 6 hours of stroke onset (Table 1, Figure 1). As a result of these trials, endovascular mechanical thrombectomy is now part of the standard of care of acute stroke patients. Stent-retrievers were used in 86–100% of patients enrolled in the mechanical thrombectomy arms of the aforementioned five clinical trials. More recently, there is also evidence to support the use of thromboaspiration catheters. These are trackable catheters with large distal inner diameters that can be hooked up proximally to a proprietary vacuum system for removal of clot via suction. Thromboaspiration, commonly referred to as the ADAPT technique (A Direct Aspiration First Pass *Technique* for Acute Stroke Thrombectomy), has been shown to be effective in achieving TICI 2B or 3 reperfusion in 78–82% of patients, and in 88–95% when used in combination with a stent-retriever (Humphries *et al.*, 2015; Turk *et al.*, 2014). The latter is known as the "Solumbra" technique.

Table 1. Summary of mechanical thrombectomy stroke trails. Adapted from Grotta & Hacke (2015) with permission from Lippincott Williams & Wilkins.

Trial N IAT+/CTL	NIHSS range		r-tPA (%)	TICI 2B/3 (%)	LSN to Groin Mdn	GA (%)	mRS 0–2 at 90 d (%)		sICH (%)		Device complications	Mortality (%)	
	CTL	IAT+					CTL	IAT+	CTL	IAT+		CTL	IAT+
MR CLEAN 500 233/267	18 (14–21)	17 (14–22)	90	59	260	38	19	33	6.4	7.7	Embol. 13	22	21
ESCAPE 315 165/150	17 (12–20)	16 (13–20)	76	72	200	9	29	53	2.7	3.6	Perf. 1	19	10
EXTEND IA 70 35/35	13 (9–19)	17 (13–20)	100	86	210	36	40	71	6	0	Perf. 1 Embol.2	20	9
SWIFT PRIME 196 98/98	17 (13–19)	17 (13–20)	98	88	224	37	36	60	3	0	SAH 4	12	9
REVASCAT 206 103/103	17 (12–19)	17 (14–20)	73	66	269	7	28	44	1.9	1.9	Perf. 5 Embol. 5	16	18

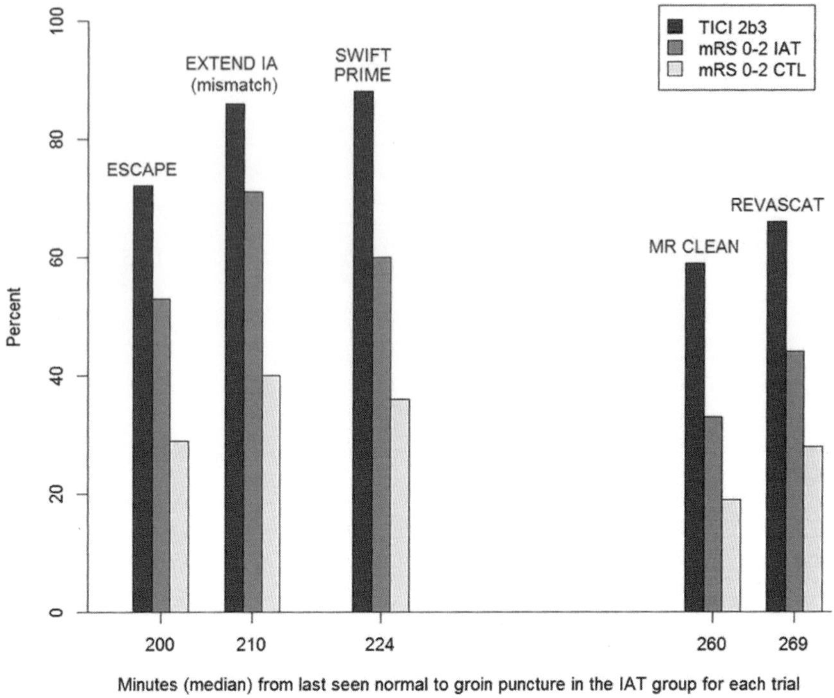

Figure 1. Relationship of percentage of patients achieving thrombolysis in cerebral infarction grade 2b or 3 (TICI2b3) flow and patients achieving modified Rankin Scale (mRS) of 0–2 in intervention (intra-arterial thrombectomy [IAT]) and control (CTL) groups (*y*-axis) versus minutes from last seen normal to groin puncture for each of the 5 studies (*x*-axis). The exact time to achieving maximal reperfusion is not available from published data for all studies, but where available is proportionate to time to groin puncture. Note the following: (1) the percentage of patients achieving good outcome is strikingly proportionate to the percentage of patients achieving TICI2b3 flow. (2) There is a consistent difference between the IAT and CTL groups in the percent achieving mRS 0 to 2 across all studies with the difference diminishing with increased time from last seen normal to groin puncture. (3) The percentage of patients achieving good outcome is roughly proportionate to the time from last seen normal to groin puncture (earlier groin puncture = higher proportion good outcome) with the exception being the EXTEND IA study, which was the only study to use advanced imaging for patient selection suggesting its use to identify responsive patients at delayed time intervals. ESCAPE indicates Endovascular Treatment for Small Core and Anterior Circulation Proximal Occlusion With Emphasis on Minimizing CT to Recanalization Times; EXTEND IA, Extending the Time for Thrombolysis in Emergency Neurological Deficits — Intra-Arterial; MR CLEAN, Multicenter Randomized Clinical Trial of Endovascular Treatment for Acute Ischemic Stroke in the Netherlands; SWIFT

(*Continued*)

Figure 1. (*Continued*) PRIME, Solitaire With the Intention for Thrombectomy as Primary Endovascular Treatment Trial; and REVASCAT, Randomized Trial of Revascularization With the Solitaire FR Device Versus Best Medical Therapy in the Treatment of Acute Stroke Due to Anterior Circulation Large Vessel Occlusion Presenting Within Eight Hours of Symptom Onset. Grotta & Hacke (2015) with permission from Lippincott Williams & Wilkins.

The THERAPY trial, another randomized control trial designed similarly to the stent-retreiver trials with the exception of the use of the ADAPT technique in place of stent-retrievers, was stopped early due to the results of the previous trials (Mocco *et al.*, 2016).

In a typical large vessel occlusion acute ischemic stroke presentation, a patient is brought into the emergency department by ambulance and a "code stroke" is activated. After evaluation by a neurologist, the patient with a suspected acute stroke will be brought to a CT scanner. According the inclusion criteria used in the clinical trials, to perform endovascular mechanical thrombectomy, there must be:

- Clinical and imaging evidence of an acute ischemic stroke (NIHSS score >5)
- No evidence of hemorrhage and without a large area (ASPECTS >6 or <70–100 cc) of brain that is not already infarcted
- Evidence of a large vessel occlusion on CT angiography (e.g., ICA, MCA, or basilar artery)
- Groin puncture can be achieved within 6 hours of stroke symptom onset
- Functionally independent (mRS<3) prior to stroke onset

After informed consent is obtained from the patient or next of kin, the patient is urgently brought to the angiography suite for the mechanical thrombectomy. Hospital and departmental protocols should be followed to ensure speedy patient transfer, physician and staff availability, and procedure suite readiness. Administration of IV tPA should occur while the patient is still on the CT scanner, regardless of whether the patient is a mechanical thrombectomy candidate or not, and concurrently with procedural informed consent and transport processes.

Femoral arterial access should be obtained as quickly as possible after sterile patient preparation. Having multiple team members working on different tasks improves efficiency. For example, one member can prepare necessary catheters and IV drips, while another obtains femoral access.

Currently, there is no consensus on the type of anesthesia a patient should receive during mechanical thrombectomy, although there is evidence supporting the use of monitored anesthesia care (MAC) instead of general anesthesia to avoid hypotension-induced expansion of the core infarct (Abou-Chebl *et al.*, 2010; McDonagh *et al.*, 2010). A retrospective multicenter study of 980 acute stroke patients demonstrated that general anesthesia was an independent predictor for poor outcome (Al-Mufti *et al.*, 2016). Potential benefits of general anesthesia include airway protection, adequate pain control, and immobilization. These benefits, however, must be weighed against the risk of hemodynamic instability during the induction phase of intubation, the risk of pulmonary aspiration, and delays in initiation of mechanical thrombectomy. MAC enables the team to maintain hemodynamic stability and to perform neurological examination intraprocedurally. Using MAC, however, has the risk of loss of airway protection and patient agitation, which may cause procedural difficulty. A metanalysis of the 5 stroke trials demonstrated that mechincal thrombectomy performed under MAC had better functional outcomes (OR, 2.08; 95% CI, 1.47–2.96) without increased risk of symptomatic intracranial hemorrhage or short-term mortality compared with general anesthesia (Ouyang *et al.*, 2016). A neurointerventionalist may consider the use of MAC in the absence of concerns about airway protection, hemodynamic stability, or inability to cooperate.

The goals of endovascular mechanical thrombectomy are to rapidly and effectively restore blood flow, avoid adjunctive treatments, avoid embolization to new territory, and avoid vascular injury. There is evidence in the literature to support the use of one or more of these three strategies:

- Stent-retriever with balloon guide catheter flow arrest during clot retrieval (Nguyen *et al.*, 2014) (Figure 2).
- Thromboaspiration using aspiration catheter and vacuum assistance (ADAPT technique) (Figure 3).
- A combination of the stent-retriever and thromboaspiration, also known as, "Solumbra" technique, whereby a stent-retriver is deployed across the entire clot, followed by advancement of an aspiration catheter to the face of a clot resulting in corking of the clot. The stent-retriever and aspiration catheter are then removed as a unit from the patient with vacuum assistance on both the aspiration catheter and guide catheter.
- IA thrombolysis (now less favored as a first-line treatment)

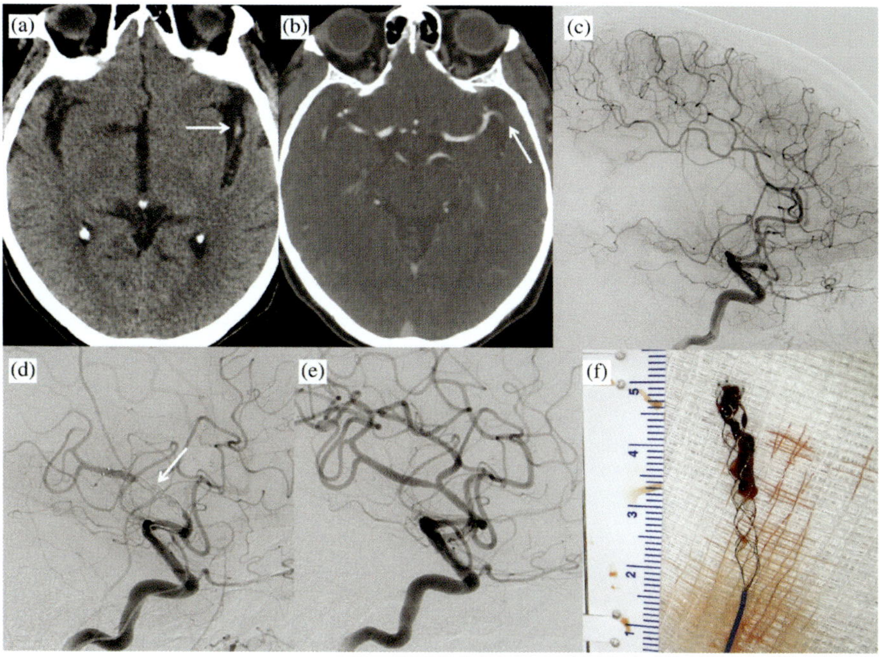

Figure 2. 73 year-old male who underwent coronary artery bypass grafting and aortic valve repair. On post-operative day 3, he developed acute-onset left gaze preference, expressive aphasia, right facial droop, and right hemiparesis (NIH stroke scale 23). A non-contrast head CT revealed a hyperdense left middle cerebral artery (MCA) (a). Given the clinical presentation and CT angiographic confirmation of a large vessel occlusion (b), the patient was taken to the neuroangiography suite for emergent mechanical thrombectomy. A cerebral angiogram (left internal carotid artery (ICA) injection), lateral views) confirmed the presence of thrombus within a left M2 segment MCA (c). A stent-treiver device was deployed across the clot (d) and the entire clot was removed from the patient (e), 193 minutes after the patient was last seen normal in the ICU. Complete recanalization and restoration of flow to the left MCA territory was seen on post-thrombectomy angiography (f). His stroke symptoms resolved over the next 48 hours.

10.2 Atherosclerotic Cerebrovascular Disease—Extracranial Carotid Disease

ICA stenoses are most commonly atherosclerotic in etiology. Extracranial ICA stenosis accounts for 15–25% of ischemic stroke, with an incidence that may be

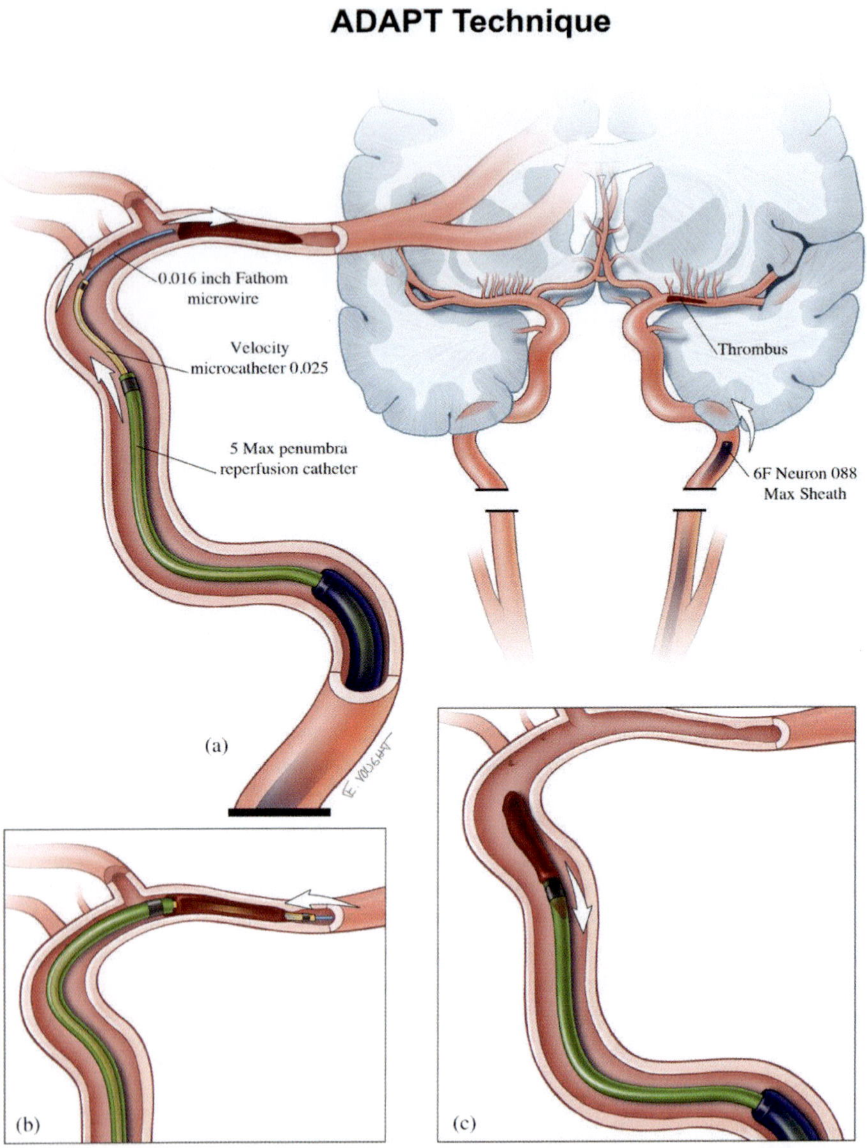

Figure 3. Illustration of the ADAPT technique in a middle cerebral artery clot. (a) The Neuron Max guide catheter is positioned as far distally as possible in the supplying ICA. Through this a 5 Max reperfusion catheter is advanced over a Velocity microcatheter with a 0.016 inch Fathom wire. (b) The Fathom wire and Velocity microcatheter are advanced through and distal to the thrombus to provide stable support for the 5 Max to be advanced to the face of the thrombus. (c) Aspiration is applied to the 5 Max until aspiration becomes occlusive and the 5 Max is then removed while maintaining aspiration to ensure the clot remains engaged in the catheter tip. Turk *et al.* (2014) with permission from BMJ Publishing Group.

as high as 10% in patients greater than 80 years of age. Ipsilateral high-grade ICA stenosis or occlusion is present in approximately 10–20% of patients presenting with acute large vessel occlusion stroke (Grau *et al.*, 2001) which further complicates endovascular access and may lead to a delay in recanalization of the target vessel occlusion. The risk for stroke in patients with extracranial ICA stenosis is associated with the degree of narrowing; for asymptomatic patients with less than 75% stenosis, the risk of stroke is less than 1%/year, but the risk increases to 2–5%/year for patients with greater than 75% stenosis. In symptomatic patients (previous TIA or stroke), the risk is considerably higher; nearly 10% in the first year and 30–35% over the next 5 years for patients with stenoses larger than 70%. Currently, the three major treatments for extracranial ICA stenosis are medical management, carotid endarterectomy, and carotid angioplasty with stenting.

10.2.1 *Management*

10.2.1.1 *Conservative*

Medical management of extracranial carotid atherosclerotic disease is by far the most commonly used treatment and includes aspirin 81–325 mg orally daily. It is the only modality considered an essential aspect of management for any patient with non-cardioembolic acute infarct and is used either alone or in conjunction with more invasive treatments. However, aspirin should not be given for the first 24 hours following treatment with IV or IA thrombolytic therapy. The use of dual antiplatelet therapy remains largely unproven, with the exception that short-term treatment with clopidogrel plus aspirin appears to be beneficial for high-risk TIA and minor stroke in Asian populations. Beyond the acute phase of an acute ischemic stroke or TIA, long-term antiplatelet therapy for secondary stroke prevention should be continued with aspirin, clopidogrel, or the combination of aspirin and extended-release dipyridamole. Major treatable atherosclerotic stroke risk factors, including hypertension, diabetes, smoking, and dyslipidemia should also be addressed.

10.2.1.2 *Surgical treatment*

Carotid endarterectomy is performed by surgical exploration of the carotid bifurcation, temporary occlusion of flow with bypass, arteriotomy, and dissection of plaque. Patching of the ICA may reduce the perioperative risk of stroke.

10.2.1.3 *Carotid angioplasty and stenting*

The placement of a stent requires administration of antiplatelet agents. In the non-emergent setting, a patient is preloaded with dual antiplatelet therapy, usually aspirin and clopidogrel, for at least 3 days prior to the procedure. In the acute setting, decision must be made whether to perform angioplasty of the stenosis without stenting, or stenting +/− angioplasty; aspirin can be administered intravenously (if available) or via per-rectal or oral routes. The only route of administration for clopidogrel is oral. Typical loading dose would be 325–650 mg of aspirin, and 300–600 mg clopidogrel. If there is concern regarding dysphagia in the acute setting, or the patient is sedated or obtunded, then a nasogastric tube can be placed through which antiplatelet agents are administered. A loading bolus of Abciximab 0.25 mg/kg IV may be administered after stent deployment as a bridge while antiplatelet agents achieve full effect. Stenting is usually performed with full systemic heparinization (after groin access, target ACT 2–2.5× baseline or greater than 250 seconds), unless contraindicated.

Compared with carotid angioplasty without stenting, insertion of a stent reduces the risk of embolization, thrombosis, carotid artery recoil, and long-term restenosis. A self-expanding stent should be used. Angioplasty by itself and balloon-mounted stents no longer play a role in extracranial carotid artery disease. Tapered stents are considered more suitable if the proximal and distal diameters of the vessel differ significantly because they reduce the radial outward force distally while still maintaining sufficient wall apposition proximally. Untapered stents, on the other hand, are more suitable for stenoses in which the proximal and distal diameters of the vessel are similar.

Stenting of a carotid artery stenosis may be performed with or without a distal embolic protection device. Proximal protection with balloon occlusion flow arrest is also an option. Pre-stenting balloon angioplasty may be necessary to allow the guide catheter to cross the stenosis. If balloon angioplasty of carotid bulb is being performed, the patient can be given prophylactic glycopyrrolate to prevent bradycardia or asystole. Zoll pads can also be placed on the carotid stent patient's chest prior to the procedure in case of asystole. Intravenous atropine can be administered for induced bradycardia or reflex asystole if needed; following stent insertion, post-stenting balloon angioplasty may be necessary to achieve adequate reduction in the degree of stenosis and improve wall apposition of the stent.

10.2.1.4 *Evidence from the literature for carotid stenting*

Several randomized controlled trials have addressed the treatment of patients with symptomatic extracranial ICA stenosis: the North American Symptomatic Carotid

Endarterectomy Trial (NASCET) (North American Symptomatic Carotid Endarterectomy Trial Collaborators, 1991), the European Carotid Surgery Trial (ECST) (1998a), and the Veterans Affairs Cooperative Studies Program Trial (Mayberg *et al.*, 1991). A pooled analysis of these three trials clearly demonstrated the benefits of surgery versus medical treatment in the group with stenosis larger than 70%; an absolute risk reduction of 16% was noted after 5 years of follow-up. Patients with moderate stenosis (50–69%) still benefitted from surgery, although the overall gains were more modest, with an absolute risk reduction of 4.6% after 5 years. In patients with mild stenosis (≤50%), the risks incurred during CEA outweighed the benefits of surgery (Rothwell *et al.*, 2003).

The benefit of CEA versus best medical treatment in asymptomatic patients with stenosis was also evaluated in three major trials: the Veterans Affairs Cooperative Studies Program Trial (Hobson *et al.*, 1993), the Asymptomatic Carotid Atherosclerosis Study (ACAS) (1995a), and the Asymptomatic Carotid Surgery Trial (ACST) (Halliday *et al.*, 2004). These studies demonstrated that CEA is beneficial for patients younger than 75 years of age with asymptomatic stenosis >70% because it approximately halves the net 5-year stroke risk from 12% down to 6%.

The previous studies have established CEA as the gold standard for the invasive management of symptomatic carotid stenosis and its role in asymptomatic patients. Several prospective randomized controlled trials have compared CEA to carotid artery stenting (CAS), with mixed results. The Carotid Artery Vertebral Artery Trial of Angioplasty and Stenting (CAVATAS) randomized 504 patients and found no difference in stroke or death at 30 days (10.3% versus 10.4%), and no significant difference in outcomes at 3 years between CEA and CAS (2001). The Stenting and Angioplasty with Protection in Patients with High Risk for Endarterectomy (SAPPHIRE) trial randomized 310 high-risk surgical patients with >50% symptomatic stenosis or >80% asymptomatic stenosis at 29 centers to CEA and CAS (Gurm *et al.*, 2008). At 30 days and 1 year, the CAS group had a lower rate of stroke, death, MI, and cranial nerve injury. A subgroup analysis of this trial also revealed significant benefit of CAS over CEA in diabetic patients (major adverse event rate of 4.8% versus 25%). The results of the SAPPHIRE trial established the role for CAS in high surgical risk patients.

Several trials have compared CAS with CEA in patients with standard surgical risk, with conflicting results. The Endarterectomy versus Angioplasty in Patients with Symptomatic Severe Stenosis Trial (EVA-3S) randomized 527 asymptomatic patients in France with >60% stenosis (Mas *et al.*, 2006). This trial was stopped early due to safety and futility concerns, with ipsilateral stroke or death occurring in 6.1% of CEA patients versus 11.7% of CAS patients. Major criticisms of this study, however, were that distal embolic protection devices were not used in the majority of cases, and there may have been inadequate training of interventionists. The Stent Supported Percutaneous Angioplasty of the Carotid Artery versus Endarterectomy Trial (SPACE) randomized 1,200 patients in Germany and Austria and found no difference between CEA and CAS at 30 days and 2 years in ipsilateral stroke or death (SPACE Collaborative Group *et al.*, 2006). Both EVA-3S and SPACE trial results were interpreted as failing to demonstrate non-inferiority of CAS versus CEA, even though the SPACE trial was stopped early because of lack of funding and did not have enough statistical power to assess non-inferiority. The interventionalist qualifications were more stringent in the SPACE trial than in EVA-3S; however, only 27% of the CAS patients in the SPACE trial were treated with distal embolic protection devices.

The UK-based International Carotid Stenting Study (ICSS) demonstrated a rather striking advantage of CEA versus CAS in 1,713 patients, with a risk of stroke or death following CAS of 8.5% versus 5.1% after CEA (Bonati *et al.*, 2015; ICSS investigators *et al.*, 2010). Once again, the use of an embolic protection device was not mandated in this study. Other criticisms included: (1) blinding in follow-up assessments was lacking, (2) the qualifications of stent operators were less rigorous than for surgeons (only 10 stents required, whereas surgeons had at least 50 cases), and (3) two inexperienced stentors had 5 major strokes in 11 patients. There is therefore valid concern that the experience of the interventionalists may have affected results of trials, with significant differences between rates of stroke or death following CAS in different studies: 4.4% for CREST, 6.9% for SPACE, 8.5% for ICSS, and 9.6% for EVA3S, compared with rather constant rates of stroke or death for CEA between 4% and 6%. In fact, a subgroup analysis of SPACE demonstrated that centers with low recruitment rates have profoundly (and negatively) affected the results for CAS.

The largest trial to date is the Carotid Revascularization Endarterectomy versus Stenting Trial (CREST), which compared CEA and CAS in 2,502 low or standard surgical risk patients with symptomatic (53% of enrolled patients) and asymptomatic (47%) stenosis at 117 sites (Brott *et al.*, 2010, 2016). Distal embolic protection devices were required in all patients in the CAS arm. In addition, a rigorous credentialing process for interventionalists was required. There was no difference between the two cohorts in the 4-year rates of primary endpoint (stroke, death, or MI). The clinical durability (rate of ipsilateral stroke) was also similar in both groups (2.4% in CEA group, and 2.0% in CAS group). CAS tended to show greater efficacy in patients <70 years of age, while CEA in those >70 years of age. There was a statistically significant difference in any stroke within 30 days in the stenting group (4.1%) versus the CEA group (1.9%, $p = 0.0019$); however, most of these were minor strokes. There was, however, a lower rate of MI (2.0% versus

3.4%, $p = 0.0387$) and cranial nerve injury (0.3% versus 4.7%) in the CAS group. In addition, the impact of MI, either clinical or by biomarkers alone, was found to be more significant than stroke on increasing risk of subsequent mortality.

Beyond 30 days, the efficacy of CEA versus CAS for stroke prevention, restenosis rates, and need for repeat revascularization was comparable in all trials. Overall, patients <70 years of age had better outcomes with CAS. CAS is associated with a better quality of life at 2 weeks and 1 month, although at 1 year, there was no difference in quality of life outcomes.

A 2012 Cochrane database meta-analysis of 11 trials pooled patients randomized to CEA versus CAS (Bonati *et al.*, 2012). In patients with symptomatic carotid stenosis at standard surgical risk, CAS was associated with a higher risk of any stroke or death at 30 days after treatment (primary safety outcome) than CEA. In patients <70 years old, however, there was no significant difference in the primary safety outcome. The rate of death or major or disabling stroke did not differ significantly between treatments. Endovascular treatment was associated with lower risks of myocardial infarction, cranial nerve palsy, and access site hematomas. The combination of death or any stroke up to 30 days after treatment or ipsilateral stroke during follow-up (the primary combined safety and efficacy outcome) also favored CEA, but the rate of ipsilateral stroke after the periprocedural period did not differ between treatments. Restenosis during follow-up was more common in patients receiving CAS than in patients assigned CEA. Among patients not suitable for surgery, the rate of death or any stroke between randomization and end of follow-up did not differ significantly between endovascular treatment and medical care.

Despite the controversy, there are well-accepted indications currently for CAS in situations where CEA is contraindicated: (1) previous anterior neck surgery (previous CEA or other neck dissection); (2) previous neck irradiation; (3) clinically significant cardiac, pulmonary, or other disease that would greatly increase the risk of anesthesia and surgery; (4) carotid artery stenosis in hyperacute stroke, a high bifurcation, a long segment of stenosis, or a tandem stenosis. Based on available data, patients with symptomatic carotid stenosis who do not have any of these conditions may be better served by treatment with CEA. In addition, the periprocedural risk of stroke and death with CAS for the operator or center should be less than 6%.

Results of the CREST trial suggest that technical aspects should be taken into account when considering CAS. Vessel tortuosity in the neck, atherosclerotic disease in the aortic arch or proximal great vessels, or even arch elongation (type III aortic arch) significantly increases the risk of the procedure, possibly preventing safe deployment of protection devices or leading to cerebral embolization, while navigating protection devices before their deployment. In these scenarios, CEA or best medical management may be the safest alternative for the patient.

Extracranial carotid stenosis is found in 10–20% of patients with hyperacute large vessel occlusion stroke. All mechanical thrombectomy trials published in 2015 allowed inclusion of patients with proximal cervical carotid stenosis, and all but one allowed inclusion of patients with complete atherosclerotic cervical ICA occlusion (SWIFT PRIME). One problem with this exclusion criterion is that differentiating complete cervical ICA occlusion from a distal ICA occlusion is often not possible using CTA or MRA. The number of patients with cervical ICA occlusion or stenosis was not consistently reported was substantial, ranging from 18.6% (REVASCAT) to 32.2% (MR CLEAN). Stenting of the underlying stenosis or occlusion was discouraged in the ESCAPE protocol. In the MR CLEAN trial, 30/75 patients with carotid stenosis or occlusion in the thrombectomy arm underwent CAS during the procedure. In the REVASCAT trial, 9/19 patients with ICA occlusion were stented. In the other trials, the management of the underlying lesion was not reported. Better outcomes for the subgroup of patients with carotid occlusion were reported in ESCAPE and MR CLEAN trials. Although thrombectomy for patients with cervical ICA occlusion is clearly indicated by these data, the optimal management of the underlying stenosis is not clear. There are several potential advantages and disadvantages for angioplasty and stenting at the time of thrombectomy. Although immediate revascularization may reduce the risk of recurrent stroke, urgent CAS generally requires dual antiplatelet therapy, which is associated with an increased incidence of intracranial hemorrhage. In addition, there is some risk for thromboembolic stroke at the time of stenting. Further studies on the definitive management of symptomatic extracranial carotid stenosis in patients undergoing emergent mechanical thrombectomy are needed.

10.3 Arteriovenous Malformations

10.3.1 *Natural history*

Current understanding of the natural history of unruptured AVMs is limited by the absence of level I or II studies (Ogilvy *et al.*, 2001). Furthermore, currently available natural history studies are limited by selection bias towards untreatable AVMs, lack of correlation to presentation, variation in analyses, and lack of long-term follow-up. The annual hemorrhage risk of an unruptured brain AVM has been reported in several hospital-based retrospective series between 2–4% (Brown *et al.*, 1988; Crawford *et al.*, 1986; Graf *et al.*, 1983; Itoyama *et al.*, 1989; Ondra *et al.*, 1990; Stapf *et al.*, 2006). For a ruptured AVM, the risk of re-hemorrhage in the first year is 7–18% in the first year (Halim *et al.*, 2004; Mast *et al.*, 1997), but decreasing to the baseline of 2% and 4% per year over time (Halim *et al.*, 2004).

The management of AVMs is dependent on occurrence of bleeding, clinical symptoms, presence of high-risk features, and treatment risk. Treatment of an AVM that has bled is often necessary because of the significant risk for rebleeding. Patients with unruptured AVMs must be appropriately worked up to determine their symptoms and whether the AVM may be the cause. Hemorrhage is not the only way an AVM may become symptomatic, however. Unruptured AVMs may cause symptoms related to venous congestion, mass effect of a venous pouch on adjacent parenchyma, or perinidal gliosis.

10.3.2 *Role of catheter angiography*

Patients with suspected brain AVMs should undergo six-vessel cerebral angiography. AVMs demonstrate arteriovenous shunting as a result of direct (fistulous and/or plexiform) connections between arteries and veins without an intervening capillary bed. The goal of angiography is also to determine the arterial territories supplying the AVM, the feeding arteries, presence of arterial stenosis, ectasias, nidus shape, size, location, venous drainage (deep, superficial, number of draining veins), and venous drainage of normal parenchyma. In addition, angiography should determine the presence of high-risk features known to increase the risk of bleed, such as feeding artery or intranidal aneurysms, venous outflow stenosis, venous ectasia (a marker for increased intranidal pressure), a diffuse nidus, a single draining vein, and deep venous drainage (Al-Shahi & Warlow, 2001; Pollock *et al.*, 1996; Stapf *et al.*, 2006). Special attention on the angiogram should be made to the presence of the number and type of feeding arteries to the AVM, whether multiple brain AVMs are present, the presence of direct arteriovenous fistulas (without an intervening nidus), and the number of draining veins (Valavanis, 1996). The brain parenchymal blush should also be analyzed for the presence of venous congestion or arterial steal.

10.3.3 *AVM treatment*

Only complete exclusion of blood flow through an AVM can completely eliminate the risk for hemorrhage. Currently, there are four management options for AVM treatment, alone or in combination:

- Microvascular neurosurgery
- Endovascular embolization
- Stereotactic radiosurgery
- Conservative (Expectant)

No randomized clinical trials have compared treatment options against each other, and therefore, optimal treatment remains controversial. In general, patients presenting with hemorrhage from an AVM should have treatment aimed at obliteration of the AVM. If this is not possible, treatment should be directed at the most probable cause of the rupture, e.g., a feeding artery aneurysm. While some feeding artery aneurysms may regress with AVM resection alone (Redekop *et al.*, 1998), this is often delayed and incomplete.

For most AVMs for which treatment is indicated, embolization is usually performed prior to surgery or radiation. Embolization performed prior to definitive microvascular neurosurgical removal improves surgical efficacy and decreases surgical morbidity and mortality (DeMeritt *et al.*, 1995; Ezura *et al.*, 2000; Jafar *et al.*, 1993; Pasqualin *et al.*, 1991; Rosenblatt *et al.*, 1998; Thompson *et al.*, 1998). At our institution, we typically perform embolization the day prior to surgery. It should also be noted that various grading systems used for estimating the risk of surgical morbidity and mortality of an AVM, such as the Spetzler–Martin scale and Lawton Supplementary scale (Lawton *et al.*, 2010; Spetzler & Martin, 1986), do not apply to endovascular management. An AVM embolization grading scale has not been developed. Embolization may also be performed prior to radiosurgery to reduce the size of the AVM nidus/target volume, as radiosurgery has been shown to be more effective for smaller AVMs (Friedman *et al.*, 1995; Gobin *et al.*, 1996; Karlsson *et al.*, 1997; Pollock *et al.*, 1998). Single-center retrospective series of AVM embolization followed by radiosurgery have yielded conflicting results, however (Back *et al.*, 2008; Gobin *et al.*, 1996). Embolization may also be performed prior to radiosurgery to occlude high-risk feeding artery or nidal aneurysms and to occlude high flow direct AVF components that are unlikely to be obliterated using radiosurgery (Gobin *et al.*, 1996). Repeat embolization may also be performed after radiosurgery (Marks *et al.*, 1993; Steinberg *et al.*, 1996).

While complete obliteration of an AVM with embolization alone is technically possible, embolization as a stand-alone therapy is currently reserved for symptomatic surgically unresectable AVMs or very small symptomatic AVMs, although the latter are typically surgically resectable with low morbidity and mortality (Panagiotopoulos *et al.*, 2009; Weber *et al.*, 2007) unless they are deep or central (Hurst *et al.*, 1995). This very small subset of all AVMs have a small, accessible nidus with no feeding arteries that are en passage vessels (i.e., supply AVM nidus and normal brain), and have a small number of feeding arteries. The use of Onyx may allow slightly larger AVMs to be curable; however, the safety and efficacy of Onyx in this setting is not well established (Panagiotopoulos *et al.*, 2009). In general, the most appropriate treatment option for each individual patient is determined by individual AVM angioarchitectural characteristics, local expertise, experience, and preference of the neurosurgical and neurointerventional teams.

For an unruptured AVM, the therapeutic risk must be lower than the natural history risk. Unfortunately, the natural history and etiology of AVMs remain unclear. The treatment of incidentally discovered brain AVMs is controversial as the efficacy and relative risks of treatments are not well established, and treatment approaches may vary widely among institutions. A Randomized Multicenter Clinical Trial of Unruptured Brain AVMs (ARUBA) showed that conservative medical management was superior to combined intervention (embolization, surgery, or radiosurgery) and medical therapy for the prevention of death or stroke in patients with unruptured brain AVMs with a mean follow-up of 33 months (Mohr *et al.*, 2014) (Table 2). The ARUBA trial is currently continuing its observational phase to establish whether the disparities will persist over an additional 5 years of follow-up. The treatment arm of the ARUBA trial included 116 patients undergoing a heterogeneous variety of interventions: neurosurgery alone (n = 5), embolisation alone (n = 30), radiotherapy alone (n = 31), or using a multimodal approach combining embolisation with either neurosurgery (n = 12) or radiotherapy (n = 15), or both (n = 1). In light of the ARUBA trial results, therefore, intervention (whether surgery, embolization, and/or radiosurgery, alone or in combination) of asymptomatic unruptured AVMs should be avoided.

Table 2. ARUBA Trial: Stroke and mortality as treated* (Mohr *et al.*, 2014).

	Interventional therapy (n = 98)	**Medical management (n = 125)**	**Risk ratio (95% CI)**
Death or stroke	36 (36.7%)	10 (8.0%)	0.22 (0.11–0.42)
Death			
All cause	3 (3.1%)	2 (1.6%)	0.52 (0.09–3.07)
AVM-related	2 (2.0%)	0	—
First stroke			
All	35 (35.7%)	8 (6.4%)	0.18 (0.09–0.37)
Haemorrhagic	24 (24.5%)	7 (5.6%)	0.23 (0.10–0.51)
Ischaemic	11 (11.2%)	1 (0.8%)	0.07 (0.10–0.54)

Notes: Data are number of patients (%). AVM = arteriovenous malformation.
*Primary outcome events only.

10.3.4 *Embolic agents*

Liquid embolic agents, such as *n*-butyl-cyanoacrylate (nBCA) and ethyl vinyl alcohol (Onyx), are most commonly used for endovascular embolization of AVMs (Howington *et al.*, 2005). NBCA is a liquid adhesive, or glue, that polymerizes rapidly when it comes into contact with ions in the bloodstream. It has been in widespread use for AVM treatment for decades and is typically mixed with Ethiodol contrast or tantalum powder for visualization under fluoroscopy (Debrun *et al.*, 1997). The concentration of nBCA is adjusted by varying the ratio of glue to ethiodol. Pure nBCA is infrequently used, usually in the setting of a high-flow AVF that cannot be reached by a coiling microcatheter. To prevent premature polymerization, microcatheters are flushed with 5% dextrose in water (D5W). Within a solidly embolized blood vessel, nBCA elicits an inflammatory response leading to vessel necrosis and fibrosis, which renders the embolization permanent (Gruber *et al.*, 1996; Howington *et al.*, 2005). Without solid embolization, intranidal capillary recruitment and recurrence can occur (Gruber *et al.*, 1996). A newer embolic agent, Onyx, is a cohesive (not adhesive) elastic copolymer comprised of ethylene vinyl alcohol (EVOH) dissolved in dimethyl sulfoxide (DMSO), and suspended micronized tantalum powder to provide contrast for fluoroscopic visualization. It does not adhere to the endothelium or microcatheter tip. A DMSO-compatible microcatheter must be used and must be primed with DMSO prior to Onyx injection. Once in contact with blood, the DMSO diffuses away and the EVOH forms a solid precipitate from the outer surface inward, usually within 5 minutes. A multicenter randomized clinical trial showed non-inferiority of Onyx compared to nBCA in achieving ≥50% nidus volume reduction with no difference in adverse events, resulting in FDA approval of Onyx for AVM embolization in 2005 (Loh *et al.*, 2010). Typical Onyx injections are much slower but easier to control than nBCA. This also means that more volume can be injected with theoretically more nidus penetration; however, care must be taken when using Onyx to avoid occlusion of outflow veins without occlusion of the nidus, as this may result in increases in intranidal pressure and subsequent hemorrhage. Extreme care is required when using embolic materials to avoid the unintentional permanent retention of portions of a catheter within a patient due to improper removal of the catheter after embolization.

Coil embolization is used in AVM treatment to occlude large direct arteriovenous fistulous components. In certain circumstances, such as when a significant distance between the microcatheter tip and the AVM nidus exists and/or distal selective catheterization is not possible, as long as there are no en passage branches to normal brain, ethanol sclerotherapy may be used to occlude the feeding artery and the AVM nidus compartment fed by that artery (Settecase *et al.*, 2016; Yakes *et al.*, 1997). The use of polyvinyl alcohol (PVA) has largely been replaced by other embolic agents.

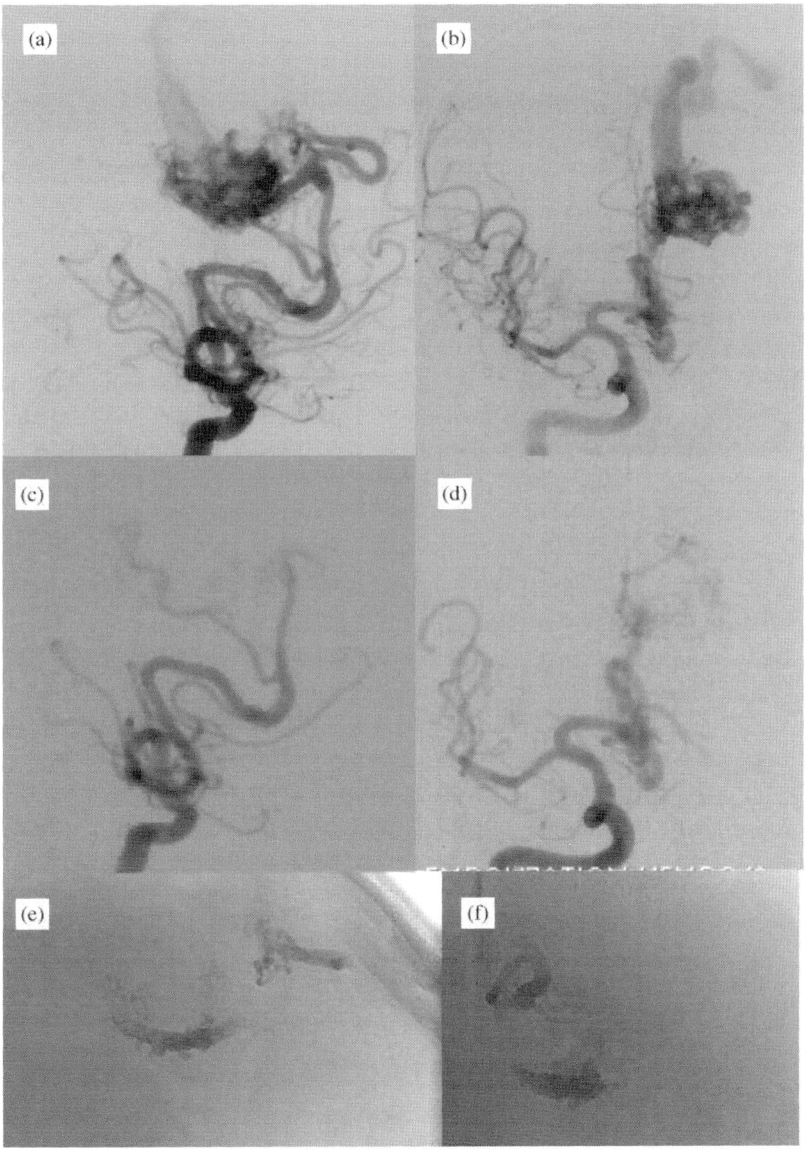

Figure 4. 53-year-old male with an unruptured AVM. (a) DSA lateral and, (b) AP views demonstrating a AVM nidus centered within the left paramedian frontal lobe supplied by the left aneterior cerebral artery with superficial venous drainage. (c) DSA lateral and, (d) AP views after embolization of the AVM with nBCA showing marked reduction in blood flow through the AVM and minimal residual AV shunting. (e) and (f) Unsubtracted lateral and AP views demonstrating the post-embolization nBCA casts which are rendered visible by the lipiodol mixed with the nBCA.

10.3.5 *Technique*

The most common and safest approach is transarterial, whereby one or more arterial feeders are occluded distally. The microcatheter is positioned distally within a feeding artery, with the tip just proximal to the AVM nidus. Flow-directed or over-the-wire catheters may be used. Permanent neurologic deficit can occur from embolization of arterial branches that supply normal brain. To prevent this, selective angiograms should be performed to look for these normal branches and injection of embolic material into them should be avoided. Unfortunately, sometimes branches to normal brain may be angiographically occult. In some instances, provocative testing may be performed by injecting a short acting barbiturate (sodium amytal) via the microcatheter followed by neurologic testing. Since this is only possible in an awake patient, appropriate anesthesia and procedural planning should take place beforehand (Debrun *et al.*, 1997; Moo *et al.*, 2002; Paulsen *et al.*, 1999; Rauch *et al.*, 1992).

The embolization goal is to occlude the most distal arterial segment, the nidus, and, if possible, the most proximal venous segment of the AV shunt. Occluding a draining vein without eliminating the arterial supply should be avoided because it will result in a marked increase in intranidal blood pressure and may lead to AVM rupture (Valavanis, 1996). Proximal feeding artery occlusion may result in inadvertent occlusion of prxominal branches to normal brain. It also ensures adequate nidal penetration that prevents the nidus from recruiting new collaterals over time.

Prior to embolization, selective microcatheter angiography should be obtained to ensure the following:

- Intravascular position of the catheter without evidence of complication, such as blood vessel perforation or re-rupture of the AVM.
- Verification that the injected vessel does in fact supply the AVM and does not supply or have branches that supply normal brain (also known as en passage or indirect branches)
- Absence of kinks in the microcatheter tip, and no proximal holes in the microcatheters. The flow microcatheters commonly used in AVM embolizations are very soft and delicate at their distal end. They can be easily damaged during preparation or during the procedure.
- Rapidity of AV shunting — one of the determinants of the appropriate concentration of nBCA that will be needed to occlude the feeding artery and nidus while avoiding early polymerization in the arterial feeder or late polymerization in the draining vein. A wedged microcatheter position allows forward flow arrest, so that flow of glue can be better controlled by the rate of injection.

With a large AVM nidus, it is recommended that no more than a third of the nidus is occluded per embolization session to minimize the risk of hemorrhage from normal perfusion pressure breakthough. Other embolization stopping points include angiographic cure, adequate reduction in nidus volume, or venous outflow stagnation or compromise (Figure 4).

10.4 Intracranial Aneurysms

Due to technical advances in cross-sectional neuroimaging and the results of the International Subarachnoid Aneurysm Trial (ISAT), the number of neurointeventional referrals for work-up and treatment of aneurysms has increased.

10.4.1 *Ruptured aneurysms*

With ruptured aneurysms, if a patient survives an initial acute subarachnoid hemorrhage (SAH), there is a high risk of a potentially fatal re-rupture of the aneurysm. The risk of re-rupture is 4% in the first 24 hours, tapering to 1.5%/day for the next 2 weeks (Winn *et al.*, 1977). Rebleeds occur in 15–20% of untreated SAH patients within the first 2 weeks, and 50% will rebleed by 6 months (Winn *et al.*, 1977). Mortality from SAH is very high, and as a result, urgent exclusion of an aneurysm from the circulation is necessary to eliminate re-rupture risk. This is commonly performed using surgical clipping or endovascular coiling.

The ISAT enrolled 2,143 patients with SAH due to aneurysmal rupture, and whose aneurysms were deemed suitable for treatment by either surgical clipping or endovascular coiling at the time of randomization. The ISAT trial demonstrated significant improvement in outcomes at 1 year with coiling (190/801, or 23.7% functionally dependent or dead) compared to clipping (243/793, or 30.6% functionally dependent or dead). A second report of the ISAT with longer follow-up published in 2005 showed that 250/1,063 (23.5%) of the coiling group were functionally dependent or dead at 1 year versus 326/1,055 (30.9%) of the surgical clipping group (Molyneux *et al.*, 2005). The risk of epilepsy was also lower in the coiling group. The most recent ISAT follow-up study, published in 2009, had mean follow-up of 9 years (minimum, 6 years; maximum,14 years) (Molyneux *et al.*, 2009). There were 24 SAH recurrences, 13 of which were related to previously treated aneurysms: 10 in the coiling group and 3 in the surgical group ($p = 0.06$). The 5-year mortality was lower in the coiling group than the surgery group (11% and 14%, respectively, $p = 0.03$). Functional independence at 5 years, however, was not different between the groups (Molyneux *et al.*, 2009). Subsequent analysis also showed that, not surprisingly, coiled aneurysms had a 6.9 times higher risk of needed retreatment due to persisting neck remnants or recurrence of the aneurysm (Campi *et al.*, 2007). Some criticisms of ISAT include: exclusion of almost 80% of screened patients, patient enrollment predominantly from European centers, over-representation of low/good SAH grade patients, and lack of minimum criteria for neurosurgeons performing clipping. Nonetheless, the results of the ISAT trial favor endovascular treatment of intracranial aneurysms for aneurysms that are amenable to surgical or endovascular treatment.

Treatment plans and the decision to coil versus clip should be individualized to each patient, taking the following into account: SAH grade; aneurysm location; size; presence of calcifications on the aneurysm (making clipping more difficult); dome-to-neck ratio (wide-necked aneurysms are difficult to coil); patient age (older patients do better with coiling); presence of significant medical co-morbidities; presence of an intraparenchymal hematoma causing mass effect in need of neurosurgical evacuation and decompression; and existence of concomitant vasospasm. In addition to a CT angiogram, a diagnostic cerebral angiogram should be performed to better evaluate the anatomy of the aneurysm and detect any other aneurysms that may be present.

10.4.2 *Unruptured aneurysms*

Unfortunately, no randomized trials on the treatment of unruptured cerebral aneurysms exist upon which to base recommendations. The lifetime rupture risk of the aneurysm must be weighed against the treatment risk.

Important information about the natural history of unruptured aneurysms comes from the International Study of Unruptured Intracranial Aneurysms (ISUIA) (1998b) (Wiebers *et al.*, 2003) and the Unruptured Cerebral Aneurysms Study (UCAS) (UCAS Japan Investigators *et al.*, 2012). The ISUIA study enrolled 2,621 patients in the US, Canada, and Europe, including 1,449 patients with 1,937 unruptured aneurysms enrolled retrospectively. The UCAS study followed 5,720 patients with 6,697 aneurysms (11,660 aneurysm-years). Both ISUIA and UCAS studies noted that aneurysm size and location were associated with the risk of rupture. More specifically, cavernous carotid artery aneurysms had the lowest rates of rupture. Anterior circulation aneurysms (anterior communicating, anterior cerebral, or ICA) had intermediate rates of rupture. Posterior circulation aneurysms (vertebrobasilar, posterior cerebral arterial system, or posterior communicating arteries) had the highest rates of rupture. Cumulative five-year rates of rupture according to aneurysm site and size at diagnosis were as follows:

- For 7–12 mm aneurysms, rupture rates for cavernous carotid, anterior circulation, and posterior circulation aneurysms were 0, 2.6, and 14.5%, respectively.

- For 13–24 mm aneurysms, rupture rates for cavernous carotid, anterior circulation, and posterior circulation aneurysms were 3, 14.5, and 18.4%, respectively.
- For ^{3}25 mm aneurysms, rupture rates for cavernous carotid, anterior circulation, and posterior circulation aneurysms were 6.4, 40, and 50%, respectively.

Other important considerations include: growth over time (with larger aneurysms more likely to continue enlarging (Burns *et al.*, 2009)), prior history of SAH (Wiebers *et al.*, 2003), family history of SAH (Broderick *et al.*, 2009; Ronkainen *et al.*, 1997), ethnic background, and presence of daughter lobule(s) (UCAS Japan Investigators *et al.*, 2012). A 5-year aneurysm rupture risk prediction score, the PHASES score (Greving *et al.*, 2014), was developed from a pooled analysis of risk factors from six prospective cohort studies (Ishibashi *et al.*, 2009; Juvela *et al.*, 2013; Sonobe *et al.*, 2010; UCAS Japan Investigators *et al.*, 2012; Wermer *et al.*, 2006; Wiebers *et al.*, 2003). Risk factors from this study were: age, hypertension, history of SAH, aneurysm size, aneurysm location, and geographical region (Table 3).

Table 3. Predictors composing the PHASES aneurysm rupture risk score.

PHASES aneurysm risk score	Points
(P) Population	
North American, European (other than Finnish)	0
Japanese	3
Finnish	5
(H) Hypertension	
No	0
Yes	1
(A) Age	
<70 years	0
≥70 years	1
(S) Size of aneurysm	
<7.0 mm	0
7.0–9.9 mm	3
10.0–19.9 mm	6
≥20 mm	10
(E) Earlier SAH from another aneurysm	
No	0

(Continued)

Table 3. (*Continued*)

PHASES aneurysm risk score	Points
Yes	1
(S) Site of aneurysm	
ICA	0
MCA	2
ACA/Pcom/posterior	4

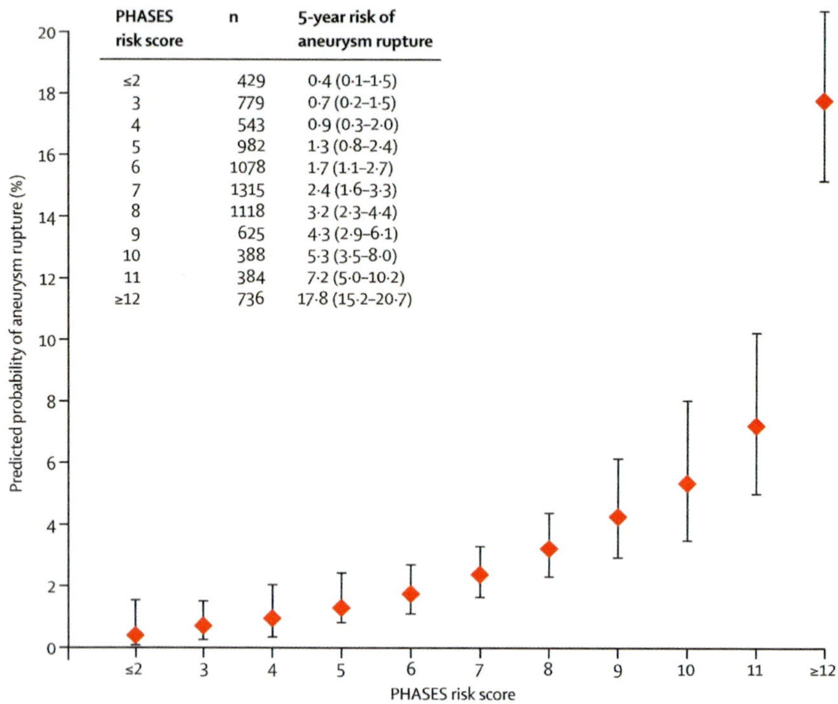

To calculate the PHASES risk score for an individual, the number of points associated with each indicator can be added up to obtain the total risk score. For example, a 55-year-old North American man with no hypertension, no previous SAH, and a medium-sized (8 mm) posterior circulation aneurysm will have a risk score of 0 + 0 + 0 + 3 + 0 + 4 = 7 points. This score corresponds to a 5-year risk of rupture of 2·4%. SAH = subarachnoid haemorrhage. ICA = internal carotid artery. MCA = middle cerebral artery. ACA = anterior cerebral arteries (including the anterior cerebral artery, anterior communicating artery, and pericallosal artery). Pcom = posterior communicating artery. Posterior = posterior circulation (including the vertebral artery, basilar artery, cerebellar arteries, and posterior cerebral artery). Greving, Wermer *et al.* (2014) with permission from Elsevier Inc.

The decision whether or not to treat an unruptured aneurysm must be based on the most efficacious and durable treatment with the lowest risk and upon patient preference. Several observational studies have demonstrated lower morbidity and mortality with endovascular treatment compared to surgical clipping (Kotowski *et al.*, 2013; Wiebers *et al.*, 2003). A systematic review and meta-analysis of 71 studies on the potential safety and efficacy of endovascular treatment of unruptured aneurysms showed an immediate satisfactory aneurysm angiographic occlusion rate of 86.1% (2,660/3,089 aneurysms) (Naggara *et al.*, 2010). During a follow-up period of 0.4 to 3.2 years, the recurrence rate was 24.4% with retreatment of 9.1% of aneurysms. SAH risk post-treatment was 0.2%. A systematic review and meta-analysis of 60 observational studies from 1990 to 2011 comprising 9,845 patients and 10,845 aneurysms found that the overall mortality associated with surgical clipping of unruptured aneurysms was 1.7%, with unfavorable outcomes occurring in 6.7% (Kotowski *et al.*, 2013). In experienced hands, the combined morbidity and mortality risk of endovascular coiling has been reported from 1.5 to 9.4% (Benes *et al.*, 2010; Murayama *et al.*, 2003; Pandey *et al.*, 2007; van Rooij & Sluzewski, 2006).

Endovascular techniques have evolved significantly over the last several decades. Currently, aneurysm treatment can be divided into three main categories: reconstructive, deconstructive, and flow diversion. The reconstructive approach is most commonly used and involves occlusion of the aneurysm by placement of one or more detachable platinum coils into an aneurysm, while keeping the parent vessel patent. The presence of coils within the aneurysm elicits a thrombotic and inflammatory reaction that causes the aneurysm to occlude (Figure 5). Some aneurysm neck configurations will not allow coils to stay within an aneurysm, with tendency for the coils to herniate into the parent vessel, which can put the patient at significant risk of thromboembolic complications. In these circumstances, coiling with balloon- or stent-assisted technique can be used. With the balloon remodeling technique, a balloon microcatheter is temporarily inflated at the neck of the aneurysm while coils are placed via a serparate microcatheter (Aletich *et al.*, 2000; Cekirge *et al.*, 2011; Lubicz *et al.*, 2008; Malek *et al.*, 2000; Mericle *et al.*, 1997; Moret *et al.*, 1997; Shapiro *et al.*, 2008). Stent-assisted coiling has represented a major advancement in the treatment of cerebral aneurysms (Benitez *et al.*, 2004; Fiorella *et al.*, 2005; Fiorella *et al.*, 2004; Lanzino *et al.*, 1999; Liang *et al.*, 2010; Maldonado *et al.*, 2011; Mericle *et al.*, 1998; Sedat *et al.*, 2009; Wajnberg *et al.*, 2009; Wanke *et al.*, 2003). Stents commonly used for stent-assisted coiling include Neuroform (Stryker, Fremont, CA) or Enterprise (Codman Neurovascular, Raynham, MA), and more recently, the Low Profile Visualized Intraluminal Support (LVIS) Device (Microvention, Aliso Viejo, CA). Stent placement, however, requires a patient to be prophylactically treated beforehand with dual

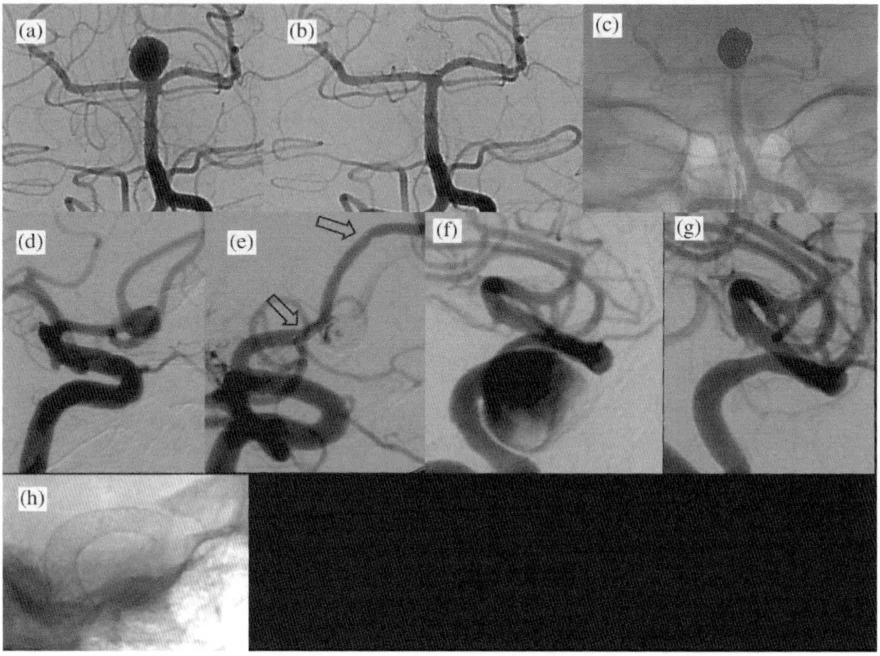

Figure 5. (a) Cerebral digital subtraction angiogram (DSA), AP view, demonstrating a basilar artery tip aneurysm. (b) DSA, after endovascular coiling of the aneurysm using detachable coils, no residual aneurysm filling is seen. (c) Unsubtracted DSA, showing coil configuration within the aneurysm. (d) DSA, lateral view, showing a wide-necked anterior communicating artery aneurysm. (e) DSA, lateral view, after Neuroform stent assisted coiling. Note only distal and proximal stent tynes are visible (arrows). (f) DSA, Schuller's view, showing a large cavernous segment ICA aneurysm. (g) After flow diversion with a PED, the aneurysm has thrombosed. (h) An unsubtracted lateral view xray of the PED.

antiplatelets (aspirin 325 po qd and Plavix 75 mg po qd) for 3–5 days to prevent thrombotic complications. As a result, stent placement is currently not recommended in patients with an acute SAH. Flow diversion is a new concept in the endovascular treatment of aneurysms, and can be considered a subtype of reconstructive approach. A flow diverting stent (e.g., Pipeline Embolization Device (PED) [ev3, Irvine, CA], Silk stent [BALT, Montmorency, France], Surpass stent [Styker, Fremont, CA]) contains more metal than a Neuroform or Enterprise stent, resulting in increased coverage of the aneurysm neck. After flow diverter placement, the changes in flow dynamics through the stented segment result in reduced arterial blood inflow to the aneurysm, leading to increased stasis of blood within the aneurysm and progressive thromboocclusion of the aneurysm while preserving patency of the parent vessel. Flow diverters render some aneurysms that are technically difficult to treat endovascularly using reconstructive techniques or inaccessible to surgery (e.g., fusiform cavernous segment ICA aneurysms) amenable to treatment (Becske *et al.*, 2013; Lubicz *et al.*, 2008; Lylyk *et al.*, 2009; Nelson *et al.*, 2011). The Pipeline was FDA-approved in 2011 for the treatment of large or giant wide-neck intracranial aneurysms in the ICA from the petrous to the superior hypophyseal segments. Use of flow diverters in other regions of the brain has not been FDA approved, although published literature has reported its use outside the ICA. Under the humanitarian device exemption regulation pathway, stent-assisted coiling systems are approved with embolic coils for the treatment of wide neck, intracranial saccular aneurysms arising from a parent vessel with a diameter of >2 mm and <4.5 mm that are not amenable to treatment with surgical clipping. The number of observational studies reporting their successful off-label use in other arteries is increasing. Dual anti-platelets must also be started prior to stent placement (although the length varies between 5 and 21 days in different centers). Although, stent-assisted coiling and flow diversion techniques have been found to be efficacious and relatively safe, unlike reconstructive endovascular coiling, their use has not been compared against other treatments in randomized controlled trials.

Deconstructive treatment of an aneurysm is also performed using coils and refers to the sacrifice of the parent vessel from which the aneurysm arises. This is performed in limited circumstances and only if surgical clipping, reconstructive endovascular coiling, or flow diversion (Figure 5g–h) are not possible (e.g., dissecting aneurysms, pseudoaneurysms, or recurrent aneurysms) (Elhammady *et al.*, 2010; Nelson *et al.*, 2011; van der Schaaf *et al.*, 2002). A balloon test occlusion should be performed beforehand in most cases to determine whether the patient has adequate collaterals that will tolerate the cessation of anterograde flow within that artery without neurologic deficits.

Acknowledgments

This research is supported by a grant from the U.S. National Institutes of Health (R01 EB012031) to S.W.H.

References

Abou-Chebl, A., Lin, R., *et al.* (2010). Conscious sedation versus general anesthesia during endovascular therapy for acute anterior circulation stroke: Preliminary results from a retrospective, multicenter study. *Stroke* **41**(6): 1175–1179.

Al-Mufti, F., Dancour, E., *et al.* (2016). Neurocritical care of emergent large-vessel occlusion: The era of a new standard of care. *J Intensive Care Med.*

Al-Shahi, R., Warlow, C. (2001). A systematic review of the frequency and prognosis of arteriovenous malformations of the brain in adults. *Brain* **124**(Pt 10): 1900–1926.

Aletich, V.A., Debrun, G.M., *et al.* (2000). The remodeling technique of balloon-assisted Guglielmi detachable coil placement in wide-necked aneurysms: Experience at the University of Illinois at Chicago. *J Neurosurg* **93**(3): 388–396.

Back, A.G., Vollmer, D., *et al.* (2008). Retrospective analysis of unstaged and staged Gamma Knife surgery with and without preceding embolization for the treatment of arteriovenous malformations. *J Neurosurg* **109**(Suppl): 57–64.

Becske, T., Kallmes, D.F., *et al.* (2013). Pipeline for uncoilable or failed aneurysms: Results from a multicenter clinical trial. *Radiology* **267**(3): 858–868.

Benes, V., 3rd, Mitchell, P., *et al.* (2010). Endovascular coiling in 131 patients with low complication rate justifies treating most unruptured intracranial aneurysms. *Cent Eur Neurosurg* **71**(1): 1–7.

Benitez, R.P., Silva, M.T., *et al.* (2004). Endovascular occlusion of wide-necked aneurysms with a new intracranial microstent (Neuroform) and detachable coils. *Neurosurgery* **54**(6): 1359–1367; discussion 1368.

Berkhemer, O.A., Fransen, P.S., *et al.* (2015). A randomized trial of intraarterial treatment for acute ischemic stroke. *N Engl J Med* **372**(1): 11–20.

Bhatia, R., Hill, M.D., *et al.* (2010). Low rates of acute recanalization with intravenous recombinant tissue plasminogen activator in ischemic stroke: Real-world experience and a call for action. *Stroke* **41**(10): 2254–2258.

Bonati, L.H., Dobson, J., *et al.* (2015). Long-term outcomes after stenting versus endarterectomy for treatment of symptomatic carotid stenosis: The International Carotid Stenting Study (ICSS) randomised trial. *Lancet* **385**(9967): 529–538.

Bonati, L.H., Lyrer, P., *et al.* (2012). Percutaneous transluminal balloon angioplasty and stenting for carotid artery stenosis. *Cochrane Database Syst Rev* (9): CD000515.

Broderick, J.P., Brown, R.D. Jr., *et al.* (2009). Greater rupture risk for familial as compared to sporadic unruptured intracranial aneurysms. *Stroke* **40**(6): 1952–1957.

Brott, T.G., Hobson, R.W. 2nd, *et al.* (2010). Stenting versus endarterectomy for treatment of carotid-artery stenosis. *N Engl J Med* **363**(1): 11–23.

Brott, T.G., Howard. G., *et al.* (2016). Long-term results of stenting versus endarterectomy for carotid-artery stenosis. *N Engl J Med* **374**(11): 1021–1031.

Brown, R.D., Jr., Wiebers, D.O., *et al.* (1988). The natural history of unruptured intracranial arteriovenous malformations. *J Neurosurg* **68**(3): 352–357.

Burns, J.D., Huston, J. 3rd, *et al.* (2009). Intracranial aneurysm enlargement on serial magnetic resonance angiography: Frequency and risk factors. *Stroke* **40**(2): 406–411.

Campbell, B.C., Mitchell, P.J., *et al.* (2015). Endovascular therapy for ischemic stroke with perfusion-imaging selection. *N Engl J Med* **372**(11): 1009–1018.

Campi, A., Ramzi, N., *et al.* (2007). Retreatment of ruptured cerebral aneurysms in patients randomized by coiling or clipping in the International Subarachnoid Aneurysm Trial (ISAT). *Stroke* **38**(5): 1538–1544.

Cekirge, H.S., Yavuz, K., *et al.* (2011). HyperForm balloon remodeling in the endovascular treatment of anterior cerebral, middle cerebral, and anterior communicating artery aneurysms: Clinical and angiographic follow-up results in 800 consecutive patients. *J Neurosurg* **114**(4): 944–953.

Crawford, P.M., West, C.R., *et al.* (1986). Arteriovenous malformations of the brain: Natural history in unoperated patients. *J Neurol Neurosurg Psychiatry* **49**(1): 1–10.

Debrun, G.M., Aletich, V., *et al.* (1997). Embolization of the nidus of brain arteriovenous malformations with n-butyl cyanoacrylate. *Neurosurgery* **40**(1): 112–120; discussion 120–121.

del Zoppo, G.J., Poeck, K., *et al.* (1992). Recombinant tissue plasminogen activator in acute thrombotic and embolic stroke. *Ann Neurol* **32**(1): 78–86.

DeMeritt, J.S., Pile-Spellman, J., *et al.* (1995). Outcome analysis of preoperative embolization with N-butyl cyanoacrylate in cerebral arteriovenous malformations. *AJNR Am J Neuroradiol* **16**(9): 1801–1807.

Elhammady, M.S., Wolfe, S.Q., *et al.* (2010). Carotid artery sacrifice for unclippable and uncoilable aneurysms: Endovascular occlusion vs common carotid artery ligation. *Neurosurgery* **67**(5): 1431–1436; discussion 1437.

Endarterectomy for asymptomatic carotid artery stenosis. (1995a). Executive committee for the asymptomatic carotid atherosclerosis study. *JAMA* **273**(18): 1421–1428.

Endovascular versus surgical treatment in patients with carotid stenosis in the Carotid and Vertebral Artery Transluminal Angioplasty Study (CAVATAS): A randomised trial. (2001). *Lancet* **357**(9270): 1729–1737.

Ezura, M., Takahashi, A., *et al.* (2000). Endovascular treatment of aneurysms associated with cerebral arteriovenous malformations: Experiences after the introduction of Guglielmi detachable coils. *J Clin Neurosci* **7**(Suppl 1): 14–18.

Fiorella, D., Albuquerque, F.C., *et al.* (2005). Usefulness of the Neuroform stent for the treatment of cerebral aneurysms: Results at initial (3–6-month) follow-up. *Neurosurgery* **56**(6): 1191–1201; discussion 1201–1202.

Fiorella, D., Albuquerque, F.C., *et al.* (2004). Preliminary experience using the Neuroform stent for the treatment of cerebral aneurysms. *Neurosurgery* **54**(1): 6–16; discussion 16–17.

Friedman, W.A., Bova, F.J., *et al.* (1995). Linear accelerator radiosurgery for arteriovenous malformations: The relationship of size to outcome. *J Neurosurg* **82**(2): 180–189.

Furlan, A., Higashida, R., *et al.* (1999). Intra-arterial prourokinase for acute ischemic stroke. The PROACT II study: A randomized controlled trial. Prolyse in Acute Cerebral Thromboembolism. *JAMA* **282**(21): 2003–2011.

Gobin, Y.P., Laurent, A., *et al.* (1996). Treatment of brain arteriovenous malformations by embolization and radiosurgery. *J Neurosurg* **85**(1): 19–28.

Goyal, M., Demchuk, A.M., *et al.* (2015). Randomized assessment of rapid endovascular treatment of ischemic stroke. *N Engl J Med* **372**(11): 1019–1030.

Graf, C.J., Perret, G.E., *et al.* (1983). Bleeding from cerebral arteriovenous malformations as part of their natural history. *J Neurosurg* **58**(3): 331–337.

Grau, A.J., Weimar, C., *et al.* (2001). Risk factors, outcome, and treatment in subtypes of ischemic stroke: The German stroke data bank. *Stroke* **32**(11): 2559–2566.

Greving, J.P., Wermer, M.J., *et al.* (2014). Development of the PHASES score for prediction of risk of rupture of intracranial aneurysms: A pooled analysis of six prospective cohort studies. *Lancet Neurol* **13**(1): 59–66.

Grotta, J.C., Hacke, W. (2015). Stroke neurologist's perspective on the new endovascular trials. *Stroke* **46**(6): 1447–1452.

Gruber, A., Mazal, P.R., *et al.* (1996). Repermeation of partially embolized cerebral arteriovenous malformations: A clinical, radiologic, and histologic study. *AJNR Am J Neuroradiol* **17**(7): 1323–1331.

Gurm, H.S., Yadav, J.S., *et al.* (2008). Long-term results of carotid stenting versus endarterectomy in high-risk patients. *N Engl J Med* **358**(15): 1572–1579.

Hacke, W., Kaste, M., *et al.* (2008). Thrombolysis with alteplase 3 to 4.5 hours after acute ischemic stroke. *N Engl J Med* **359**(13): 1317–1329.

Halim, A.X., Johnston, S.C., *et al.* (2004). Longitudinal risk of intracranial hemorrhage in patients with arteriovenous malformation of the brain within a defined population. *Stroke* **35**(7): 1697–1702.

Halliday, A., Mansfield, A., *et al.* (2004). Prevention of disabling and fatal strokes by successful carotid endarterectomy in patients without recent neurological symptoms: Randomised controlled trial. *Lancet* **363**(9420): 1491–1502.

Hobson, R.W., 2nd, Weiss, D.G., *et al.* (1993). Efficacy of carotid endarterectomy for asymptomatic carotid stenosis. The veterans affairs cooperative study group. *N Engl J Med* **328**(4): 221-227.

Howington, J.U., Kerber, C.W., *et al.* (2005). Liquid embolic agents in the treatment of intracranial arteriovenous malformations. *Neurosurg Clin N Am* **16**(2): 355–363, ix–x.

Humphries, W., Hoit, D., *et al.* (2015). Distal aspiration with retrievable stent assisted thrombectomy for the treatment of acute ischemic stroke. *J Neurointerv Surg* **7**(2): 90–94.

Hurst, R.W., Berenstein, A., *et al.* (1995). Deep central arteriovenous malformations of the brain: The role of endovascular treatment. *J Neurosurg* **82**(2): 190-195.

International Carotid Stenting Study investigators, Ederle, J., *et al.* (2010). Carotid artery stenting compared with endarterectomy in patients with symptomatic carotid stenosis (International Carotid Stenting Study): An interim analysis of a randomised controlled trial. *Lancet* **375**(9719): 985–997.

International Study of Unruptured Intracranial Aneurysms Investigators. (1998b). Unruptured intracranial aneurysms — risk of rupture and risks of surgical intervention. International study of unruptured intracranial aneurysms investigators. *N Engl J Med* **339**(24): 1725–1733.

Ishibashi, T., Murayama, Y., *et al.* (2009). Unruptured intracranial aneurysms: Incidence of rupture and risk factors. *Stroke* **40**(1): 313–316.

Itoyama, Y., Uemura, S., *et al.* (1989). Natural course of unoperated intracranial arteriovenous malformations: Study of 50 cases. *J Neurosurg* **71**(6): 805–809.

Jafar, J.J., Davis, A.J., *et al.* (1993). The effect of embolization with N-butyl cyanoacrylate prior to surgical resection of cerebral arteriovenous malformations. *J Neurosurg* 78(1): 60–69.

Jansen, O., von Kummer, R., *et al.* (1995). Thrombolytic therapy in acute occlusion of the intracranial internal carotid artery bifurcation. *AJNR Am J Neuroradiol* **16**(10): 1977–1986.

Jovin, T.G., Chamorro, A., *et al.* (2015). Thrombectomy within 8 hours after symptom onset in ischemic stroke. *N Engl J Med* **372**(24): 2296–2306.

Juvela, S., Poussa, K., *et al.* (2013). Natural history of unruptured intracranial aneurysms: A long-term follow-up study. *Stroke* **44**(9): 2414–2421.

Karlsson, B., Lindquist, C., *et al.* (1997). Prediction of obliteration after gamma knife surgery for cerebral arteriovenous malformations. *Neurosurgery* **40**(3): 425–430; discussion 430–431.

Khatri, P., Yeatts, S.D., *et al.* (2014). Time to angiographic reperfusion and clinical outcome after acute ischaemic stroke: An analysis of data from the Interventional Management of Stroke (IMS III) phase 3 trial. *Lancet Neurol* **13**(6): 567–574.

Kotowski, M., Naggara, O., *et al.* (2013). Safety and occlusion rates of surgical treatment of unruptured intracranial aneurysms: A systematic review and meta-analysis of the literature from 1990 to 2011. *J Neurol Neurosurg Psychiatry* **84**(1): 42–48.

Lanzino, G., Wakhloo, A.K., *et al.* (1999). Efficacy and current limitations of intravascular stents for intracranial internal carotid, vertebral, and basilar artery aneurysms. *J Neurosurg* **91**(4): 538–546.

Lawton, M.T., Kim, H., *et al.* (2010). A supplementary grading scale for selecting patients with brain arteriovenous malformations for surgery. *Neurosurgery* **66**(4): 702–713; discussion 713.

Lees, K.R., Bluhmki, E., *et al.* (2010). Time to treatment with intravenous alteplase and outcome in stroke: An updated pooled analysis of ECASS, ATLANTIS, NINDS, and EPITHET trials. *Lancet* **375**(9727): 1695–1703.

Liang, G., Gao, X., *et al.* (2010). Neuroform stent-assisted coiling of intracranial aneurysms: A 5 year single-center experience and follow-up. *Neurol Res* **32**(7): 721–727.

Lima, F.O., Furie, K.L., *et al.* (2014). Prognosis of untreated strokes due to anterior circulation proximal intracranial arterial occlusions detected by use of computed tomography angiography. *JAMA Neurol* **71**(2): 151–157.

Loh, Y., Duckwiler, G.R., *et al.* (2010). A prospective, multicenter, randomized trial of the Onyx liquid embolic system and N-butyl cyanoacrylate embolization of cerebral arteriovenous malformations. Clinical article. *J Neurosurg* **113**(4): 733–741.

Lubicz, B., Lefranc, F., *et al.* (2008). Balloon-assisted coiling of intracranial aneurysms is not associated with a higher complication rate. *Neuroradiology* **50**(9): 769–776.

Lylyk, P., Miranda, C., *et al.* (2009). Curative endovascular reconstruction of cerebral aneurysms with the pipeline embolization device: The Buenos Aires experience. *Neurosurgery* **64**(4): 632–642; discussion 642–643; quiz N6.

Maldonado, I.L., Machi, P., *et al.* (2011). Neuroform stent-assisted coiling of unruptured intracranial aneurysms: Short- and midterm results from a single-center experience with 68 patients. *AJNR Am J Neuroradiol* **32**(1): 131–136.

Malek, A.M., Halbach, V.V., *et al.* (2000). Balloon-assist technique for endovascular coil embolization of geometrically difficult intracranial aneurysms. *Neurosurgery* **46**(6): 1397–1406; discussion 1406–1407.

Marks, M.P., Lane, B., *et al.* (1993). Endovascular treatment of cerebral arteriovenous malformations following radiosurgery. *AJNR Am J Neuroradiol* **14**(2): 297–303; discussion 304–305.

Mas, J.L., Chatellier, G., *et al.* (2006). Endarterectomy versus stenting in patients with symptomatic severe carotid stenosis. *N Engl J Med* **355**(16): 1660–1671.

Mast, H., Young, W.L., *et al.* (1997). Risk of spontaneous haemorrhage after diagnosis of cerebral arteriovenous malformation. *Lancet* **350**(9084): 1065–1068.

Mayberg, M.R., Wilson, S.E., *et al.* (1991). Carotid endarterectomy and prevention of cerebral ischemia in symptomatic carotid stenosis. Veterans Affairs Cooperative Studies Program 309 Trialist Group. *JAMA* **266**(23): 3289–3294.

McDonagh, D.L., Olson, D.M., *et al.* (2010). Anesthesia and sedation practices among neurointerventionalists during acute ischemic stroke endovascular therapy. *Front Neurol* **1**: 118.

Mericle, R.A., Lanzino, G., *et al.* (1998). Stenting and secondary coiling of intracranial internal carotid artery aneurysm: Technical case report. *Neurosurgery* **43**(5): 1229–1234.

Mericle, R.A., Wakhloo, A.K., *et al.* (1997). Temporary balloon protection as an adjunct to endosaccular coiling of wide-necked cerebral aneurysms: Technical note. *Neurosurgery* **41**(4): 975–978.

Mocco, J.O. Zaidat, O., *et al.* (2016). Aspiration thrombectomy after intravenous alteplase versus intravenous alteplase alone. *Stroke* **47**(9): 2331–2338.

Mohr, J.P., Parides, M.K., *et al.* (2014). Medical management with or without interventional therapy for unruptured brain arteriovenous malformations (ARUBA): A multicentre, non-blinded, randomised trial. *Lancet* **383**(9917): 614–621.

Molyneux, A.J., Kerr, R.S., *et al.* (2009). Risk of recurrent subarachnoid haemorrhage, death, or dependence and standardised mortality ratios after clipping or coiling of an intracranial aneurysm in the International Subarachnoid Aneurysm Trial (ISAT): long-term follow-up. *Lancet Neurol* **8**(5): 427–433.

Molyneux, A.J., Kerr, R.S., *et al.* (2005). International subarachnoid aneurysm trial (ISAT) of neurosurgical clipping versus endovascular coiling in 2143 patients with ruptured intracranial aneurysms: A randomised comparison of effects on survival, dependency, seizures, rebleeding, subgroups, and aneurysm occlusion. *Lancet* **366**(9488): 809–817.

Moo, L.R., Murphy, K.J., *et al.* (2002). Tailored cognitive testing with provocative amobarbital injection preceding AVM embolization. *AJNR Am J Neuroradiol* **23**(3): 416–421.

Moret, J., Cognard, C., *et al.* (1997). The "Remodelling Technique" in the treatment of wide neck intracranial aneurysms. Angiographic results and clinical follow-up in 56 cases. *Interv Neuroradiol* **3**(1): 21–35.

Murayama, Y., Nien, Y.L., *et al.* (2003). Guglielmi detachable coil embolization of cerebral aneurysms: 11 years' experience. *J Neurosurg* **98**(5): 959–966.

Naggara, O.N., White, P.M., *et al.* (2010). Endovascular treatment of intracranial unruptured aneurysms: Systematic review and meta-analysis of the literature on safety and efficacy. *Radiology* **256**(3): 887–897.
National Institute of Neurological Disorders and Stroke rt-PA Stroke Study Group. (1995b). Tissue plasminogen activator for acute ischemic stroke. The National Institute of Neurological Disorders and Stroke rt-PA Stroke Study Group. *N Engl J Med* **333**(24): 1581–1587.
Nelson, P.K., Lylyk, P., *et al.* (2011). The pipeline embolization device for the intracranial treatment of aneurysms trial. *AJNR Am J Neuroradiol* **32**(1): 34–40.
Nguyen, T.N., Malisch, T., *et al.* (2014). Balloon guide catheter improves revascularization and clinical outcomes with the Solitaire device: Analysis of the North American Solitaire Acute Stroke Registry. *Stroke* **45**(1): 141–145.
Nogueira, R.G., Lutsep, H.L., *et al.* (2012). Trevo versus Merci retrievers for thrombectomy revascularisation of large vessel occlusions in acute ischaemic stroke (TREVO 2): A randomised trial. *Lancet* **380**(9849): 1231–1240.
North American Symptomatic Carotid Endarterectomy Trial Collaborators. (1991). Beneficial effect of carotid endarterectomy in symptomatic patients with high-grade carotid stenosis. *N Engl J Med* **325**(7): 445–453.
Ogilvy, C.S., Stieg, P.E., *et al.* (2001). Recommendations for the management of intracranial arteriovenous malformations: A statement for healthcare professionals from a special writing group of the Stroke Council, American Stroke Association. *Circulation* **103**(21): 2644–2657.
Ondra, S.L., Troupp, H., *et al.* (1990). The natural history of symptomatic arteriovenous malformations of the brain: A 24-year follow-up assessment. *J Neurosurg* **73**(3): 387–391.
Ouyang, F., Chen, Y., *et al.* (2016). Selection of patients and anesthetic types for endovascular treatment in acute ischemic stroke: A meta-analysis of randomized controlled trials. *PLoS One* **11**(3): e0151210.
Panagiotopoulos, V., Moller-Hartmann, W., *et al.* (2009). Onyx embolization as a first line treatment for intracranial dural arteriovenous fistulas with cortical venous reflux. *Rofo* **181**(2): 129–138.
Pandey, A.S., Koebbe, C., *et al.* (2007). Endovascular coil embolization of ruptured and unruptured posterior circulation aneurysms: Review of a 10-year experience. *Neurosurgery* **60**(4): 626–636; discussion 636–637.
Pasqualin, A., Scienza, R., *et al.* (1991). Treatment of cerebral arteriovenous malformations with a combination of preoperative embolization and surgery. *Neurosurgery* **29**(3): 358–368.
Paulsen, R.D., Steinberg, G.K., *et al.* (1999). Embolization of rolandic cortex arteriovenous malformations. *Neurosurgery* **44**(3): 479–484; discussion 484–486.
Pollock, B.E., Flickinger, J.C., *et al.* (1996). Factors that predict the bleeding risk of cerebral arteriovenous malformations. *Stroke* **27**(1): 1–6.
Pollock, B.E., Flickinger, J.C., *et al.* (1998). Factors associated with successful arteriovenous malformation radiosurgery. *Neurosurgery* **42**(6): 1239–1244; discussion 1244–1247.

Randomised trial of endarterectomy for recently symptomatic carotid stenosis: Final results of the MRC European Carotid Surgery Trial (ECST). (1998a). *Lancet* **351**(9113): 1379–1387.

Rauch, R.A., Vinuela, F., *et al.* (1992). Preembolization functional evaluation in brain arteriovenous malformations: The superselective Amytal test. *AJNR Am J Neuroradiol* **13**(1): 303–308.

Redekop, G., TerBrugge, K., *et al.* (1998). Arterial aneurysms associated with cerebral arteriovenous malformations: Classification, incidence, and risk of hemorrhage. *J Neurosurg* **89**(4): 539–546.

Rha, J.H., Saver, J.L. (2007). The impact of recanalization on ischemic stroke outcome: A meta-analysis. *Stroke* **38**(3): 967–973.

Riedel, C.H., Zimmermann, P., *et al.* (2011). The importance of size: Successful recanalization by intravenous thrombolysis in acute anterior stroke depends on thrombus length. *Stroke* **42**(6): 1775–1777.

Ronkainen, A., Hernesniemi, J., *et al.* (1997). Familial intracranial aneurysms. *Lancet* **349**(9049): 380–384.

Rosenblatt, S., Lewis, A.I., *et al.* (1998). Combined interventional and surgical treatment of arteriovenous malformations. *Neuroimaging Clin N Am* **8**(2): 469–482.

Rothwell, P.M., Eliasziw, M., *et al.* (2003). Analysis of pooled data from the randomised controlled trials of endarterectomy for symptomatic carotid stenosis. *Lancet* **361**(9352): 107–116.

Saver, J.L., Goyal, M., *et al.* (2015). Stent-retriever thrombectomy after intravenous t-PA vs. t-PA alone in stroke. *N Engl J Med* **372**(24): 2285–2295.

Saver, J.L., Jahan, R., *et al.* (2012). Solitaire flow restoration device versus the Merci Retriever in patients with acute ischaemic stroke (SWIFT): A randomised, parallel-group, non-inferiority trial. *Lancet* **380**(9849): 1241–1249.

Sedat, J., Chau, Y., *et al.* (2009). Endovascular occlusion of intracranial wide-necked aneurysms with stenting (Neuroform) and coiling: Mid-term and long-term results. *Neuroradiology* **51**(6): 401–419.

Settecase, F., Hetts, S.W., *et al.* (2016). Superselective intra-arterial ethanol sclerotherapy of feeding artery and nidal aneurysms in ruptured cerebral arteriovenous malformations. *AJNR Am J Neuroradiol* **37**(4): 692–697.

Shapiro, M., Babb, J., *et al.* (2008). Safety and efficacy of adjunctive balloon remodeling during endovascular treatment of intracranial aneurysms: A literature review. *AJNR Am J Neuroradiol* **29**(9): 1777–1781.

Sonobe, M., Yamazaki, T., *et al.* (2010). Small unruptured intracranial aneurysm verification study: SUAVe study, Japan. *Stroke* **41**(9): 1969–1977.

SPACE Collaborative Group, Ringleb, P.A., *et al.* (2006). 30 day results from the SPACE trial of stent-protected angioplasty versus carotid endarterectomy in symptomatic patients: A randomised non-inferiority trial. *Lancet* **368**(9543): 1239–1247.

Spetzler, R.F., Martin, N.A. (1986). A proposed grading system for arteriovenous malformations. *J Neurosurg* **65**(4): 476–483.

Stapf, C., Mast, H., *et al.* (2006). Predictors of hemorrhage in patients with untreated brain arteriovenous malformation. *Neurology* **66**(9): 1350–1355.

Steinberg, G.K., Chang, S.D., *et al.* (1996). Surgical resection of large incompletely treated intracranial arteriovenous malformations following stereotactic radiosurgery. *J Neurosurg* **84**(6): 920–928.

Thompson, R.C., Steinberg, G.K., *et al.* (1998). The management of patients with arteriovenous malformations and associated intracranial aneurysms. *Neurosurgery* **43**(2): 202–211; discussion 211–212.

Turk, A.S., Frei, D., *et al.* (2014). ADAPT FAST study: A direct aspiration first pass technique for acute stroke thrombectomy. *J Neurointerv Surg* **6**(4): 260–264.

UCAS Japan Investigators, Morita, A., *et al.* (2012). The natural course of unruptured cerebral aneurysms in a Japanese cohort. *N Engl J Med* **366**(26): 2474–2482.

Valavanis, A. (1996). The role of angiography in the evaluation of cerebral vascular malformations. *Neuroimaging Clin N Am* **6**(3): 679–704.

van der Schaaf, I.C., Brilstra, E.H., *et al.* (2002). Endovascular treatment of aneurysms in the cavernous sinus: A systematic review on balloon occlusion of the parent vessel and embolization with coils. *Stroke* **33**(1): 313–318.

van Rooij, W.J., Sluzewski, M. (2006). Procedural morbidity and mortality of elective coil treatment of unruptured intracranial aneurysms. *AJNR Am J Neuroradiol* **27**(8): 1678–1680.

Wajnberg, E., de Souza, J.M., *et al.* (2009). Single-center experience with the Neuroform stent for endovascular treatment of wide-necked intracranial aneurysms. *Surg Neurol* **72**(6): 612–619.

Wanke, I., Doerfler, A., *et al.* (2003). Treatment of wide-necked intracranial aneurysms with a self-expanding stent system: initial clinical experience. *AJNR Am J Neuroradiol* **24**(6): 1192–1199.

Weber, W., Kis, B., *et al.* (2007). Endovascular treatment of intracranial arteriovenous malformations with onyx: Technical aspects. *AJNR Am J Neuroradiol* **28**(2): 371–377.

Wermer, M.J., van der Schaaf, I.C., *et al.* (2006). Yield of short-term follow-up CT/MR angiography for small aneurysms detected at screening. *Stroke* **37**(2): 414–418.

Wiebers, D.O., Whisnant, J.P., *et al.* (2003). Unruptured intracranial aneurysms: Natural history, clinical outcome, and risks of surgical and endovascular treatment. *Lancet* **362**(9378): 103–110.

Winn, H.R., Richardson, A.E., *et al.* (1977). The long-term prognosis in untreated cerebral aneurysms: I. The incidence of late hemorrhage in cerebral aneurysm: A 10-year evaluation of 364 patients. *Ann Neurol* **1**(4): 358–370.

Yakes, W.F., Krauth, L., *et al.* (1997). Ethanol endovascular management of brain arteriovenous malformations: Initial results. *Neurosurgery* **40**(6): 1145–1152; discussion 1152–1154.

11 Stem Cell-mediated Biobridge: Crossing the Great Divide Between Bench and Clinic in Translating Cell Therapy for Stroke

Trenton Lippert, Marci Crowley, M. Grant Liska, and Cesar V. Borlongan

Department of Neurosurgery and Brain Repair,
University of South Florida, Morsani College of Medicine, Tampa, Florida, USA

The utilization of stem cells as a neurotherapeutic agent has been demonstrated through the stimulation of endogenous stem cells (Barha *et al.*, 2011; Borlongan, 2011; Jaskelioff *et al.*, 2011; Wang *et al.*, 2011) and transplantation of exogenous cells adjacent to the damaged brain tissues (Andres *et al.*, 2011; Borlongan *et al.*, 2006; Hargus *et al.*, 2010; Lee *et al.*, 2010; Liu *et al.*, 2011; Mazzocchi-Jones *et al.*, 2009; Mezey, 2011; Yasuda *et al.*, 2011). Many endogenous and exogenous cell-based therapeutics for brain disorders are tailored to arrest the secondary cell death triggered by cerebrovascular insults. A chief tenet of regenerative medicine for neurological disorders primarily involves the mobilization of both endogenous and exogenous stem cells toward the injured brain following insults, such as stroke (Andres *et al.*, 2011; Barha *et al.*, 2011; Borlongan *et al.*, 2006, 2011; Hargus *et al.*, 2010; Jaskelioff *et al.*, 2011; Lee *et al.*, 2010; Liu *et al.*, 2011; Mazzocchi-Jones *et al.*, 2009; Mezey, 2011; Wang *et al.*, 2011; Yasuda *et al.*, 2011; Yasuhara *et al.*, 2008). Despite positive outcomes of stroke therapies in pre-clinical laboratory studies, there has been negligible success in translating these novel treatments into clinical applications (Pollock *et al.*, 2006; Seol *et al.*, 2011; Yasuhara *et al.*, 2009). The success of these treatments would likely broaden the therapeutic window for stroke treatment, which has remained limited at ~4.5 hours.

Here, we discuss the current status of stem cell therapy for stroke, expanding upon the concept of transplanted stem cell-mediated biobridge as a therapeutic mechanism underlying stroke recovery. Additionally, we explore major intricacies of stem cell therapies in stroke such as safety, efficacy, dosing, and the possibilities offered by combination therapies. Finally, the issues of public opinion and the morality of stem cell use will be addressed. Improving our understanding of these topics will be instrumental in providing a practical cell therapy option for stroke.

11.1 Benefits of Stem Cell Treatment in Stroke

The regeneration associated with stem cell therapy treatment was traditionally hypothesized to occur through the mechanism of direct replacement of necrotizing cells, or the indirect repair of injured tissues via secretion of trophic factors by the transplanted stem cells (Lee *et al.*, 2007; Redmond *et al.*, 2007). For the last 25 years, regenerative processes initiated by direct cell replacement and growth factor secretion have been focal points of the stem cell research attempting to transition cell therapy from the laboratory to the clinic. However, the paradoxically-robust functional recovery observed in stroke animals, despite poor graft survival rates of the stem cells and acute upregulation of growth factors in the brain, can only be partially explained by these two established mechanisms of action. The concept of biobridge formation — whereby transplanted stem cells create a cellular migratory pathway in an effort to promote recruitment of endogenous stem cells — was proposed as another brain repair mechanism facilitating the biological connection of the neurogenic niches of the brain to the damaged regions within the cortex (Sanai *et al.*, 2011; Tajiri *et al.*, 2013). The significance of this biobridge mechanism is that in mobilizing endogenous stem cells to mount the brain repair, graft persistence, and prolonged growth factor upregulation will not a be a prerequisite for neuroregeneration. Accordingly, the observation of biobridge formation addresses the conundrum of why many transplanted studies have detected robust functional recovery even with low graft survival and mediocre growth factor upregulation.

11.2 Secondary Cell Death in Stroke as a Therapeutic Target for Stem Cell Transplantation

Blood flow restriction to a region of the brain results in stroke. This interruption in blood flow within the brain is commonly referred to as ischemic stroke, whereas hemorrhagic stroke is defined by blood vessel bleeding or rupture. An area of dead neuronal cells called the infarct core, and a surrounding region of injured and dying tissue known as the penumbra, is the signature pathological characteristic of stroke. As a consequence of the vascular disablement associated with ischemic stroke, regional deficiency of glucose and oxygen develops and leads to the progression of an elaborate cascade of events, referred to as secondary cell death processes (Merson & Bourne, 2014). We focus on stem cell therapy in ischemic stroke as it offers a more applicable use of stem cells than hemorrhagic stroke. That ischemic stroke subsequently manifests a secondary cell death cascade, which presents with a wide therapeutic window lasting for days, weeks, months, and potentially years, makes it a reasonable target for delayed therapeutic intervention. The primary necrotic core that evolves within minutes after stroke becomes fixed and unsalvageable. Conversely, the secondary cell death that arises in the tissue that perimeters the core corresponds to the developing ischemic penumbra (Broughton *et al.*, 2009), and, with proper treatment intervention such as stem cell therapy, may be revived.

A better prognosis has been indicated from the arrest of secondary cell death that presents as subacute or even chronic progression of the penumbra in contrast to the supracute window in the formation of the ischemic core. Minimization of the necrotic core and limitation of the evolving penumbra requires that sufficient circulation be re-established within a short time frame, preferably less than 4.5 hours. In this regard, the only FDA-approved drug to combat stroke, tPA, must be administered no later than the 4.5 hours after the initial stroke onset, and beyond this acute window, the risk of hemorrhage is too high to indicate its use (Graham, 2003). In comparison, when targeting the progressive penumbra, treatment interventions over a much protracted window post-stroke may allow for the sequestration of secondary cell death events. Indeed, penumbra-associated functional and neurostructural deficits have been shown to diminish and even reverse upon activation of endogenous neurogenesis, angiogenesis, and neuroplasticity.

11.3 Inciting the Self-regenerative Mechanism in Stroke

Despite the long-standing generalization that the adult mammalian brain lacks the capacity for regenerative processes, this belief has been confronted with scientific evidence that neurogenesis occurs in distinct regions of the adult brain, specifically in the subventricular zone (SVZ) of the lateral ventricles and the subgranular zone (SGZ) of the dentate gyrus in the hippocampus (Gage, 2000; Ming & Song, 2011). This endogenous regeneration that reaches into adulthood (Merson & Bourne, 2014) has been observed to undergo further stimulation as a result of injury, indicating that the adult brain has the capacity to support a reparative process against stroke. Unfortunately, these efforts are largely insufficient in that they are unlikely to yield meaningful improvements after stroke due to a number of factors, including meager proliferation levels of newly formed cells, restriction in their neural lineage commitment, and poor migratory capacity. In view of these limitations of spontaneous host regeneration, the field of regenerative medicine has sought to find a treatment strategy to enhance endogenous neurogenesis. Treatment regimens that modify the host microenvironment to cater to newly generated neurons have been explored in hopes of assisting proliferation, migration, and differentiation of these neural progenitor cells. Despite the use of neurogenic drugs, to date, the extent of host neurogenesis remains relatively restrained within the neurogenic niches, namely the SVZ and SGZ, and their neighboring tissues; therefore, the applications of this endogenous repair mechanism are relegated within these brain regions proximal to the neurogenic tissues (Gould, 2007; Ming & Song, 2011). Because cerebrovascular diseases, such as stroke, entails primary injury as well as secondary cell death, with many of the damaged tissues likely remote from the neurogenic niches, purely targeting the endogenous neurogenesis niches to stimulate brain repair will likely result in limited functional outcomes.

After an ischemic stroke, upregulation of specific molecules that stimulate neurogenic niches has been observed to improve neural progenitor cell survival, proliferation, migration, and differentiation (Merson & Bourne, 2014). These mobilized neural progenitor cells (although a minor population) have been shown to travel from the SVZ to the post-ischemic striatum in rat stroke models (Yamashita *et al.*, 2006; Zhang *et al.*, 2009). Furthermore, it has been documented that within the hippocampus and cerebral cortex following stroke, resident stem cells that remain dormant during homeostasis are activated. When examined in the clinical setting, however, post-mortem analysis of the stroke brain at 3 days to 13 years after stroke onset has not presented any indications of newly generated cortical neurons (Huttner *et al.*, 2014). By rough estimation 0.1%, or 1 out of every 1,000 neurons is a newly born cell resulting from host neurogenesis after stroke (Huttner *et al.*, 2014), implying that relying merely on the scarcity of newly proliferated cells and their limited ability to migrate from the neurogenic niche to the damaged area may not be practical when considering endogenous neurogenesis as a regenerative process to counter stroke.

11.4 Regenerative Medicine Circumvents the Restricted Therapeutic Window for Stroke

As noted above, regenerative medicine for CNS disorders primarily engages endogenous and exogenous stem cells for repair of the aging and diseased brain (Androutsellis-Theotokis *et al.*, 2006; Lindvall & Kokaia, 2006; Muotri & Gage, 2006).

Transplantation of stem cells has been evaluated in various models of neurological disorder, generating promising results signifying its role as a stroke treatment (Boltze *et al.*, 2014; Borlongan, 2016; Chen *et al.*, 2003). Currently, the accessibility of the thrombolytic agent tissue plasminogen activator (tPA) for stroke, its limited treatment period, and its related unfavorable effects, has not ameliorated the disease as a primary cause of morbidity and mortality around the world. We alluded above to stem cell therapy's extended time frame to initiate intervention — namely the subacute and chronic stages of stroke — that drastically prolongs the window of effective treatment, allowing for treating more patients to take advantage of this therapy. Various categories of transplantable cells have been tested in animal models of neurological disorders, with initiation of clinical trials over the last decade. The stem cells examined in stroke models include embryonic stem cells, neural stem/progenitor cells, fetal cells, NT2N cells, CTX0E3, umbilical cord blood, adipose, amnion and induced pluripotent stem cells, which are further discussed below (De La Peña *et al.*, 2015; Hara *et al.*, 2008; Kaneko *et al.*, 2011; Li *et al.*, 2008; Liu *et al.*, 2014; Stroemer *et al.*, 2009). Largely as a result of a reliable safety profile in blood-borne diseases, the leading choice for transplantable donor cell in the laboratory and clinic is the bone marrow and its cellular derivatives in particular (Borlongan *et al.*, 2011; Steinberg *et al.*, 2016). Peripheral transplantation, such as intra-arterial, intravenous and intranasal, and direct intracerebral transplantation, has demonstrated the functional recovery using bone marrow-derived stem cells (Acosta *et al.*, 2015; Borlongan, 2011; Borlongan *et al.*, 2004, 2011; Prasad *et al.*, 2014). Initial clinical trials have begun, and initial data have exhibited safety; however, efficacy warrants further studies (Steinberg *et al.*, 2016). We discuss the numerous stem cell profiles and origins, specifically the adult tissue-derived mesenchymal stem cells and their application in cell transplantation, translational impediments, and anticipated necessity for adjunctive therapies. Additionally, we discuss the existing general public views on stem cell therapy. Our objective in the subsequent sections is to evaluate the science supporting regenerative medicine that caters to validating the safety, efficacy, and clinical application of stem cell therapy in stroke.

11.5 Stem Mobilization as Standalone and in Combination Therapy

11.5.1 *Directing stem cells under the influence of drugs*

As a consequence of the difficulty in extending tPA's therapeutic window for drug treatments, many pharmaceutical companies have abandoned the research and development of new drugs for stroke. In recent years, however, there appears to be a resurgence in stroke drug investigations. The emergence of two novel classes of drugs that target the subacute and chronic phases of stroke pathology has reinvigorated the search for stroke therapies — namely biologic factors that promote neurogenesis, angiogenesis, and synaptogenesis and compounds that attenuate harmful inflammation (Chen *et al.*, 2015; Hosaka & Hoh, 2014; Xiong & Yang, 2015). Included in these are candidates for stroke therapeutics being examined in the laboratory, and a few early clinical trials such as statins (Chen *et al.*, 2003; Reuter *et al.*, 2015; White *et al.*, 2000), G-CSF (De La Peña *et al.*, 2015; Peña & Borlongan, 2015; Yamauchi *et al.*, 2014), erythropoietin (Ding *et al.*, 2010; Gonzalez *et al.*, 2013; Souvenir *et al.*, 2014; Wang *et al.*, 2004), candesartan (Chen *et al.*, 2003; White *et al.*, 2000), metformin (Abdelsaid *et al.*, 2015; Ishrat *et al.*, 2013), minocycline (Fagan *et al.*, 2011), apurinic/apyrimidinic site-repairing enzyme endonuclease-1 (Leak *et al.*, 2015), Isoxazole-9 (Koh *et al.*, 2015), arachidonic acid metabolites such as Lipoxin A4 (Guo *et al.*, 2016), and recombinant T-cell receptor ligands (Zhu *et al.*, 2014, 2015).

When contemplating these new stroke drugs, the idea of combination therapy with stem cell therapy becomes appealing because of overlapping therapeutic targets in the ischemic brain. For example, MSCs have been shown to promote neurogenesis and suppress inflammation (Lee *et al.*, 2007; Massensini *et al.*; 2015, Skop *et al.*, 2014), implying that combining MSCs with drugs that act on these two regenerative pathways may provide additive effects. In fact, stem cell adjunct therapies for stroke have ignited much interest in the stroke therapy field (De La Peña *et al.*, 2015). In addition to pharmacological approaches, non-pharmacological methods of promoting neurogenesis and dampening harmful inflammation include remote ischemic conditioning (Borlongan *et al.*, 2008; Oh *et al.*, 2011; Park *et al.*, 2009; Savitz *et al.*, 2011a, 2014; Stem Cell Therapies as an Emerging Paradigm in Stroke, 2009) and localized cerebral hypothermia (Borlongan *et al.*, 2010). Unique combination therapy options may exist, but would require rigorous preclinical studies in order to translate these into the clinical realm.

11.5.2 *Nourishing stem cells in a microenvironment favorable for regeneration*

The pathology of the ischemic brain produces an area of parenchymal and supporting tissue that is not conducive to the survival of endogenous and exogenous stem cells. This poses an innate limit to the effectiveness of both transplanted stem cells and endogenous reparative stem cells. To circumvent this hostile stroke microenvironment, the use of biomaterials, such as hydrogels, may aid in remodeling neural tissue, making it more conducive for stem cells to survive, to integrate with the host tissue, and to exert functional effects, altogether nurturing an improved regenerative process (Ghuman *et al.*, 2016). Cell-free extracellular matrix (ECM) has been shown to promote the movement of host brain cells into the injury site while sequestering the necrotic debris (Ghuman *et al.*, 2016). Hydrogel treatment could provide dual benefits, recruiting an endogenous stem cell repair effort (Ghuman *et al.*, 2016) and modulating the microenvironment to facilitate the survival of exogenous transplanted stem cells (Stabenfeldt *et al.*, 2010). Specifically, hydrogels act as scaffolding that mediates interactions between the transplanted cells and the damaged host tissue, occupies the tissue cavity formed by the necrotic core, and regulates the localized delivery of stem cell-secreted therapeutic molecules (Ghuman *et al.*, 2016; Massensini *et al.*, 2015). Thus, this material can alter the stroke brain microenvironment to be more apt to promote transplanted stem cell survival, in addition to supporting the endogenous stem cells that migrate from the neurogenic niches into the damaged brain tissue (Skop *et al.*, 2014). Comparable to the drugs discussed above, the use of biomaterial may serve as adjunct therapies to stem cell transplantation, but will require significant translational research in order to optimize the material's safety and efficacy in promoting neural regeneration in the stroke brain.

11.6 Multitude of Options, but Transplantable Stem Cells Need to be Identified

An equally key factor in cell transplantation outcomes — along with optimizing cell route, dosage amount, and time of delivery — is the identification of a well-defined set of stem cells (Savitz *et al.*, 2011b, 2014; Stem Cell Therapies as an Emerging Paradigm in Stroke, 2009). Arising from initial studies using fetal stem cells in stroke models, notable levels of cellular and functional recovery have now been demonstrated using NT2N, CTX0E3, embryonic stem cells, hematopoietic stem cells, neural stem cells, adult tissue-derived stem cells, and induced pluripotent stem cells (Borlongan *et al.*, 2006, 2008; De La Peña *et al.*, 2015; Hara *et al.*, 2008; Oh *et al.*, 2011; Park *et al.*, 2009). In addition to the scientific merits and clinical practical factors dictating the preferred cell type, ethical and logistical concerns have also contributed to the search for optimal stem cells (Borlongan *et al.*, 2010; Lo & Parham, 2009). In particular, the harvest of fetal and embryonic stem cells has experienced resistance due to ethical issues. In tandem, generating sufficient quantities of cells possessing high levels of "stemness" has posed logistical issues. Bone marrow-derived stem cells have emerged as an attractive candidate for the treatment of CNS disorders, such as stroke, due to their adult tissue origins and their possession of necessary stem cell phenotypic qualities (Tang *et al.*, 2007). From bone marrow, specialized populations of cells can be isolated or engineered into stem/progenitor cells, including MSCs, endothelial progenitor cells (EPCs), SB623, multipotent adult progenitor cells (MAPCs), and Multilineage-differentiating stress-enduring (Muse), among others (Gnecchi & Melo, 2009; Jellema *et al.*, 2015; Li *et al.*, 2015; Tate *et al.*, 2010; Uchida *et al.*, 2016). As we noted above, MSCs in particular have been the subject of extensive research in stroke models. Of interest, MSCs have demonstrated adult multipotency characteristics by differentiating toward numerous cell lineages both *in vitro* and *in vivo* (Borlongan, 2011; Borlongan *et al.*, 2011; Eckert *et al.*, 2013; Joyce *et al.*, 2010; Kocsis & Honmou, 2012; Rowart *et al.*, 2015). When transplanted in stroke models, MSCs resulted in functional recovery of motor and cognitive performance, supported by histological evidence of a reduction in brain damage (Van Velthoven *et al.*, 2014). Mechanisms of action such as cell replacement, growth factor secretion, and activation of endogenous brain repair processes have been suggested. Possible means of endogenous repair may include neurogenesis, angiogenesis, and synaptogenesis via engrafted MSC secretions (Anderson *et al.*, 2016; Doeppner & Hermann, 2010; Duffy *et al.*, 2009; Ishikawa *et al.*, 2013; Xiong *et al.*, 2010; Xiong & Yang, 2015).

Stem cells are biologics in nature, yet initial study of their clinical applications seemed to reveal ligand-receptor mechanisms more consistent with drugs. As a result, the search for a single, overly simplistic regenerative pathway ensued attempting to reveal and exploit the underlying mechanisms of cell therapy in stroke. Fortunately, our improved understanding of the complex secondary cell death pathways initiated after stroke has allowed the multiple regenerative processes of stem cells to be fully appreciated (Puyal *et al.*, 2013; Schweizer *et al.*, 2013; Sozmen *et al.*, 2012). This complex stroke pathology also dictates the necessity for combination therapies, such as co-treatment with effective drugs or biomaterials, which, as we pointed out in the preceding section, can act synergistically to facilitate improved functional outcomes over standalone stem cell therapies.

11.7 Safety of MSCs is Only Half the Story, with their Clinical Efficacy Unproven

A limited number of clinical trials have tested the safety and efficacy of intravenous delivery of autologous MSCs. When 100 million autologous MSCs (SH-2 and SH-4 positive) were administered to five patients 4 weeks post-stroke, recovery of neurological function was seen via both Barthel Index and Rankin Score, although the benefits seemed to diminish by 12 months post-treatment (Bang *et al.*, 2005). Importantly, no negative side effects were observed or reported in the randomized trial (Bang *et al.*, 2005). Another study delivering 7–10 million mononuclear cells (MNCs) per kilogram in stroke patients delivered at 24 and 72 hours after ischemia conferred greater improvements in Barthel Index, modified Rankin Score, and the National Institute of Health's Stroke Scale (NIHSS) over a 6-month time frame (Savitz *et al.*, 2011b). Again, no adverse effects were reported in this open-labeled study (Savitz *et al.*, 2011b). In response to this promising MNC trial, a phase II, multicenter, parallel group, randomized trial with blinded outcome assessment was conducted in India with 120 patients (Prasad *et al.*, 2014). The data from this experiment revealed that stroke patients who received an average of 280.75 million MNCs (n = 58) at a median of 18.5 days post-stroke did not demonstrate significant differences in the Barthel Index, modified Rankin scale shift analysis, NIHSS score, or infarct volume when compared to non-transplanted stroke patients after 6 months (Prasad *et al.*, 2014). These results suggest that while safe, intravenous injection of MNCs is not an effective treatment for subacute stroke. Instead of general MNCs, a subpopulation of bone marrow MNCs that are CD34+ was used in a separate trial, delivering 100 million autologous cells intra-arterially within 7 days of ischemia onset in five stroke patients suffering from severe anterior circulation ischemic stroke (NIHSS score of ≥8) (Banerjee *et al.*, 2014). These patients exhibited increases in modified Rankin Scores and NIHSS scores, in addition to decreases in lesion volume at a 6-month follow-up. Again, the treatment was well-tolerated, with no patients experiencing adverse side effects from the transplantation. Thus, this open-label study revealed intra-arterial transplantation of CD34+ MNCs to be safe (Banerjee *et al.*, 2014).

The evidence from these clinical trials supports the notion that the transplantation of MSCs and their derivatives, such as MNCs, are safe in stroke patients. However, the efficacy of these treatments has not been demonstrated conclusively due to small sample sizes and the open-label nature of most of the studies. Additionally, inconsistencies between the individual studies make cross-study evaluation difficult. The discrepancy between clinical and laboratory transplant regimens is also problematic. Differences in the donor cells used for transplantation among these clinical protocols likely influenced the varying functional recoveries seen. Bang *et al.* for example, used SH-2 and SH-4 as the criteria for their cells, while Savitz and colleagues employed an extensive panel of markers (CD3, CD14, CD16, CD19, CD20, CD34, CD45, CD56, Lin 1, CD133-2) to define their MNCs via flow cytometry. Prasad *et al.* also used flow cytometry to characterize their MNCs but limited their criteria to CD34 and CD45 (Prasad *et al.*, 2014), while Banerjee and colleagues used magnetic cell isolation to isolate CD34+ cells

(Banerjee *et al.*, 2014). Based solely upon the phenotypic variations of the donor cells, comparing the outcomes of these trials would be inconclusive. Additionally, the time points of intervention were significantly different across the trials, ranging from 1 day to 4 weeks post-stroke. The route of administration also varied, with Bang, Savitz, and Prasad delivering cells intravenously, and Banerjee administering the cells intra-arterially. The doses used in these trials also pose concern, as the effective dose for intravenous administration has been estimated at approximately 840 million cells for a 75-kg human (based on preclinical studies demonstrating that 4 million cells as effective dose for a 250-g rat), yet amounts far below this cell dose were used in three of the clinical trials. Savitz's trial was the only one to utilize doses that closely approximated the preclinical estimations. While this trial did result in patients showing clinical improvements, the efficacy of the treatment should be approached cautiously as the study was open-label. Moreover, the literature supporting MNCs is scarce, raising the question on the basis for proceeding with these clinical trials, in that limited scientific data support the safety, efficacy, and mechanisms of action for the donor cells used. Except for Savitz's clinical trial, the other studies did not conform to the lab-to-clinic translation guidelines established by Stem Cell Therapeutics as an Emerging Paradigm for Stroke (STEPS) (Savitz *et al.*, 2014). The STEPS recommendations serve as a guide for laboratory research to form the basis of clinical trial designs, thereby increasing the likelihood of success of translating stem cell-based therapies to the clinic.

11.8 Biobridge: Stem Cell-paved Path Between Endogenous Neurogenic Niche and Remote Injured Brain Areas

As mentioned previously, cell replacement therapy and modification of the microenvironment via trophic factors are the two commonly accepted mechanisms for transplanted stem cells' therapeutic effects (Lee *et al.*, 2007; Redmond *et al.*, 2007). The novel biobridge concept was first reported in a study utilizing the transplantation of modified MSCs in a traumatic brain injury (TBI) model, in which adult rats were subject to a controlled cortical impact (CCI) and subsequently received MSC transplantation (Tajiri *et al.*, 2013). The MSCs provided a pathway through which host endogenous stem cells could travel from the neurogenic niche to the damaged cortex via their formation of a biobridge that was later viewed using immunohistochemistry and laser capture assay (Tajiri *et al.*, 2013). Imaging analysis revealed that significantly higher levels of extracellular matrix metalloproteinases (MMPs) were present, specifically within the region of the biobridge. This indicates the potential role of MMP in the facilitation of mobilizing neural progenitor cells following brain injury, an event that would likely release particular chemoattractant signals from the area of insult, further suggesting that an inflammation (*albeit* cytokines, chemokines, MMPs), enriched mechanism closely mediates that biobridge formation. Notably, grafted MSCs are subsequently replaced by host cells after a period of time (Tajiri *et al.*, 2013), which is a welcome translational safety outcome because tumor formation has been associated with prolonged, unregulated survival and proliferation of stem cells. Furthermore, long-term graft survival is not required for robust and steady stroke recovery supporting the observed strong functional improvements in several studies despite noting an inferior engraftment.

A series of experiments designed to provide additional characterization of the biobridge determined that MMPs played a key role in generating the biobridge (Tajiri *et al.*, 2013). The transplanted stem cells appeared to provide MMPs by themselves that molded the initial biobridge at the acute stages following transplantation. Once the biobridge is established, transplanted cells are replaced by endogenous stem cells, originating from the neurogenic niche (Tajiri *et al.*, 2013). This biobridge then behaved as a pathway for endogenous stem cells to travel from the SVZ to the TBI insult area. The stem cells proceeded to spread throughout the peri-impact area and exerted their benefits to the surrounding region.

The development of a biobridge creates a secondary pathway to augment the repair mechanism of stem cell therapy, along with the release of growth factors for regenerative processes and cell replacement. A biobridge generates an advantageous microenvironment for axonal growth and cell migration. The transplanted injured brain displaying biobridge formation exhibited elevated levels of proliferation and differentiation, which were detected in the peri-impact area; a series of cells marked with DCX and nestin, indicators of maturing neurons, was observed (Tajiri *et al.*, 2013). On the other hand, animals who were administered the vehicle did not display substantial degrees of differentiation or migration. In addition, the MMP-9 levels of manifestation and movement were upregulated nine-fold and correlated with endogenous stem cell migration, in comparison with animals that were administered the vehicle transplantation (Tajiri *et al.*, 2013). Current investigations have exhibited the importance of ECMs and MMPs within stroke pathology (Borlongan, 2011; Yasuhara *et al.*, 2008), which provides additional support for the significance of forming biobridges in stem cell therapy for treating stroke. With this in mind, we foresee that grafted stem cells through comparable biobridges could stimulate the movement of stem cells from the neurogenic niche to the infarct region and stroke core.

Stem cells isolated from the adult brain, peripheral blood and umbilical cord blood can alter the behavior and levels of ECMs and MMPs (Barkho *et al.*,

2008; Lin *et al.*, 2013; Sobrino *et al.*, 2012). As a result of ECMs and MMPs being involved in the pathology of stroke, the capacity of stem cells to produce a biobridge in stroke necessitates further studies. A recent study has demonstrated that restriction of MMPs minimized the movement of endogenous stem cells from the niche to the stroke site (Zhao *et al.*, 2006), possibly a result of diminished neurovascular network remodeling. In contrast, the generation of a biobridge could incite the manufacturing of MMPs, which can improve transport of host endogenous stem cells and the repair of the neurovascular network. Moreover, the physiochemical homeostasis between cell deformation, migration rates, and tissue barrier restrictions coincides with limiting factors that determine the efficiency of cell migration. Sequentially, this physiochemical homeostasis is normalized by the capacity to breakdown the ECM using proteolytic enzymes such as integrin, MMPs and actomyosin-mediated mechanocoupling (Wolf *et al.*, 2013).

This unique concept of biobridge formation as a result of stem cell transplantation is not restricted only to treatment in TBI but may also extend to stroke. The affected regions of the brain in stroke are generally more widespread in comparison to TBI, which presents with more localized brain injury; therefore, it may prove more difficult for endogenous stem cells to migrate toward the injured site. Nonetheless, adjusting the cell dose, route of delivery, and timing of transplantation may allow a more effective biobridge that can resolve the long distances of stem cell migration from the neurogeneic niches to the ischemic brain areas. In the end, the biobridge formation provides a direct transport of neurogenic niche-derived cells to the remote infarcted areas via an ECM where the endogenous stem cells may attach and mobilize to the remote regions (Tajiri *et al.*, 2013). Accordingly, we posit that a comparable biobridge will inevitably be achieved following an intracerebral stem cell treatment into the stroke brain. This biobridge will likewise create a path for the endogenous stem cells to transport from the endogenous niches, such as the SGZ and SVZ, to the infarct areas within the stroke brain, leading to functional recovery (Stone *et al.*, 2013). Laboratory investigations have demonstrated that transplantation using MSCs assists in recovery from stroke, by modulating immune and inflammatory response (Yoo *et al.*, 2013). Of note, transplanted stem cells have exhibited the ability to produce transforming growth factor-beta (TGF-β) that can reduce the dispersal of the inflammatory MPC-1 cells (Yoo *et al.*, 2013). In this regard, transplanted MSCs may establish a biobridge by producing immunomodulatory or anti-inflammatory cytokines, termed MSC secretome (Hsieh *et al.*, 2013). The MSC secretome can involve glial cell-line derived neurotrophic factor (GDNF), insulin-like growth factor (IGF)-1, and stromal cell derived factor (SDF-1α) (Drago *et al.*, 2013), which are all growth factors that are upregulated in the stroke brain following intracerebral treatment of MSCs (Drago *et al.*, 2013). These MSC-secreted growth factors along the biobridge will likely not only guide the migration of endogenous stem cells but may directly afford brain repair by rescuing dying host cells in both neurogenic and non-neurogenic ischemic tissues. As we discussed earlier, it may very well be that stem cell therapy will achieve its optimal benefits not as a standalone therapy but via combined or multipronged regenerative processes.

11.9 Translational Caveats of Biobridge Formation from Bench to Clinic

Although the manifestation of the biobridge in enhancing stem cell therapy is evident, there are impediments in its translation to the clinic. Further investigations are required to adequately describe the biobridge mediation of endogenous stem cell migration. Specifically, the creation of an ensuing biobridge following the peripheral transplantation of stem cells should be investigated as a minimally invasive technique that could be favored in contrast to direct intracerebral transplantation when intervening in the subacute phase of stroke.

Another hindrance of the biobridge is that the ability to generate itself is contingent on the severity of the injury. The chronic stage of stroke seemingly exhibits more significantly damaged tissue that could be a factor in the creation of biobridges. The pathology of stroke is complex and varies from patient to patient as a result of the severity of the damage, demanding specific stem cell regimens, e.g., timing of initial cell transplantation, quantity of transplantations over the extended phase of secondary cell death, along with brain-region-specific factors that could affect the formation of a biobridge. Furthermore, the creation of a biobridge may be affected by the non-conducive surrounding of the brain's microenvironment, bringing to light the necessity to intervene in severe cases of stroke during the acute stage to promptly arrest the advancement of the disease and limit the progression of a non-ideal host microenvironment.

The inflammatory reactions associated with stroke also require further research, aimed to uncover how biobridge creation is affected by the secondary cell death phase. On one hand, sustained inflammation has negative effects; however, acute inflammation could assist the brain and biobridge formation. There is no consensus timeline for when inflammation yields positive effects and then shifts into a detrimental effect. Perhaps, treatment at a time closer to post-insult could aid the creation of the biobridge and may elevate the beneficial rather than the detrimental effects of inflammation. Further investigations are necessary to discern the optimal time point to maximize the therapeutic effects of inflammation on stem cell transplantation and its regulation of biobridge creation.

11.10 Outstanding Issues Relevant to Stem Cell Therapy in Stroke

Approximately 1 in every 18 deaths in the U.S. is a result of a stroke-related incident, with nearly 800,000 Americans suffering from stroke annually at a mortality rate of 17% (Lloyd-Jones *et al.*, 2010). Those who survive stroke exhibit enduring disabilities varying from modest to severe (Lloyd-Jones *et al.*, 2010). Every year, it is expected that $18.8 billion will be spent on healthcare expenses as a consequence of disabilities caused by stroke (Lloyd-Jones *et al.*, 2010). This estimation would rise to nearly $34.3 billion if the further limited productivity and premature deaths, $15.5 billion, were included (Centers for Disease Control and Prevention, 2012). Regardless of the fact that stroke mortality rate has diminished 33.5% from 1996 to 2006, the expenses of healthcare accompanying stroke have increased exponentially in recent decades (Lloyd-Jones *et al.*, 2010).

With healthcare expenses continuing to rise, the necessity for more efficient treatments for neurological ailments, such as stroke, is apparent. Stem cell therapy remains an experimental treatment option for stroke (Acosta *et al.*, 2014; Chiang *et al.*, 2001; Pastori *et al.*, 2008; Snyder *et al.*, 1997). Rehabilitation regimens are the standard treatment for stroke patients; however, the majority of clinical trials for stem cell therapies has disregarded this established form of treatment and fails to include rehabilitation as a control in their experimental design (Langhorne *et al.*, 2011; Ploughman *et al.*, 2015). To truly understand the possible benefits and shortcomings of cell therapies in both stroke animal models and stroke patients, it is essential that rehabilitation therapy be used as a standard for comparison.

With the demonstration of the therapeutic potential of stem cell therapies, public concerns have been raised about stem cell sources and their commercialization (Lo & Parham, 2009). The ethical controversy has been propagated by media outlets that focus on fetal and embryonic-derived stem cells, contributing to the misinformation and misunderstanding among the public of the true scientific potential of stem cells and their origins. In addition, the financial rewards associated with patents on certain stem cells, coupled with the rather expensive price tags for cell therapies, have also garnered ethical issues. Connecting the public with stem cell experts via open dialogue may resolve many of the ethical concerns that surround the field of stem cell therapy (Borlongan *et al.*, 2010). The establishment of consortia that place the public as main stakeholder will also allow a healthy forum to discuss the hopes and hypes of stem cell therapy (Borlongan *et al.*, 2010).

11.11 Conclusion

The damaged and/or aging brain has the capacity to regenerate even into adulthood. The process of neuroregeneration that occurs spontaneously after stroke is largely insufficient to mount robust and stable brain repair. Therapeutic intervention is required to produce any noticeable clinical improvements after stroke. Stem cell transplantation has emerged as a key strategy of regenerative medicine for neurological disorders. While traditionally considered as a therapy to replace dead or dying brain cells, and subsequently as a biological vehicle to deliver growth factors into the CNS, recent evidence indicates the ability of stem cells to form a biobridge that facilitates the migration of endogenous stem cells from host neurogenic niches towards the site of injury.

Following encouraging preclinical studies, stem cell therapy for stroke has proceeded to limited clinical trials. MSCs in particular have received substantial interest and promise as a therapeutic option due to their solid safety profiles in blood-borne diseases. More current and rigorous laboratory studies have supported the use of MSC transplantation in stroke patients. Importantly, recent data from completed and ongoing clinical trials indicate that while safe, stem cell therapies have not demonstrated significant efficacy — likely due to deviations of clinical protocols from the ideal cell doses, routes of delivery, and transplantation timing established by laboratory studies. In fact, cell doses in the clinic are often as much as eight-fold lower than the doses recommended from stroke studies in animals. Thus, using the laboratory studies as a guide to clinical trial designs will likely increase the effectiveness of stem cell therapy in stroke. Additionally, using rehabilitation therapy as a control measure in stem cell clinical trial will provide a more accurate assessment of this regenerative medicine therapy. As we move forward with transplant regimen optimization studies in the laboratory and efficacy trials in the clinic, it is imperative that we continue to address the ethical concerns that may arise with the large-scale use of stem cell therapy in stroke. A better understanding of the scientific basis, the practical issues in the clinic, and the public concerns associated with stem cell transplantation will guide its translation as a safe and effective stroke therapeutic.

Acknowledgments

C.V.B. is funded by grants from the U.S. National Institutes of Health (R01 NS071956, R01 NS090962, R21 NS089851, R21 NS094067), the U.S. Department of Defense (TATRC W811XWH-11-1-0634), the U.S. Veterans Affairs (BX001407-01), and the James and Esther King Biomedical Research Program (09KB-01-23123, 1KG01-33966).

References

Abdelsaid, M., Prakash, R., *et al.* (2015). Metformin treatment in the period after stroke prevents nitrative stress and restores angiogenic signaling in the brain in diabetes. *Diabetes* **64**(5): 1804–1817.

Acosta, S.A., Tajiri, N., *et al.* (2014). Combination therapy of human umbilical cord blood cells and granulocyte colony stimulating factor reduces histopathological and motor impairments in an experimental model of chronic traumatic brain injury. *PLoS One* **9**(3): e90953.

Acosta, S.A., Tajiri, N., *et al.* (2015). Intravenous bone marrow stem cell grafts preferentially migrate to spleen and abrogate chronic inflammation in stroke. *Stroke* **46**(9): 2616–2627.

Anderson, J.D., Johansson, H.J., *et al.* (2016). Comprehensive proteomic analysis of mesenchymal stem cell exosomes reveals modulation of angiogenesis via nuclear factor-kappaB signaling. *Stem Cells* **34**(3): 601–613.

Andres, R.H., Horie, N., *et al.* (2011). Human neural stem cells enhance structural plasticity and axonal transport in the ischaemic brain. *Brain* **134**(Pt 6): 1777–1789.

Androutsellis-Theotokis, A., Leker, R.R., *et al.* (2006). Notch signalling regulates stem cell numbers *in vitro* and *in vivo*. *Nature* **442**(7104): 823–826.

Banerjee, S., Bentley, P., *et al.* (2014). Intra-arterial immunoselected CD34+ stem cells for acute ischemic stroke. *Stem Cells Transl Med* **3**(11): 1322–1330.

Bang, O.Y., Lee, J.S., *et al.* (2005). Autologous mesenchymal stem cell transplantation in stroke patients. *Ann Neurol* **57**(6): 874–882.

Barha, C.K., Ishrat, T., *et al.* (2011). Progesterone treatment normalizes the levels of cell proliferation and cell death in the dentate gyrus of the hippocampus after traumatic brain injury. *Exp Neurol* **231**(1): 72–81.

Barkho, B.Z., Munoz, A.E., *et al.* (2008). Endogenous matrix metalloproteinase (MMP)-3 and MMP-9 promote the differentiation and migration of adult neural progenitor cells in response to chemokines. *Stem Cells* **26**(12): 3139–3149.

Boltze, J., Lukomska, B., *et al.* (2014). Mesenchymal stromal cells in stroke: Improvement of motor recovery or functional compensation? *J Cereb Blood Flow Metab* **34**(8): 1420–1421.

Borlongan, C.V. (2011). Bone marrow stem cell mobilization in stroke: A "bonehead" may be good after all! *Leukemia* **25**(11): 1674–1686.

Borlongan, C.V. (2016). Age of PISCES: Stem-cell clinical trials in stroke. *Lancet* **388**(10046): 736–738.

Borlongan, C.V., Chopp, M., *et al.* (2008). Potential of stem/progenitor cells in treating stroke: The missing steps in translating cell therapy from laboratory to clinic. *Regen Med* **3**(3): 249–250.

Borlongan, C.V., Fournier, C., *et al.* (2006). Gene therapy, cell transplantation and stroke. *Front Biosci* **11**: 1090–1101.

Borlongan, C. ., Glover, L.E., *et al.* (2011). The great migration of bone marrow-derived stem cells toward the ischemic brain: Therapeutic implications for stroke and other neurological disorders. *Prog Neurobiol* **95**(2): 213–228.

Borlongan, C.V., Lind, J.G., *et al.* (2004). Intracerebral xenografts of mouse bone marrow cells in adult rats facilitate restoration of cerebral blood flow and blood–brain barrier. *Brain Res* **1009**(1–2): 26–33.

Borlongan, C.V., Mcwhirter, C., *et al.* (2010). The case for an ethics research consortium for emerging technologies: Public perception of stem cell research and development. *Technol Innov* **12**: 21–28.

Broughton, B.R., Reutens, D.C., *et al.* (2009). Apoptotic mechanisms after cerebral ischemia. *Stroke* **40**(5): e331–339.

Centers for Disease Control and Prevention (2012). Prevalence of stroke — United States, 2006–2010. *MMWR Morb Mortal Wkly Rep* **61**(20): 379–382.

Chen, J., Zhang, Z.G., *et al.* (2003). Statins induce angiogenesis, neurogenesis, and synaptogenesis after stroke. *Ann Neurol* **53**(6): 743–751.

Chen, S., Yang, Q., *et al.* (2015). An update on inflammation in the acute phase of intracerebral hemorrhage. *Transl Stroke Res* **6**(1): 4–8.

Chiang, Y., Morales, M., *et al.* (2001). Fetal intra-nigral ventral mesencephalon and kidney tissue bridge transplantation restores the nigrostriatal dopamine pathway in hemiparkinsonian rats. *Brain Res* **889**(1–2): 200–207.

De La Peña, I., Sanberg, P.R., *et al.* (2015). G-CSF as an adjunctive therapy with umbilical cord blood cell transplantation for traumatic brain injury. *Cell Transplant* **24**(3): 447–457.

Ding, G., Jiang, Q., *et al.* (2010). Cerebral tissue repair and atrophy after embolic stroke in rat: A magnetic resonance imaging study of erythropoietin therapy. *J Neurosci Res* **88**(14): 3206–3214.

Doeppner, T.R., Hermann, D.M. (2010). Mesenchymal stem cells in the treatment of ischemic stroke: Progress and possibilities. *Stem Cells Cloning* **3**: 157–163.

Drago, D., Cossetti, C., *et al.* (2013). The stem cell secretome and its role in brain repair. *Biochimie* **95**(12): 2271–2285.

Duffy, G.P., Ahsan, T., *et al.* (2009). Bone marrow-derived mesenchymal stem cells promote angiogenic processes in a time- and dose-dependent manner *in vitro*. *Tissue Eng Part A* **15**(9): 2459–2470.

Eckert, M.A., Vu, Q., *et al.* (2013). Evidence for high translational potential of mesenchymal stromal cell therapy to improve recovery from ischemic stroke. *J Cereb Blood Flow Metab* **33**(9): 1322–1334.

Fagan, S.C., Cronic, L.E., *et al.* (2011). Minocycline development for acute ischemic stroke. *Transl Stroke Res* **2**(2): 202–208.

Gage, F.H. (2000). Mammalian neural stem cells. *Science* **287**(5457): 1433–1438.

Ghuman, H., Massensini, A.R., *et al.* (2016). ECM hydrogel for the treatment of stroke: Characterization of the host cell infiltrate. *Biomaterials* **91**: 166–181.

Gnecchi, M., Melo, L.G. (2009). Bone marrow-derived mesenchymal stem cells: Isolation, expansion, characterization, viral transduction, and production of conditioned medium. *Methods Mol Biol* **482**: 281–294.

Gonzalez, F.F., Larpthaveesarp, A., *et al.* (2013). Erythropoietin increases neurogenesis and oligodendrogliosis of subventricular zone precursor cells after neonatal stroke. *Stroke* **44**(3): 753–758.

Gould, E. (2007). How widespread is adult neurogenesis in mammals? *Nat Rev Neurosci* **8**(6): 481–488.

Graham, G.D. (2003). Tissue plasminogen activator for acute ischemic stroke in clinical practice: A meta-analysis of safety data. *Stroke* **34**(12): 2847–2850.

Guo, Z., Hu, Q., *et al.* (2016). Lipoxin A4 reduces inflammation through formyl peptide receptor 2/p38 MAPK signaling pathway in subarachnoid hemorrhage rats. *Stroke* **47**(2): 490–497.

Hara, K., Yasuhara, T., *et al.* (2008). Neural progenitor NT2N cell lines from teratocarcinoma for transplantation therapy in stroke. *Prog Neurobiol* **85**(3): 318–334.

Hargus, G., Cooper, O., *et al.* (2010). Differentiated Parkinson patient-derived induced pluripotent stem cells grow in the adult rodent brain and reduce motor asymmetry in Parkinsonian rats. *Proc Natl Acad Sci USA* **107**(36): 15921–15926.

Hosaka, K., Hoh, B.L. (2014). Inflammation and cerebral aneurysms. *Transl Stroke Res* **5**(2): 190–198.

Hsieh, J.Y., Wang, H.W., *et al.* (2013). Mesenchymal stem cells from human umbilical cord express preferentially secreted factors related to neuroprotection, neurogenesis, and angiogenesis. *PLoS One* **8**(8): e72604.

Huttner, H.B., Bergmann, O., *et al.* (2014). The age and genomic integrity of neurons after cortical stroke in humans. *Nat Neurosci* **17**(6): 801–803.

Ishikawa, H., Tajiri, N., *et al.* (2013). Vasculogenesis in experimental stroke after human cerebral endothelial cell transplantation. *Stroke* **44**(12): 3473–3481.

Ishrat, T., Pillai, B., *et al.* (2013). Candesartan reduces the hemorrhage associated with delayed tissue plasminogen activator treatment in rat embolic stroke. *Neurochem Res* **38**(12): 2668–2677.

Jaskelioff, M., Muller, F.L., *et al.* (2011). Telomerase reactivation reverses tissue degeneration in aged telomerase-deficient mice. *Nature* **469**(7328): 102–106.

Jellema, R.K., Ophelders, D.R., *et al.* (2015). Multipotent adult progenitor cells for hypoxic-ischemic injury in the preterm brain. *J Neuroinflammation* **12**: 241.

Joyce, N., Annett, G., *et al.* (2010). Mesenchymal stem cells for the treatment of neurodegenerative disease. *Regen Med* **5**(6): 933–946.

Kaneko, Y., Hayashi, T., *et al.* (2011). Human amniotic epithelial cells express melatonin receptor MT1, but not melatonin receptor MT2: A new perspective to neuroprotection. *J Pineal Res* **50**(3): 272–280.

Kocsis, J.D., Honmou, O. (2012). Bone marrow stem cells in experimental stroke. *Prog Brain Res* **201**: 79–98.

Koh, S.H., Liang, A.C., *et al.* (2015). Differential effects of isoxazole-9 on neural stem/progenitor Cells, oligodendrocyte precursor cells, and endothelial progenitor cells. *PLoS One* **10**(9): e0138724.

Langhorne, P., Bernhardt, J., *et al.* (2011). Stroke rehabilitation. *Lancet* **377**(9778): 1693–1702.

Leak, R.K., Li, P., *et al.* (2015). Apurinic/apyrimidinic endonuclease 1 upregulation reduces oxidative DNA damage and protects hippocampal neurons from ischemic injury. *Antioxid Redox Signal* **22**(2): 135–148.

Lee, H.S., Bae, E.J., *et al.* (2010). Foxa2 and Nurr1 synergistically yield A9 nigral dopamine neurons exhibiting improved differentiation, function, and cell survival. *Stem Cells* **28**(3): 501–512.

Lee, J.P., Jeyakumar, M., *et al.* (2007). Stem cells act through multiple mechanisms to benefit mice with neurodegenerative metabolic disease. *Nat Med* **13**(4): 439–447.

Li, Y.F., Ren, LN., *et al.* (2015). Endothelial progenitor cells in ischemic stroke: An exploration from hypothesis to therapy. *J Hematol Oncol* **8**: 33.

Li, Z., Mckercher, S.R., *et al.* (2008). Myocyte enhancer factor 2C as a neurogenic and antiapoptotic transcription factor in murine embryonic stem cells. *J Neurosci* **28**(26): 6557–6568.

Lin, C.H., Lee, H.T., *et al.* (2013). Role of HIF-1alpha-activated Epac1 on HSC-mediated neuroplasticity in stroke model. *Neurobiol Dis* **58**: 76–91.

Lindvall, O., Kokaia, Z. (2006). Stem cells for the treatment of neurological disorders. *Nature* **441**(7097): 1094–1096.

Liu, S. P., Fu, R. H., *et al.* (2014). Mouse-induced pluripotent stem cells generated under hypoxic conditions in the absence of viral infection and oncogenic factors and used for ischemic stroke therapy. *Stem Cells Dev* **23**(4): 421–433.

Liu, Z., Li, Y., *et al.* (2011). Bone marrow stromal cells promote skilled motor recovery and enhance contralesional axonal connections after ischemic stroke in adult mice. *Stroke* **42**(3): 740–744.

Lloyd-Jones, D., Adams, R.J., *et al.* (2010). Heart disease and stroke statistics — 2010 update: A report from the American Heart Association. *Circulation* **121**(7): e46–e215.

Lo, B., Parham, L. (2009). Ethical issues in stem cell research. *Endocr Rev* **30**(3): 204–213.

Massensini, A.R., Ghuman, H., *et al.* (2015). Concentration-dependent rheological properties of ECM hydrogel for intracerebral delivery to a stroke cavity. *Acta Biomater* **27**: 116–130.

Mazzocchi-Jones, D., Döbrössy, M., *et al.* (2009). Embryonic striatal grafts restore bi-directional synaptic plasticity in a rodent model of Huntington's disease. *Eur J Neurosci* **30**(11): 2134–2142.

Merson, T.D., Bourne, J.A. (2014). Endogenous neurogenesis following ischaemic brain injury: Insights for therapeutic strategies. *Int J Biochem Cell Biol* **56**: 4–19.

Mezey, E. (2011). The therapeutic potential of bone marrow-derived stromal cells. *J Cell Biochem* **112**(10): 2683–2687.

Ming, G.L., Song, H. (2011). Adult neurogenesis in the mammalian brain: Significant answers and significant questions. *Neuron* **70**(4): 687–702.

Muotri, A.R., Gage, F.H. (2006). Generation of neuronal variability and complexity. *Nature* **441**(7097): 1087–1093.

Oh, J.S., Kim, K.N., *et al.* (2011). Cotransplantation of mouse neural stem cells (mNSCs) with adipose tissue-derived mesenchymal stem cells improves mNSC survival in a rat spinal cord injury model. *Cell Transplant* **20**(6): 837–849.

Park, D.H., Borlongan, C.V., *et al.* (2009). Human umbilical cord blood cell grafts for brain ischemia. *Cell Transplant* **18**(9): 985–998.

Pastori, C., Librizzi, L., *et al.* (2008). Arterially perfused neurosphere-derived cells distribute outside the ischemic core in a model of transient focal ischemia and reperfusion *in vitro*. *PLoS One* **3**(7): e2754.

Peña, I., Borlongan, C.V. (2015). Translating G-CSF as an adjunct therapy to Stem cell transplantation for stroke. *Transl Stroke Res* **6**(6): 421–429.

Ploughman, M., Austin, M.W., *et al.* (2015). The effects of poststroke aerobic exercise on neuroplasticity: A systematic review of animal and clinical studies. *Transl Stroke Res* **6**(1): 13–28.

Pollock, K., Stroemer, P., *et al.* (2006). A conditionally immortal clonal stem cell line from human cortical neuroepithelium for the treatment of ischemic stroke. *Exp Neurol* **199**(1): 143–155.

Prasad, K., Sharma, A., *et al.* (2014). Intravenous autologous bone marrow mononuclear stem cell therapy for ischemic stroke: A multicentric, randomized trial. *Stroke* **45**(12): 3618–3624.

Puyal, J., Ginet, V., *et al.* (2013). Multiple interacting cell death mechanisms in the mediation of excitotoxicity and ischemic brain damage: A challenge for neuroprotection. *Prog Neurobiol* **105**: 24–48.

Redmond, D.E., Bjugstad, K.B., *et al.* (2007). Behavioral improvement in a primate Parkinson's model is associated with multiple homeostatic effects of human neural stem cells. *Proc Natl Acad Sci USA* **104**(29): 12175–12180.

Reuter, B., Rodemer, C., *et al.* (2015). Effect of simvastatin on MMPs and TIMPs in human brain endothelial cells and experimental stroke. *Transl Stroke Res* **6**(2): 156–159.

Rowart, P., Erpicum, P., *et al.* (2015). Mesenchymal stromal cell therapy in ischemia/reperfusion injury. *J Immunol Res* **2015**: 602597.

Sanai, N., Nguyen, T., *et al.* (2011). Corridors of migrating neurons in the human brain and their decline during infancy. *Nature* **478**(7369): 382–386.

Savitz, S.I., Chopp, M., *et al.* (2011a). Stem Cell Therapy as an Emerging Paradigm for Stroke (STEPS) II. *Stroke* **42**(3): 825–829.

Savitz, S.I., Cramer, S.C., *et al.* (2014). Stem cells as an emerging paradigm in stroke 3: Enhancing the development of clinical trials. *Stroke* **45**(2): 634–639.

Savitz, S.I., Misra, V., *et al.* (2011b). Intravenous autologous bone marrow mononuclear cells for ischemic stroke. *Ann Neurol* **70**(1): 59–69.

Schweizer, S., Meisel, A., *et al.* (2013). Epigenetic mechanisms in cerebral ischemia. *J Cereb Blood Flow Metab* **33**(9): 1335–1346.

Seol, H.J., Jin, J., *et al.* (2011). Genetically engineered human neural stem cells with rabbit carboxyl esterase can target brain metastasis from breast cancer. *Cancer Lett* **311**(2): 152–159.

Skop, N.B., Calderon, F., *et al.* (2014). Improvements in biomaterial matrices for neural precursor cell transplantation. *Mol Cell Ther* **2**: 19.

Snyder, E.Y., Yoon, C., *et al.* (1997). Multipotent neural precursors can differentiate toward replacement of neurons undergoing targeted apoptotic degeneration in adult mouse neocortex. *Proc Natl Acad Sci USA* **94**(21): 11663–11668.

Sobrino, T., Perez-Mato, M., *et al.* (2012). Temporal profile of molecular signatures associated with circulating endothelial progenitor cells in human ischemic stroke. *J Neurosci Res* **90**(9): 1788–1793.

Souvenir, R., Flores, J.J., *et al.* (2014). Erythropoietin inhibits HIF-1alpha expression via upregulation of PHD-2 transcription and translation in an *in vitro* model of hypoxia-ischemia. *Transl Stroke Res* **5**(1): 118–127.

Sozmen, E.G., Hinman, J.D., *et al.* (2012). Models that matter: White matter stroke models. *Neurotherapeutics* **9**(2): 349–358.

Stabenfeldt, S.E., Munglani, G., *et al.* (2010). Biomimetic microenvironment modulates neural stem cell survival, migration, and differentiation. *Tissue Eng Part A* **16**(12): 3747–3758.

Steinberg, G.K., Kondziolka, D., *et al.* (2016). Clinical outcomes of transplanted modified bone marrow-derived mesenchymal stem cells in stroke: A Phase 1/2a study. *Stroke* **47**(7): 1817–1824.

Stem Cell Therapies as an Emerging Paradigm in Stroke (STEPS). (2009). Stem Cell Therapies as an Emerging Paradigm in Stroke (STEPS): Bridging basic and clinical science for cellular and neurogenic factor therapy in treating stroke. *Stroke* **40**(2): 510–515.

Stone, L.L., Grande, A., *et al.* (2013). Neural repair and neuroprotection with stem cells in ischemic stroke. *Brain Sci* **3**(2): 599–614.

Stroemer, P., Patel, S., *et al.* (2009). The neural stem cell line CTX0E03 promotes behavioral recovery and endogenous neurogenesis after experimental stroke in a dose-dependent fashion. *Neurorehabil Neural Repair* **23**(9): 895–909.

Tajiri, N., Kaneko, Y., *et al.* (2013). Stem cell recruitment of newly formed host cells via a successful seduction? Filling the gap between neurogenic niche and injured brain site. *PLoS One* **8**(9): e74857.

Tang, Y., Yasuhara, T., *et al.* (2007). Transplantation of bone marrow-derived stem cells: A promising therapy for stroke. *Cell Transplant* **16**(2): 159–169.

Tate, C.C., Fonck, C., *et al.* (2010). Human mesenchymal stromal cells and their derivative, SB623 cells, rescue neural cells via trophic support following *in vitro* ischemia. *Cell Transplant* **19**(8): 973–984.

Uchida, H., Morita, T., *et al.* (2016). Transplantation of unique subpopulation of fibroblasts, muse cells, ameliorates experimental stroke possibly via robust neuronal differentiation. *Stem Cells* **34**(1): 160–173.

Van Velthoven, C.T., Gonzalez, F., *et al.* (2014). Stem cells for neonatal stroke — the future is here. *Front Cell Neurosci* **8**: 207.

Wang, L., Chopp, M., *et al.* (2011). Tumor necrosis factor α primes cerebral endothelial cells for erythropoietin-induced angiogenesis. *J Cereb Blood Flow Metab* **31**(2): 640–647.

Wang, L., Zhang, Z., *et al.* (2004). Treatment of stroke with erythropoietin enhances neurogenesis and angiogenesis and improves neurological function in rats. *Stroke* **35**(7): 1732–1737.

White, H.D., Simes, R.J., *et al.* (2000). Pravastatin therapy and the risk of stroke. *N Engl J Med* **343**(5): 317–326.

Wolf, K., Te Lindert, M., *et al.* (2013). Physical limits of cell migration: control by ECM space and nuclear deformation and tuning by proteolysis and traction force. *J Cell Biol* **201**(7): 1069–1084.

Xiong, X.Y., Yang, Q.W. (2015). Rethinking the roles of inflammation in the intracerebral hemorrhage. Transl Stroke Res **6**(5): 339–341.

Xiong, Y., Mahmood, A., *et al.* (2010). Angiogenesis, neurogenesis and brain recovery of function following injury. *Curr Opin Investig Drugs* **11**(3): 298–308.

Yamashita, T., Ninomiya, M., *et al.* (2006). Subventricular zone-derived neuroblasts migrate and differentiate into mature neurons in the post-stroke adult striatum. *J Neurosci* **26**(24): 6627–6636.

Yamauchi, T., Saito, H., *et al.* (2014). Platelet lysate and granulocyte-colony stimulating factor serve safe and accelerated expansion of human bone marrow stromal cells for stroke therapy. *Transl Stroke Res* **5**(6): 701–710.

Yasuda, A., Tsuji, O., *et al.* (2011). Significance of remyelination by neural stem/progenitor cells transplanted into the injured spinal cord. *Stem Cells* **29**(12): 1983–1994.

Yasuhara, T., Hara, K., *et al.* (2008). Intravenous grafts recapitulate the neurorestoration afforded by intracerebrally delivered multipotent adult progenitor cells in neonatal hypoxic-ischemic rats. *J Cereb Blood Flow Metab* **28**(11): 1804–1810.

Yasuhara, T., Matsukawa, N., *et al.* (2009). Notch-induced rat and human bone marrow stromal cell grafts reduce ischemic cell loss and ameliorate behavioral deficits in chronic stroke animals. *Stem Cells Dev* **18**(10): 1501–1514.

Yoo, S.W., Chang, D.Y., *et al.* (2013). Immune following suppression mesenchymal stem cell transplantation in the ischemic brain is mediated by TGF-beta. *Neurobiol Dis* **58**: 249–257.

Zhang, R.L., Chopp, M., *et al.* (2009). Patterns and dynamics of subventricular zone neuroblast migration in the ischemic striatum of the adult mouse. *J Cereb Blood Flow Metab* **29**(7): 1240–1250.

Zhao, B.Q., Wang, S., *et al.* (2006). Role of matrix metalloproteinases in delayed cortical responses after stroke. *Nat Med* **12**(4): 441–445.

Zhu, W., Casper, A., *et al.* (2015). Preclinical evaluation of recombinant T cell receptor ligand RTL1000 as a therapeutic agent in ischemic stroke. *Transl Stroke Res* **6**(1): 60–68.

Zhu, W., Libal, N.L., *et al.* (2014). Recombinant T cell receptor ligand treatment improves neurological outcome in the presence of tissue plasminogen activator in experimental ischemic stroke. *Transl Stroke Res* 5(5): 612–617.

Index